Viruses

More Friends Than Foes

Viruses

More Friends Than Foes

Karin Moelling

University of Zurich, Switzerland &
Max Planck Institute for Molecular Genetics,
Berlin, Germany

World Scientific

NEW JERSEY · LONDON · SINGAPORE · BEIJING · SHANGHAI · HONG KONG · TAIPEI · CHENNAI · TOKYO

Published by

World Scientific Publishing Co. Pte. Ltd.

5 Toh Tuck Link, Singapore 596224

USA office: 27 Warren Street, Suite 401-402, Hackensack, NJ 07601

UK office: 57 Shelton Street, Covent Garden, London WC2H 9HE

Library of Congress Cataloging-in-Publication Data
Mölling, Karin, 1943–
    Viruses : More Friends Than Foes / by Karin Moelling, University of Zurich,
Switzerland, and Max Planck Institute for Molecular Genetics, Berlin, Germany.
        pages cm -- New Jersey : World Scientific, 2016.
    Includes bibliographical references and index.
    ISBN 9789813147812 (hard cover : alk. paper) -- ISBN 9789813147829 (pbk. : alk. paper)
    1. Viruses. 2. Microbiology. 3. Well-being.
    QR360 .M575 2016
    616.9/101   2016034785

British Library Cataloguing-in-Publication Data
A catalogue record for this book is available from the British Library.

Originally published in German as *Supermacht des Lebens: Reisen in die erstaunliche Welt der Viren*,
C.H. Beck, Munich, 2015.

Printed in Singapore by Mainland Press Pte Ltd.

Preface

We stand on the shoulders of giants. Standing next to giants, I have over the decades made the acquaintance of some of the most prominent researchers of my generation — listening to and watching them, mostly with high admiration and respect, and sometimes contributing myself. On this basis, but also to some extent in opposition to what I saw, this book was initiated as an "anti-virus book". Viruses as we usually encounter them are dangerous, terrifying creatures; they are destructive, threatening and downright nasty. However, they contribute to our existence, to our environment, to the development of life and to evolution. They are parts of our genes! This positive side of viruses and microorganisms is almost always ignored, and they deserve more credit for it than they are generally given. Here the reader will encounter this other world of viruses, and I hope that our journey through it will be entertaining, not sinister, not too scientific, always a bit relaxing, sometimes provocative, scientifically up to date, and sometimes a little futuristic.

Every reader will find occasion to be surprised about the viruses: where they are active on our planet, in the oceans, in our gardens and trees, inside and outside our body — including guts, brain, or birth canal. The viruses influence our well-being, our souls, fear or courageousness, depression, freedom and decision-making — and to give an example: even obesity. Imagine, HIV-like viruses made egg-laying obsolete for us humans since millions of years. I hope that readers will be fascinated more than once — as indeed I was too during writing.

I was studying and teaching about the disease-causing viruses for more than 40 years — which are not the main topic here — so, I know about HIV/AIDS — but I believe that most viral diseases are man-made due to poverty, lack of hygiene, mobility, or habits.

The reader will set off on a journey into the innermost part of what makes up our world. Today, Goethe's or Marlowe's Dr. Faustus would quite probably be a molecular biologist — perhaps even a virologist, because the "Viro-sphere" includes the whole world, even the universe! Perhaps Goethe would have honored the viruses if he knew about their existence and their importance.

Indeed, the "Homunculus", which Goethe designed, places Dr. Faustus close to our molecular world. How did life start? And how does it end? How do progress and innovation come about — of course by viruses and their famous "sloppiness". And the "noisy lemurs, which fall head-over-tail into their graves" can also be found here, even though they are not quite like those in Goethe's "Faust" — in fact, they are very strange creatures, which since 13 million years carry HIV-like viruses!

The book does not need to be read from beginning to end, but rather by choosing, picking, jumping, turning over pages, and skipping passages if they seem too difficult or contain too much scientific detail. A glossary and list of references is provided for further information. However, the end, the last chapter, is intended for everybody, as it condenses and summarises many details, with voices combined in a final *tutti* as at the end of a fugue by my hero, Johann Sebastian Bach.

I take the risk of reflecting on how science progresses, and on what forces drive the everyday activities of research scientists. I consider myself as a witness, a spectator, who can describe some of my own experiences — though with the idea and the hope that they will be taken as representative and general, and not appear too personal. At some points my comments will be quite critical, but I write without bitterness or resentment — rather, slightly amused about what happened, what I missed, and what today still keeps me going. So some parts are detective story-like for entertainment. Other parts are rather philosophical, so you, dear reader, can choose.

A colleague characterized the German version of this book as three volumes in one: a detective story about scientists, a popular account of science spanning many decades, and a philosophical work. I am not a philosopher, but science leads us to wonder and to reflect. Another apposite comment comes from the German writer and film-producer Alexander Kluge in his recent "Chronicle", in which he describes me as story-teller of "Goodnight Stories", just like the ones he liked to listen to by his nanny as a 5-year-old boy. There are two options, then, either you fall asleep or you find the stories amusing and simple enough to listen. I am writing not only for academic colleagues from similar or related fields, but also, and perhaps primarily, for students and for lay people of all kinds — readers who can simply ignore some of the more specialized scientific remarks. Read about Cesarean sections or Dutch Famine and their consequences for the newborns. Did you know that viruses can "see"? Or look at the chapter on tulips, the first financial crisis ever — with viruses as the cause! Thus, economists may also learn some surprising facts.

Whom do I have to thank? All those whom I have met — not only the giants, because every human being is inspiring and has something to hand on to others. I have always liked to listen to the "little shots".

So many people and organisations have contributed so much to my life; these contributions are interwoven to such an extent that they cannot be separated any more. Many have supported me: my parents, school, universities, funding organisations, scholarships, research organisations, as well as society as a whole. They have all accompanied me along the path and allowed me to find my way. Not least, they have financed my research, the most expensive hobby on earth: on viruses and cancer.

Along the way, I have always had to interact with young people, many of whom it was my responsibility to motivate and support — this I did not only as a duty, but also with great pleasure and enjoyment, sometimes with success They kept me young and flexible. And, last but not least: strong enemies made me strong. This last challenge was harder to meet than it sounds, and I had no idea how to cope with it — so I took a colleague's advice: "Hide in a sewer, close the cover, and try to produce good papers — that will be your life-saver!"

The foundation *Studienstiftung* showed great generosity and patience in supporting my unexpected transition from physics to molecular biology at Berkeley, USA, at a time when student riots were raging across the campus; in fact, the *Studienstiftung* made that transition possible. The decision to dive into the unknown, the field of molecular biology, at a time when nobody could even explain to me what molecular biology was about, was one of the hardest I have ever had to take. Then I became a researcher and scientist against everybody's advice. The Max Planck Institute for Molecular Genetics, Berlin, supported me so that I was able to perform independent research for twenty years, during which time I had the good luck to obtain several novel results on viruses and cancer, that carried me on into the future, even though this turned out to be much more difficult than anticipated. All this happened at times when women were as rare in the Max Planck Society "as in the Berlin Philharmonic Orchestra under Herbert von Karajan, or in the Hierarchy of the Catholic Church". At the time they had one each: Sabine Meyer, the then very young clarinettist, and Saint Mary, mother of Jesus — as Heinz Schuster, then Director of the Max Planck Institute, stressed in a public lecture. I was lucky and profited from his support.

I am very grateful to the University of Zürich for allowing me to perform my research over many years and for taking the courageous decision to appoint me, as a non-medical doctor, to the Medical Faculty — as the only female Director in the "pre-gender" era. "Keep your mouth shut in public" was a serious advice from a friend, who knew the scene and his colleagues — so obviously not everybody was equally convinced. I should have been more obedient.

I am grateful to Manfred Eigen, at the Max Planck Institute for Biophysical Chemistry in Göttingen, for his numerous generous invitations to the RNA world of his renowned annual Winter Seminars in Klosters, Switzerland, where some of the ideas discussed in this book were initiated. I was admitted as a retrovirologist — since Eigen regarded these viruses as a great model for evolution, as the reader will see. I admired Eigen's broad knowledge and visions — and how his often simple calculations about reaction kinetics, numbers, interaction parameters made some speakers get lost.

I also thank the Institutes for Advanced Study at Berlin and at Princeton for their invitations and their support, for their stimulating atmospheres where ideas could be created and grow in a milieu of discussion, and for the fantastic invitation to enter very distant new worlds of thought. I hanker after the use of blackboard and chalk, as these evoke memories of the most lively and spontaneous discussions at Princeton. John Hopfield or Freeman Dyson at the blackboard will always be unforgettable for me. Thinking about neuronal networks or quantum mechanics they asked completely different questions than brain-washed insiders do. Dyson warned me about writing a book on viruses: "People want to know about people, people do not want to know about genetics!" I tried to follow his advice while writing this book.

My special thanks go to my former students, co-workers and co-authors on so many papers, whose work I refer to in various chapters, and who kept me young and energetic. Most of them were so enthusiastic about science that they often forgot about career planning and to think about their future. We spent many enjoyable hours and a significant part of my life together.

My special thanks go to Felix Broecker for his critical reading of the German and the English versions of the book from the point of view of a young scientist, and for verifying the many updates, numbers, facts, credits, names, and references to other publications. I am grateful to Ulrike Kahle-Steinweh for her comments from the point of view of a lay person and her never-failing encouragement. Also Alfred Pingoud, who sadly enough has already passed away, deserved many thanks, not least because he was not always of the same opinion as I was. My thanks also go to Stefan Bollmann — who was this book's editor at the German publisher C.H. Beck, and who suffered from but tolerated my style — and to Paul Woolley, who edited the first draft of my English manuscript.

I have had to experience the limits of cancer research twice in a sad way. I dedicate this book to the memory of Heinz Schuster, one of the founders of the Max Planck Institute for Molecular Genetics in Berlin. He was enthusiastic and supportive of my work, and a close and generous friend. I dedicate this book also to Paul Gredinger, in Zurich — a non-scientist, who regretted knowing so little about science, he would study in his "next life". He was known as an innovative creative operator of an Art and Think Tank. He gave me the unforgettable advice: If someone steals your results, it is an honor for you. And most importantly: Do not attend the meetings of your peers but of your non-peers; that is much more innovative. He was so right. Both of them I was allowed to accompany, to share ideas and thoughts with. They inspired me, they made my life happy — and often they had more confidence in me than I myself had.

Twice I had to learn to say goodbye forever — farewells that I cannot forget. I have tried to by writing this book.

Berlin/Zurich, 2016   Karin Moelling

Contents

Viruses

More Friends Than Foes

1 Viruses — not as you pictured them

Viruses — a success story

The word "virus" often provokes disgust: "Ugh, keep that out of here" — or "Watch out, they're contagious, they make you ill"! Yet here comes a book about the opposite: Viruses are better than their reputation. Much better. This is a surprising reverse side of the viruses, which will be the topic of this book. Viruses as friends, not foes!

Everything in virology is new. Unnoticed, a paradigm shift has taken place: the focus is no longer upon diseases, but rather on the positive side of viruses: viruses as drivers of evolution, viruses leading to innovation, viruses at the origin of life — or at least their presence from the very beginning. Throughout evolution viruses have been our "bodybuilders" or gene modulators. What is a virus? Where do viruses come from? Are they alive or not? Why and when do they make us ill? Whether you, the reader, want to continue swimming in the sea — whether babies' dummies from the Far East may cause cancer and should be avoided — whether you should stop enjoying salad because of the many plant viruses — these are questions that you can decide upon for yourself after having read this book.

While reading it, you will learn something about life — the innermost parts of your cells and your genes; you will find out how viruses contribute to our adaptation to environmental conditions; you will be confronted with the question of whether they may contribute to human free will; you will discover the extent to which we are related to bacteria and worms, and how sex can be replaced by viruses; you will find out that viruses "invented" all immune systems and supplied cells with antiviral defense mechanisms. This is much easier to understand than you may expect, and it reflects my own thinking. What part do viruses play in cancer development and gene therapy? Do "jumping genes", which are "locked-in viruses",

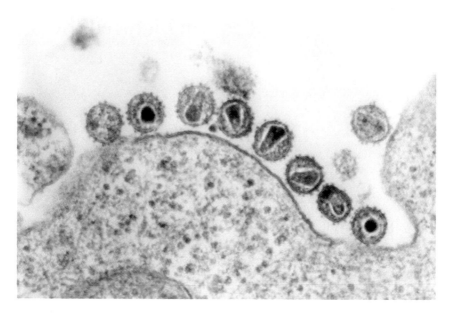

HIV particles aligned at the cell surface, electron microscopic picture.

create geniuses? Do you know that viruses can "see"? Almost they do, the color blue! You will read about efforts to save chestnut trees, about where stripes on tulips and the white patterns on blue balcony petunias come from (viruses of course) and how viruses caused the first financial crisis, known as "tulipomania". Finally, one-third of mankind may want to know how to control body mass or obesity. (By viruses? Yes, indeed.) The viruses have contributed, and are still contributing, to all of these. And now, let's start by reading about the success story of the viruses.

In 2009, Darwin's bicentennial year, I had lunch with colleagues from the Institute for Advanced Study in Berlin and, in conversation, asked them how they thought life started. They included philosophers, historians, sociologists and lawyers — so what did they suppose? The Big Bang? Certainly not Adam and Eve, they said — disdaining creationism. But their general response was one of perplexity and helplessness. If *you* are asking this question, one of the Fellows concluded, then the answer must have something to do with viruses. Yes, that is exactly what I think: Viruses were there at the beginning, or at least they contributed from very early on.

The history of medicine has led to a one-sided picture of the viruses, portraying them as causing various diseases. That is indeed how we came to know about them. Most viral diseases are incurable — no treatments are available against them — which contributes to the bad reputation of viruses. For centuries, people were helpless against viral infections. Polio, measles, pox and influenza have destroyed cultures, decided the outcome of wars, ruined cities and depopulated whole landscapes. Viral infections could not be distinguished from bacterial ones, in fact, that is not even necessary, because according to the newest research results *pestis* bacteria became as deadly as they are by modification through a phage, a virus that resides in bacteria.

There are commemorative "plague columns" in many cities — such as Vienna, where the *Pestsäule* is today a popular meeting-point. The church of Santa Maria della Salute in Venice reminds us of the Black Death in 1347, and the fear of infections and the gratitude of the survivors, even today. With colorful gondola parades, Venice commemorates the plague every year — even though that city never completely recovered from the high death toll. The Spanish *Conquistadores* owed their victory over the Mayas in Mexico mainly to measles, which was not known there and was therefore deadly for the local population. The Mayas avenged themselves by giving syphilis to their conquerors, and so to Europe. The outcome of World War One was decided at least in part by the influenza epidemic, with possibly up to 100 million deaths. Since 1981 HIV/AIDS has killed some 37 million people worldwide, and every year two million more become infected.

It has only been for the last 100 years that we have been able to distinguish viruses from bacteria. The easiest way is by their size: viruses are in most cases — we shall hear about exceptions — smaller than bacteria, and at least today they depend on cells for their replication, including mammalian or plant or bacterial cells. Bacteria, in contrast to viruses, can replicate autonomously. Today's viruses cannot do that, but they may well have been able to do so in the distant past. Both bacteria and viruses can cause diseases. Antibiotics will destroy bacteria but not viruses. If doctors nonetheless prescribe antibiotics for viral infections, then this is to protect the patients against any possible bacterial superinfection. There are very many books about viruses describing them as causes of disease. For many years I taught exactly this to medical students at the Universities of Berlin and Zurich. But that is not what I am writing about here.

No, the opposite needs some attention. Thanks to new technologies, virology has changed completely since the beginning of this century. If viruses were once regarded as enemies of humans and animals, even of all life forms, we now recognize that viruses contributed to the beginning of life and have positively contributed to its development from then on, to this day. In the last ten years or so, our perception of all microbes — viruses as well as bacteria — has changed completely. New methods, new experimental approaches and new, sensitive detection methods have revealed that viruses are by no means only pathogenic germs. Shouldn't we be surprised that viruses do not spread faster, leading to many more infections, when there are three billion flights worldwide each year, carrying about 300 billion passengers — despite the fact that the air in airplanes is only circulated around and not purified by expensive sterilizing filters. Most viruses and other microorganisms are harmless for their hosts — that is something to remember.

Viruses are everywhere. They are the oldest biological entities on our planet, as will be shown later. And they are also by far the most abundant. We were born as humans into a world which had existed for billions of years before us. We are latecomers, and have populated our planet for less than a few hundred thousand years. Those who were unable to cope with the pre-existing microorganisms died, and the others established a co-existence with them. We do not know how many populations have died out because of diseases — was Neandertal Man one of them? It is important to note: Diseases occur when a balance is disturbed and changes of environmental conditions occur through poor hygiene, traveling, overpopulated cities, disappearing forests, water reservoirs, pollution or close contact with other species that carry viruses unfamiliar to us (zoonosis). Microbes not known to an organism may cause diseases without affecting those who are used to them. Most of our human diseases are self-made — a strong statement! A simple example is that of catching a proverbial "cold" — which means a temperature change allowing some viruses to replicate better than before, leading to diseases such as rhinitis or influenza. "Catching a cold" — summarizes virology in a nutshell! We are normally in a well-balanced equilibrium with our environment, and diseases arise only if the balance gets out of control or conditions are unfamiliar. This gives viruses an opportunity to replicate and make us ill.

The new millennium started with a surprise. Two scientific publications changed our view of the world. One showed that viruses make up half of our genetic material, our genome, all of our genes, and the other revealed to us the dominance of microorganisms in our body and around us. These publications were both based on a new technology which became available toward the end of the last century: sequencing, the determination of the sequences of large genomes such as the human genome. The first of the two papers, in 2001, described the determination of our genes consisting of 3.2 billion building blocks, the nucleotides. This was the result of a gigantic effort, with a multimillion-dollar input. Nobody could have imagined what our genome is predominantly made of. The answer is: viruses. Around half of the human genome consists of viruses — or at least virus-related sequences or truncated viruses, or viral fossils that have inhabited our genome for millions of years. Other organisms may even harbor viral sequences that constitute up to 85% of their genes. Where is the limit? 100% — ? We will discuss that. More surprising is the fact that these virus-like elements can move, they can jump, our genomes are constantly changing. And yet another surprise is that all genomes of all species on our planet are interrelated. We are all relatives at the genetic level: flies and other insects, algae or plankton, worms, even baker's yeast, bacteria, plants, fungi all the way to humans — and the viruses anyway — because they supplied many of the genes.

The consortium that had set out to compare many genomes in the Human Genome Project (HGP) presented this result, in one of the longest publications I have ever seen in the journal *Nature*.

Recently, new methods have allowed an estimate of the numbers of viruses on our planet. There are more viruses on Earth than stars in the sky: 10^{33} viruses, 10^{31} bacteria, "only" 10^{25} stars and soon about 10^{10} humans. We are the invaders in a world of microorganisms, and not the other way round. A gigantic number of microorganisms, bacteria, archaea, viruses and fungi populate our body and dominate in our environment. Bacteria and viruses are present in kilogram amounts in our intestines — yet without causing diseases. On the contrary, they help us to digest various — even essential — nutrients, which we otherwise would not be able to consume. They also cover our skin, mouth, vagina, toes, nails and birth canal, all with site-specific bacterial and viral compositions. This very surprising observation of the ubiquitous presence of microoganisms is the result of a recent large-scale analysis, sequencing of the human microbiome, in the Human Microbiome Project (HMP). In a way it is a follow-up of the Human

Genome Project (HGP). "Microbiome" is a new word that means the sequence of all microorganisms combined, without knowledge of the individual ones present in a given sample. "Do them all" is the principle. This second epoch-making paper was published in 2010, and since then the microbiome and its role in the human gut, in nutrition, in health and disease including such urgent issues as obesity and even autism and, surprisingly, depression and even anxiety have attracted much attention. An associated question is that of our food — what is healthy food? Is it the same for Japanese as for Italians? We do not even know! And the viruses come in here, too: Viruses also exist in astronomical numbers in the oceans, every salad dish is full of viruses — and they cannot be washed off, as they are inside the cells; yet they are harmless. Viruses are everywhere, and so are bacteria and probably also all other microorganisms — and this is not in the context of diseases. All this knowledge is new; it goes back to the beginning of our century, and we owe it to the new technology, the sequencing of genes — which has become millions of times cheaper and faster within the past ten years.

Humans are a superorganism, a complete ecosystem. Healthy humans comprise 10^{13} cells which are authentically human, our "self", and in addition we host about 10^{14} bacteria and, in addition, at least ten to a hundred times more viruses. Our genome, consisting of about 20,000 to 22,000 genes is augmented by more than several millions of genes, 350 times higher than the number of our genuinely human cells. The microorganisms reside in our guts and populate our skin — we may ask whether they should be removed by a daily shower — I would say no! They are useful and protect us from foreign ones.

Viral and bacterial sequences have even entered our genomes. This seems unbelievable. How much is left of us as humans? The challenge of explaining this to as many people as possible was the reason for me to write this book.

Bacteria have been called our second genome. This is generally accepted. We shall have to add the viruses as our third genome. And then there are also millions of fungi. Are they our fourth genome? And what about the archaea? Yes, they certainly contribute as well.

This ecosystem is not characterized by a constant war — not by killing, but rather by a ping-pong game, by a well-balanced co-existence, by co-evolution. The "war" vocabulary has to be abandoned. Things only get dangerous when we destroy the balance. In most cases humans cause diseases themselves. Viruses and bacteria behave opportunistically. They take advantage of unusual situations, of weaknesses of their hosts. This description is as far as I would go; the "war" vocabulary I do not accept.

Another novelty has been the discovery of giant viruses, the mimiviruses, the biggest viruses ever encountered, bigger than many bacteria. These viruses can even play host to smaller viruses. They also have some properties in which they resemble bacteria. They were not detected "mimicking" bacteria, which gave them their name "mimiviruses". Thus the borderline between viruses and cells is not a sharp one, and the world of viruses and bacteria is a continuum. All known definitions of viruses have become obsolete. How are we to define a virus? What about the transition between living and non-living matter?

After the Big Bang

The Big Bang occurred almost 14 billion years ago. It was the beginning of our universe — not yet of life! Since then the universe has been expanding. 4.5 billion years ago the sun emerged and our solar system arose through asteroids, rocks populating the universe. The rocks and gas gathered under the force of gravity, and its pressure fused them to form heavier elements. They collided and clumped together, forming the planets that circle around the sun. Today, most small asteroids do not survive the passage through our atmosphere and can be seen at night as shooting stars,

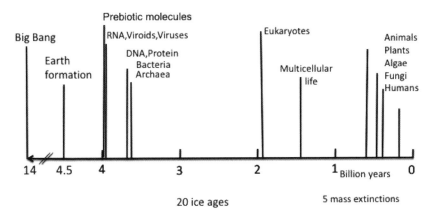

Timeline of the earth.

glowing and dying. Only the bigger ones made it all the way to the surface of our planet. Those asteroids that reached the earth delivered some of the elements known to us from the Periodic Table of the Elements. We are made of "stardust", which sounds quite poetic.

Oxygen and iron cannot have been synthesized so easily, just by collision; they needed an energy supply from the explosions of supernovae. I always wonder how iron got into the center of our haemoglobin — our "elixir of life"! As a real curiosity, I keep a piece of an iron meteorite in my *Wunderkammer* (a collection of curiosities including fossils such as ammonites, shells, petrified trees, amber and corals; unfortunately, I cannot collect fossil viruses, though I would love to do so). The iron meteorite once hit the Campo del Cielo (what a meaningful name!) in the north of Venezuela. The piece of iron is dark and very heavy, so I take it on trust that it is a meteorite (the material on earth arising from asteroids) and that it is really 500 million years old; however, I cannot prove that. Some people think it is magic and can cure diseases — I don't! Recently I added a can of Coca Cola to my curiosity collection — why? You will see.

We learn the names of the planets at school: Mercury, Venus, Earth, and Mars. Big Jupiter is further away, and its gravity prevented the clumping of asteroids, so that thousands of residual asteroids formed the Asteroid Belt. This belt is called a "habitable zone", because some scientists expect that there will be extraterrestrial life there. One such asteroid is especially interesting, Ceres, as in 2014 the space agencies detected clouds on it, suggesting it consists of water on the inside surrounded by ice. Water is an essential component of life as we know it today. So could there be life on Ceres? If we could find out, we would learn more about our own beginnings. RNA, viruses, bacteria — that is what I expect and would like to look for.

Where did the Earth's water come from? That is an interesting and still largely open question. Did "dirty snowballs" like Ceres bring water from extraterrestrial space to our planet, with the dirty dust helping the water to crystallize into ice? Can snowballs be so big as to account for all the water on our "blue planet", which looks blue from spaceships, because 2/3 of it is water? The astrophysicist Anna Frebel has calculated that each Coca Cola can today contains 5% of fossil water dating all the way back to the Big Bang, 14 billion years ago. I therefore added the drink to my collection of curiosities.

About 4 billion years ago, a star the size of Mars hit the sun and stripped off the moon, at least according to the leading theory. Since then the moon has been drifting away from the Earth by 3.8 cm each year. The moon determines our day and night rhythm, which has increased from 6 to 24 hours. The moon was essential at the beginning of life. Mars is too far away from the sun, while Venus is too close to the sun and therefore too hot. Jupiter is our bodyguard, because it attracts — with its 300 fold greater mass — the debris stones from the Universe and keeps them away from the earth. We have been lucky — so far. There is always the threat that one of these asteroids could hit the earth. In fact, in the year 2029 we are expecting Apophys to hit the earth. Some most recent calculations predict that it will hopefully just miss us!

Life on Earth began about 3.8 billion years go. That was about 10 billion years later than the Big Bang. The crust of the Earth differed in shape from today's and drifted on top of the Earth's molten interior. Until today, the tectonic plates have continued to move, and the distance between America and Europe increases by 2.5 cm annually. In Asia colliding plates built the Himalayas, and they are still doing so, causing terrible earthquakes even today. The plates can be detected around the world as the "rings of fire" of active volcanoes. At the contact points between the continents, the bottom of the oceans broke open, volcanoes formed and spouted out magma, which solidified quickly in the water at the bottom of the ocean. From there high chimneys formed, the hydrothermal vents or "black smokers". Black dust and smoke were released from these volcanoes at the bottom of the oceans. Water there can reach 400°C, because of the high pressure at these depths. Somewhere there, life started. That is almost generally accepted today. In the oceans, 200 meters deep and beyond, there is no sunlight any more, so it is not the sun that supplied the energy: rather, it was chemical reactions that provided the energy for early life. The beginning of life, without sunlight was based on chemical energy production. That was the motor for life. In the opinion of many scientists, it was there that the first biomolecules, such as RNA, arose. The building-blocks of RNA, the nucleotides, are rather complex. How could they have been first, was the counter argument of some scientists. One can imitate their generation in the laboratory today. Only recently a chemist in England, John D. Sutherland, produced all of the three major building-blocks of life (nucleotides for nucleic acids, amino acids for proteins and fatty acids for fat) in a single reaction vessel from very simple ingredients and chemicals: hydrogen cyanide, hydrogen sulphide, phosphate,

Hydrothermal vents or black smokers are volcanoes in the deep sea, where life began.

water (HCN, H_2S, P, H_2O) and energy in the form of ultraviolet light in this case. Some minerals speeded up the multistep "one pot" reaction, as stated by the authors. The prebiotic chemical reactions were supported by zinc and copper as catalysts. Thus, biological material could have formed even all at once. The beginning must have been simple — that is what I believe. This sounds simple.

Instead of Adam and Eve

Early life created its own growth conditions. It did not adjust to the existing environment, but created the necessary conditions itself, especially oxygen. Light contributed (by photosynthesis) to the increase of oxygen, which displaced the then dominant methane. The oxygen concentration increased and became the basis of breathing of mammals. About 2.2 billion years ago the earth froze to ice. Yet life already existed. Where did it hide? Next to the stove, the hot hydrothermal submarine vents? Did they protect life from dying out? It is also conceivable that life started not in the warm but in the cold; early life may have existed in ice crystals with fluid passing through channels, and some molecules may have led to biomolecules such as RNA. This theory has some adherents, because RNA has been found within ice crystals with fluid streams inside. Model RNAs are even under investigation after storage in deep freezers in the laboratory — does it change, mutate or evolve? In the proximity of the black smokers, life progressed rapidly: 2 billion years ago there were oceans with bacteria and unicellular organisms, colonies of algae grew, and 600 million years ago sponges, jellyfish and worms developed; 500 million years ago there were hard shells, corals and teeth. Then there came the so-called Cambrian explosion, the dramatic increase in numbers of all forms of life. The first vertebrates appeared. 300 million years ago the mottled fish *Coelacanth* appeared, which was thought to have become extinct but was then discovered accidentally by a museum's director in a fisherman's net in South Africa in 1999. Its genome was sequenced, and this made it possible to reconstruct the transition from fins to legs. This is a conclusion drawn from the genes! 250 million years ago the earth consisted of a supercontinent, Pangea, and a first mass extinction occurred which makes me wonder whether it could happen again.

65 million years ago a catastrophe killed up to 90% of all living species. This was due to an asteroid, a large meteorite, as big as the Himalaya mountains, which hit the peninsula of Yucatan in Mexico: Samples from drilling in the bottom of oceans, as collected by the marine research center "MARE" in Bremen, supplied the proof of this collision. A 15-cm-broad dark layer represents fire and ashes. This layer was designated as the K-T layer, indicating the transition period between the Earth's Cretaceous and Tertiary ages. This K-T extinction layer can be found all over the world. The sudden increase in amounts of the element iridium, termed the "iridium anomaly", is also regarded as proof of the hit by an extraterrestrial carrier. The asteroid of Yucatan resulted in a tsunami, 100 m high; the sun was darkened by dust, and the dinosaurs became extinct. Their death opened up the chance for other living animals, small ones, similar to mice, which were omnivores and ate everything without preference. That helped them to survive. They could hide underneath the surface in protective holes. Finally, we arrived on the scene! Five million years ago monkeys developed. Chimpanzees are our closest relatives, with a genetic content that is 98.4% identical to ours. What distinguishes us from them? Not only a bigger brain, but also special — combinatorial — features of our genes has increased

our complexity; this will be discussed later. 3.2 million years ago, Lucy lived in Eastern Africa, the oldest human ancestor. 1.7 million years ago, *Homo erectus* appeared, who came out of Africa. Perhaps he was able to make and use fire. Could he speak? He became extinct. Then, 200,000 years ago, *Homo sapiens* appeared, again out of Africa, and colonized Eurasia until 60,000 years ago. The Neanderthals lived from about 200,000 years ago until they died out 40,000 years ago. Did they die of hunger, pathogens, diseases, cold or climate changes? They left behind wall paintings in caves and some genes in our genomes. Here I dare to include a comment: I am not so surprised about this as some newspaper-columnists have been. Yet I do not believe reports that the Neanderthals suffered from depression which some humans still suffer from today. Who knows the sequence of a depression gene(s)?

12,000 years ago hunters and gatherers settled and became farmers. Then our ancestors started drinking milk, or at least the majority of them did. Some, like me in fact, have never adjusted to milk up to this very day. A mutation is required for its digestion. Milk protected humans against vitamin D deficiency — a selective advantage for those who live in darker Northern regions of the world. About 12,000 years ago the last ice age ended, and humans could walk on foot from Siberia to America. That was the beginning of our civilization.

Proximity between livestock and humans led to diseases transmitted from animals to humans, by the mechanism referred to as zoonosis. This remains until today the most frequent cause of infectious diseases: "bush meat" with monkey viruses ended up in the food chain of humans, causing HIV, and Ebola originates from fruit bats. During the past 2.5 million years there have been about 20 ice ages, and the warm intervals were more like intermediate periods between ice ages. One cause of these is the change of the orbit of the earth around the sun every 100,000 years. It makes it easy to remember the dates of the ice ages: 450,000, 350,000, 150,000 and 50,000 years ago. The last one ended about 12,000 years ago. The Earth's axis turns in cycles of 24,000 years and by doing so influences our climate. The next change, with an ice age, will come in 39,000 years. Who thinks of ice ages — the opposite, global warming is our present discussion. There are no known models in the history of our planet to explain rapid changes in the weather — but since there have been ice ages, global warming must have occurred before.

In the beginning were viruses

"A beginning must be small and simple." That seems logical, but it is also just a belief. Bacteria or archaea reach genomic sizes of around one million nucleotides — that is far and away too large for a beginning. This at least is what I think. It must have been smaller, much smaller, more primitive, perhaps a mixture of several molecules, biomolecules in Darwin's often-cited "warm little pond". There RNA, as the first biomolecule, may have been the beginning of the tree of life. The discoverers of the giant viruses strongly support the view that giant viruses were on the scene there first, but they may be biased — these viruses are too big for a start. The first RNA is already somehow a naked virus — more exactly, a viroid. This kind of RNA continues to haunt our cells today. (Details will come later.)

Viruses have been around since early on; no pipette in the world and no infection could have caused such a worldwide spread of viruses — they pervade every single living being, without exception. After all, there are about 1.8 million different species known on our planet and about ten times more are unknown, not even including bacteria. There are no exceptions — even though researchers, including myself, keep on looking for them. I have not been able to find a virus-free biological system — or perhaps a single one, the worm *C. elegans*? It has the strongest antiviral defense! Therefore, it must at least have experienced viruses in the past, because antiviral defense comes from viruses! As you will see....

What did Darwin think about the origin of life? He was careful, and mentioned in a letter to a friend — the botanist Joseph D. Hooker, Director of Kew Gardens in London — dated 1871: "It is often said that all the conditions for the first production of a living organism are present, which could ever have been present. But if (and oh! what a big if!) we could conceive in some warm little pond, with all sorts of ammonia and phosphoric salts, light, heat, electricity, etc., present, that a protein compound was chemically formed ready to undergo still more complex changes, at the present day such matter would be instantly devoured or absorbed, which would not have been the case before living creatures were formed." Thus the beginning cannot easily be repeated today. The environmental conditions on the primordial Earth are not known. But Darwin also said that we cannot exclude the possibility that all living creatures on earth have one single origin. As a former physicist, I share this simple thought with enthusiasm.

Recently I published an article with the title: "Are viruses our oldest ancestors?" (*EMBO Report* 2012). With the question mark at the end, because one cannot really be sure, the article was put into the category "Opinion". A reader soon sent me an e-mail, pointing out Félix d'Herelle, who together with his colleague John Burdon Sanderson Haldane will be mentioned in more detail later; both published essays, already in the 1920s about viruses as the possible origin of life. At that time d'Herelle had just discovered the phages, the viruses of bacteria. Immediately, the two researchers speculated that viruses, which can self-reproduce, are the "primordial origin" as Darwin called it. I share their vision. However, their contemporaries rejected this idea vehemently.

Every year the New York agent and publicist John Brockman asks leading scientists an annual question, "The Edge", about the knowledge of tomorrow. "What we believe but cannot prove: Today's leading thinkers on science in the age of uncertainty" was the question in the year 2005. My answer would have been: Viruses got here first!

This requires a definition of what viruses exactly are.

Looking back

What is a virus? First of all, the word "virus" in Latin means sap, slime, or poison.

A colleague — Eckard Wimmer from Stony Brook in the USA — studies one of the smallest human virus particles, the one that causes polio, which contains 3,326,552 carbon atoms, 492,288 hydrogen atoms, 1,131,196 oxygen atoms, 98,245 nitrogen atoms, 7501 phosphorus atoms and 2340 sulphur atoms. Because it is possible to set out this molecular-style description, he designates the virus as a chemical, at least as long as it resides outside of the cell. Inside the cell it is not really only a chemical any longer, since it replicates itself and multiplies. Such a separation into two life forms is quite unique. Are then humans also only chemicals? That cannot be the answer.

It was just 120 years ago that viruses were first transmitted experimentally, causing diseases. The filtrate of sick tobacco leaves was transferred to healthy ones, which in turn became infected. The discoverer was the Russian botanist Dmitri Ivanovsky in 1892. However, he always believed that he was looking at something related to bacteria. It was therefore the Dutch microbiologist Martinus Beijerinck who is credited with the discovery of viruses, even though he himself acknowledged Ivanovsky's work. Beijerinck coined the word "virus" to distinguish them from the larger bacteria, which cannot pass through the filters, the so-called Chamberland filters, where only the small viruses run through. In animals, Friedrich Loeffler and Paul Frosch discovered almost at the same time — in 1898 — a transmissible small agent causing foot-and-mouth disease in cattle. The virus infects cows and is extremely contagious, and for that reason the first research institute for studying this virus was founded on a peninsula in the Baltic Sea. However, the wind was enough to spread the virus even from there. This research institute, named after the two aforementioned pioneers the LFI, is the biggest of its kind in Europe. Its reopening a few years ago attracted so many curious people that finally nobody could get there because of the traffic jam. The sterilization chambers (autoclaves) there are big enough to disinfect cadavers of whole cows.

Until recently it was taken for granted that all viruses are small, are nanoparticles, can only be detected in the electron microscope, cannot be kept back by filters, and contain either RNA or DNA often within symmetrical protein structures such as icosahedra; they do not replicate by themselves, they are parasites, they need cells within which to replicate, they cannot perform protein synthesis and they need energy from the cell. They are mostly specialized to living in certain hosts, and are sometimes covered by a coat that can be derived from the host cell and which often also carries receptors for binding to specific host cells. They are pathogens, cause diseases, are dangerous, steal their genes from the host, betray and abuse their host cells for the benefit of their own progeny, use disguises and hide in Trojan horses. In short: Viruses are enemies.

In recent years we have found out that almost all of this is wrong. Viruses are not *only* small; they can be bigger than many bacteria. Viruses themselves can be hosts of viruses, can be much bigger than nanoparticles — or even much smaller; in fact, they are not always particles! They can have sizes varying by a factor of 10,000 — a very broad range — they have very different morphologies, about a dozen different types of genomes, and a variety of totally different replication strategies. The number of genes that a virus can have reaches from zero (!) to 2500 — for comparison, humans have 20,000, only ten times more. "Zero genes" are present in viroids — though these are not generally accepted to be viruses. There are viruses that consist only of nucleic acids, without proteins, or (the other way round) only proteins without nucleic acids. The latter are the prions, which are often not considered as viruses either, but I would like to include them as well. There are viruses that have only foreign genes and none of their own, such as the very exotic plant viruses, poly-DNA-viruses (or polydnaviruses PDVs), a fact that may tell us something about evolution. Then there are the endogenous viruses, which never leave their host cell, and the rudimentary viruses that jump around in our genomes. These two types of virus do not have coats and are therefore locked-in viruses, unable to move between cells.

Viruses are mobile (genetic) elements — is that perhaps a useful definition? Viruses need energy, yes, but not necessarily from a host cell. Chemical energy will do, and that can come from around the black smokers, where life may have started and where no sunshine ever reaches. Viruses need niches, compartments, clay as catalysts — Darwin's warm little pond — so that the concentration of components can be high. Such a first kind of containment could have been lipid bags, and one can ask whether this was an early virus or an early cell. There was initially no sharp boundary between viruses and cells — rather, they together make up a continuum. Especially the newly discovered giant viruses break taboos, because they are almost bacteria; they even have a hallmark normally only assumed to exist in bacteria: components for protein synthesis. This is often used as a definition of life, the ability to synthesize proteins. Thus these "almost-bacteria" represent a transition between viruses and bacteria, between lifeless and living. The discovery of giant viruses has revolutionized our view of viruses and has shifted viruses more towards "life" than had previously been assumed. A minimalistic definition of viruses includes their inability to perform protein synthesis, which is regarded as a hallmark of life. But the giant viruses can "almost" synthesize proteins — after all!

Viruses are found wherever life is. Viruses can take up and deliver genes, can mutate, recombine, insert, delete, or mix genes. Their replication is error-prone and therefore innovative for the virus and the host. Tumor viruses can pick up genes from the cell and mutate them during replication, which can increase their oncogenicity. But the opposite is also true: they can deliver genes to the cell, supplying new features, sometimes beneficial and sometimes detrimental. They can bring oncogenes into a cell and cause cancer, or they can introduce genes to cure cancer. More genes go into cells than come out. Viruses do not cause "wars" or lead to "crossing swords" or "arms races"; these negative descriptions are inadequate. They play ping-pong with their host. Horizontal gene transfer, between microorganisms and all other living hosts have led to complicated genomes. This is how our genome became such a colorful mixture of genes from very different other organisms and other genes. Every organism has a complex number of genes taken over from many other organisms, most frequently in fact from viruses. The viruses have by far the largest repertoire of genes, the largest sequence space available on earth — most of it is not even used. Viruses have a higher variety of genes than cells have, supporting the assertion that viruses were first on the scene, earlier than cells (more about that later).

How far back do we know about viruses? Let's go backwards. 35 years ago HIV started to invade the human population, so far causing more than 37 million deaths. 100 years ago the influenza pandemic during World War I killed perhaps up to 100 million people. Measles killed the Mayas after being imported from Europe by the *conquistadores*. During the Middle Ages plague bacteria killed one-third of all Europeans, about 25 million people. Some 600 years earlier, in 542, the "Plague of Justinian" devastated Rome and spread as a pandemic around the Mediterranean Sea to Constantinople, at its height killing 6000 people there each day. Thucydides described an unknown disease in Athens during the Peloponnesian War in about 400 B.C. which could have been caused by Ebola, pox, measles or other viruses, or pesti bacteria. An Egyptian Pharaoh must have suffered from polio virus 3500 years ago as can be judged from a crippled leg shown on a gravestone. Retrovirus-like elements existed in Neanderthal Man, who lived between 250,000 and 30,000 years ago, after which the Neanderthals became extinct. Then there is a gap. A great surprise was the detection of an HIV-like virus in rabbits, RELIK, dating from 12 million years ago.

Other HIV-like viruses can be dated back 4.2 million years, in lemurs (relatives of monkeys) on the island of Madagascar. Nobody had anticipated that HIV-like viruses had been around for so long and can even be inherited.

A new field of science, paleovirology, has been a hot research topic at Princeton and in London throughout the last ten years. Sequences from Ebola virus 50 million years old have been discovered in the genomes of bats, pigs and monkeys, while Bornavirus sequences have been found in humans but not in horses. Only the horses get sick with Bornavirus, while humans do not. Thus, endogenous sequences and their products protect an organism against the corresponding viral diseases. These RNA viruses should not normally be integrated into DNA at all, but they are — by "illegitimate" mechanisms using some cellular-molecular tricks such as a foreign reverse transcriptase. Even our human placenta we owe to relatives of HIV, the human endogenous retrovirus, HERV-W, from about 30 million years ago. Human endogenous retroviruses, which can be found in our human genome, are estimated to date back 35 to 100 million years. Some of them are intact viruses, which can form particles, yet normally are no longer infectious. Endogenous viruses are probably much older than we can judge, because they cannot be recognized as viruses any more. A dinosaur, now in the Natural History Museum in Berlin, suffered 150 million years ago from a virus infection caused by a paramyxovirus similar to measles virus, *osteodystrophia deformans*, which led to bone deformations, a disease still in existence and known as Paget's syndrome.

Back to about 200 million years ago we can witness viral footprints, but there our journey into the past ends. Virus information disappears in the genetic "background noise" due to mutations. Endogenous retroviral fossils can be detected as proof of viruses. The newly rediscovered fish *Coelacanth*, which was assumed to have become extinct, has been around for the last 300 million years; it is genetically surprisingly stable and it also harbors retroviral fossils.

There are tricks however, that lead to even older clues to early viruses. The giant viruses can be found in today's amoebae, but also in macrophages, two lineages which diverged from each other 800 million years ago and developed independently, and which are therefore both thought to have been infected already before they diverged. Further evidence going farther back than 800 million years is almost unobtainable. Yet that leaves an enormous gap back to the origin of life, about 3.8 billion years ago. Viruses probably belong to the oldest biological fossils known. A real surprise are the viroids, which are virus-like structures and present till today — not only as such, but also as ribozymes or relatives of circular RNA in all our present-day human cells. They date back to the epoch when there was no genetic code — maybe 3.5 billion years ago. In a scientific publication I once tried to reconstruct the evolution of life on the basis of today's viruses. The article's title was "What contemporary viruses tell us about evolution" — and the editor added "A Personal View", to be on the safe side! (*Archives of Virology*, 2013)

When the human genome was sequenced for the first time and published 15 years ago, the *Frankfurter Allgemeine Zeitung* (FAZ), printed a whole page filled with only 4 letters: A, T, G and C, the alphabet of life, without any interruption, no words, no sentences, no paragraphs. The page was awarded a prize. It indicates exactly what we know about our genes, just the letters! Almost all the rest is still waiting to be understood. The "text analysis" is ongoing. What do the letters mean? There are about 3.2 billion of them in the human genome, corresponding to 20,000 genes; however, the genes are encoded in only 2% of the whole. What is the "rest" for, the majority of the letters? Is it also genetic information, or is it the often-quoted "junk DNA", or what? I will already let the cat out of the bag as to what the rest is: mostly information for regulating the expression of the genes themselves. To understand the details will keep scientists busy for perhaps the next 50 years. The project is known as "ENCODE": Encyclopedia of DNA Elements.

Here are a few numbers worth remembering: Viruses such as HIV have about 10 genes, phages have 70, bacteria 3000, humans 20,000 to 22,000 and bananas have 32,000 — what, more than humans? Yes, surprisingly! Yet, bananas are not smarter than we are. This was even once called a paradox: the sizes of the genomes or the number of genes do NOT correlate with the complexity of a species. Humans do not have the greatest number of genes, but they have the longest genes and most importantly, these genes can be much better recombined (by splicing; see the next section) to increase their overall complexity, overtaking in this respect all the other known species. Finally: one gene of a virus is made up of about 1000 nucleotides.

Before we go on, every reader has to learn two words, or at least their abbreviations: RNA and DNA. One can just memorize them, together with some extra information. RNA and DNA are large molecules, the carriers of genetic information organized in regions as genes. The primary genetic information is normally encoded in DNA; only viruses can also use RNA, or even mixtures of RNA and DNA, as primary genetic information. DNA is called the molecule of life. It is known to everybody as the double helix, resembling as it does a circular staircase with two handrails (strands) connected by horizontal bars like stairs (stacked bases). This structure was discovered by James D. Watson and Francis Crick in 1953, then young, adventurous and ambitious scientists in Cambridge, UK, "never in a modest mood" — at least that is the first sentence that Watson used to describe

Crick in his famous book *The Double Helix*. They wanted to win the Nobel prize and they did. Important information also came from Rosalind Franklin, who produced the structural data by X-ray diffraction pattern analysis, which she tried to hide. Did she really tell them that their model was wrong, that it had to be turned around, inside out? Watson himself describes the discovery in his book, a bestseller. A new theatre play deals with Franklin's picture, *Photograph 51*, a detective story by the American playwright Anna Ziegler on how Franklin's X-ray picture contributed to the discovery without her ever knowing about her important contribution. She died of cancer as a consequence of her experiments with X-rays, sadly so, as she was still young. Less often mentioned is Franklin's head of department, whom she did not accept as such, and who shared the Nobel prize: Maurice Wilkins. He received the starting material, lots of pure DNA, from a Swiss colleague who handed it out generously. Wilkins used it as an essential source for crystallization. Later, he came under the political spotlight for possible involvement in communism, and is less widely recognized.

DNA is double-stranded, whereas RNA is single-stranded, more flexible, rope-like; it undergoes variations more easily and is very important for new sequence discoveries by viruses. Crick formulated the "central dogma of molecular biology": from "DNA to RNA to protein" describing the flow of genetic information inside the cell. Some people say that Crick was not dogmatic in his thinking but anyway his name has become attached to the dogma. DNA dominated the thinking of molecular biologists for half a century, but now RNA is catching up in importance. RNA came earlier in evolution, before DNA, so the reverse of the dogma is also true: RNA can turn into DNA. This is what we have learnt from the viruses. So, dear reader: more molecular biology is — almost — not needed. Many details can be skipped and some more details are listed in the Glossary.

A sailor and splicing

While I went on a sailing trip across the Baltic Sea on the three-masted schooner *Lily Marleen*, a sailor surprised me one morning with a present: a "spliced" rope. He had connected two ropes by "splicing" them together as a demonstration-piece for the students in my virology lectures. Splicing requires some skill, he said, and it kept him entertained during the boring night watches. A book entitled *Splices and Knots* teaches the art of these

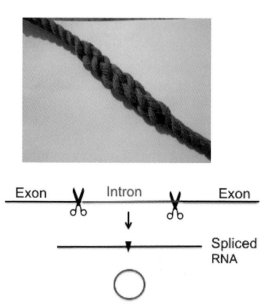

Knot-free splicing (cleaving and joining) of RNA allows combinations of exons to build new genes with high complexity, intron is deleted.

as used in sailing. A book with the same title could also be important in molecular biology. A splice is in a sense the opposite of a knot. Only when a rope is spliced can a sailor pass the connected pieces, quickly and without any hindrance, through the pulleys to pull up the sail; a knot would provide an obstacle to this. The same principle holds true in the molecular world. Double-stranded DNA is transcribed into single-stranded RNA, a flexible copy of the stiff DNA. The RNA — like a rope — can then be shortened by splicing: the rope is cut, a piece removed, and the ends that this produces are connected without knots — by splicing. The length of the pieces removed can differ, and now I can let another cat out of the bag: humans have the highest numbers of such removable portions per gene. That makes us more complicated than any other living animal. In this lies our uniqueness; perhaps we are after all the "crowning achievement of creation".

Our genes are split, consisting of *exons* — regions that can be translated into proteins — and *introns* — regions in between, which are removed by cutting and splicing (rejoining) of the RNA ends. Human genes have on average approximately seven to nine exons, interspaced by introns. Think of a garden fence, with its planks and the spaces between them: these would correspond to the exons and introns respectively, although exons and introns are less regularly spaced. Furthermore, some introns can be very large (imagine a gate in the fence).

One can combine the various possible connections between introns and exons, whereby the introns are normally the parts deleted. The introns are not really "empty", as gaps in the fence are. What information do they contain? They do not code for proteins, but direct and regulate the production of the proteins; they harbor regulatory information, determining where and when the proteins are to be produced. Thus the introns control the exons. Scientists used to think that the exons are more important, but today it is recognized that they are even the dependents. According to the scientific definition, exons "code" for proteins, whereas introns are "noncoding" (nc). The catalogue of known ncRNAs is a rapidly growing family of very important regulatory RNAs — almost a dozen of them were discovered recently. Therefore, dear reader, please remember "ncRNA". The corresponding DNA is called ncDNA, which is transcribed to give the ncRNA. I find it particularly interesting that humans and the viruses are the joint world champions in splicing! Why the viruses? Because they make so much out of so little! (Our ancestors!) The sailor did not know how complex the background was behind the spliced rope that he gave me as a present.

As an example of generating many "messages" from a single all-embracing "message", think of how many words you can generate from combined exons after splicing across any gaps: *supercalifragilisticexpialidocious* — not the longest English word, but the best-known for its length — just by deleting (but not rearranging) letters and thus created: super, supercilious, perfidious, precious, serious, superficial, fragile, pallid, series, focus; there are many more, and you may care to try for yourself. Viruses splice only within a word, as their minimalistic equipment comprises exons and no introns. In these different ways, we and the viruses have learnt to make the best possible use of our genetic material. This complexity lies at the heart of the complexity of the human organism, and that of the viruses, which is why I was doubly grateful for the sailor's present.

Viruses — dead or alive?

Viruses are not lifeless — at least not as lifeless as a stone or a crystal. One can over-simplify the issue and say that anything smaller than a virus is lifeless, and what is larger is alive. So viruses are at the borderline: dead or alive or both. I do not see a singularity, no point, no sharp borderline, but rather a continuous transition from individual biomolecules all the way up to the cell. At the origin of life RNA viruses were around as the largest biomolecules, and from then on they have always been present.

"What is life?" This question was asked in 1944, in the title of the famous book by the physicist Erwin Schrödinger, a thesis that mobilized a whole generation of physicists into doing biological research. Life follows the laws of thermodynamics and energy conservation. Living cells are characterized by negative entropy based on organized structures, whereby entropy is often simply described as a "measure of disorder". For example, left to itself my desk becomes more and more untidy; however, if I can muster enough energy to clear up, then it becomes ordered and tidy. Life and the second law of thermodynamics follow this rule: nutrition and energy allow an orderly life. Admittedly, Schrödinger was asking about the laws of life, not the origin of life.

I think that NASA must have a good definition of life, because NASA is trying to find life outside our planet. They surely know what they are looking for. Jerry Joyce, then at the Salk Institute in California, may have contributed to their definition when he succeeded in producing self-replicating RNA in a test-tube, RNA that was also capable of mutating and evolving. This was his approach to "repeating the origin of life". He may have inspired the US space agency NASA, which formulated a definition of life as a self-replicating system containing genetic information and able to evolve. (*I would even omit the word "genetic", because structural information able to evolve would also be possible. I am thinking of viroids.*)

Viruses can be compared to apples. An apple on a table cannot duplicate itself and turn into two apples — and a virus cannot do that either. An apple needs earth to become an apple tree and thereby to produce new apples. Are apples alive? What then of viruses? Can Charles Darwin help? He pictured a "warm little pond" as the place where life possibly originated, and imagined that the beginning was simple — but he predicted no more than that. A virus needs a pond, or at least a test-tube — an environment with nutrients for replication and for the production of progeny. And viruses are simple. So viruses are more alive than stones and only stones are really dead. Surprisingly, some viruses can aggregate and form symmetrical quasi-crystalline structures, which are extremely stable and heat-resistant, and in that respect they do indeed resemble stones. Crystals with malformations can even perpetuate their misfolding in a manner that looks almost like replication. Some protein aggregates in the brain can behave like this — prions for example; so are they then also something like viruses? I suggest that, as we shall see!

Bacteria are generally accepted as living microorganisms: they can divide and thus replicate themselves, and — most importantly — they can synthesize proteins. Protein synthesis is accepted as an important borderline marker distinguishing living from non-living matter. Bacteria also need food from the outside; yet they are not completely independent entities either. And bacteria are not simple! There is no such thing as a biological "perpetual-motion machine", something that can get away without an energy supply. Yet energy does not necessarily have to come from a cell. Energy can be chemical energy, without any sunshine around, as in the case of the black smokers in the depths of the ocean.

Most surprisingly, the recently discovered giant viruses contain components for protein synthesis. So they are very close to the living bacteria, they are "quasi-bacteria". Accordingly, the giant viruses are also called "mimiviruses", because they seem to mimic bacteria. Just like bacteria, these giant viruses are hosts for smaller viruses, which replicate inside the bigger ones. All this was extremely irritating for classical virologists, because no previous convention or definition of viruses fitted for these giant viruses. Their discovery in 2013 was commented upon in the journal *Nature* to the effect that viruses now qualify to take a seat at the table where life is being debated, and that they should be placed at the bottom of the tree of life, this is what the discoverers of mimiviruses hoped! At the bottom there were no cells yet, and there are no mimiviruses — both are huge in comparison with viruses, therefore both cannot represent the origin of the tree of life. Probably, the early viruses did not need cells. This is a risky idea, and the only complication in my speculation that "viruses came first". Today's viruses need cells, but that could have been the result of long evolution. Indeed, there are the viroids, naked RNAs, which can replicate and evolve and may initially not have depended on cells as they do today. They are able to do all this in Joyce's test-tube — no cell. They could be termed "naked viruses".

Viruses are inventors and suppliers of genetic innovation. They built up our genomes. This is what I believe, and I shall mention it more than once, my credo, my *ceterum censeo*.

Viruses have certainly contributed to building cells. This is indeed a hard fact and no speculation. Today they are parasites and depend on cells. A parasite can delegate functions to its host and get away with fewer genes than if it is alone or has to survive outside of a host cell. Today, we detect only cell-dependent parasitic viruses. Evolution has not only proceeded from simple to complicated structures; it has also gone in the opposite direction. Complicated systems can become simpler: they can lose genes, delegate functions and become specialized. Depending on the environment, abilities can be acquired or given up and lost. Mitochondria are an example. Wait for the last chapter of the book!

How do viruses interact with their host cells? There are host cells without nuclei, the prokaryotes, comprising the bacteria and the archaea, and cells that contain nuclei, the eukaryotes, comprising insects, worms, plants, mammals, etc. These all harbor viruses, whereby the bacterial viruses have an extra name, bacteriophages or just phages. However, there is no need to differentiate between the viruses and the phages. They behave very similarly in their host cells. There are a few possible characteristics of their "life cycles" — or replication cycles: viral entry for infection, whereafter the virus can persist, integrate, replicate and/or lyse. Persistence — often unnoticed by the host — is a chronic or long-lasting state. Herpesviruses hide this way in neurons, where they can rest for years. Many plant viruses persist for ever, never acquiring a coat, never becoming active (or virulent), and always propagated together with the plant cell. Phages can persist as integrated prophages in cells; this is referred to as a lysogenic state. Also retroviruses and some other DNA viruses can integrate into the DNA of the host genome. The host then owns a few more genes. However, integration can also cause genotoxic or mutagenic effects and be harmful to the cell. Phages and viruses can lyse their host cell, setting free thousands of viral progeny, often as a response to stress, which is similar to our experience of stress: no space, no food! A dentist's appointment can have the same effect, then herpeviruses crawl from their niche to the lips and cause a lesion.

Remember this general rule: foreign invaders can lead to integration or destruction linked by stress — this is true even for societies!

Will viruses destroy their hosts, and lead to the end of mankind by killing us all? No, that is a fairy tale and will not happen. It would be nonsense from the point of view of evolution, because then viruses would eliminate the basis for their own existence or "survival" and die out themselves. When most host cells have disappeared, then they are so scarce that the virus will never find the last ones. So if there is a shortage of hosts then viruses adjust to new types of hosts. This is the dangerous "zoonosis", in which humans become infected by animal viruses that they have not encountered before. Before all hosts are eliminated, the viruses will find an arrangement. This is a transition from parasitic behavior to a co-existence — often mutualistic, that is, benefiting both the virus and the host. If the virus supports survival of the host then it increases the chances of surviving both for itself and for its progeny. Co-evolution can lead to less aggressive or less virulent behavior. That can happen in two ways: either the host develops greater resistance, or the virus becomes harmless. The latter can be achieved by endogenization of viral sequences into the host genome — our genome is full of them, a graveyard of former viruses. Endogenization will be discussed below.

Many viruses have become less harmful for their hosts during evolution. Ebola viruses, for example, have developed such an arrangement with bats (their principal host), and SARS viruses have done the same. Similarly, the equivalent of HIV in monkeys (SIV) does not cause diseases in monkeys any longer. So, if we wait long enough, shall we also have a friendly relationship with HIV? My prediction is yes, but we would have to wait for so long.

2 Viruses — how they make us ill

Viruses wrote history

One of the biggest success stories in medicine was the elimination of poxviruses. Poxvirus outbreaks should not be possible any longer, since humans have learnt how to vaccinate against the virus: cowpox prevents smallpox — that was the discovery of Edward Jenner, who first tested this vaccine with his son in 1796 and saved more lives than anybody else in the world has done. Jenner discovered the principle about how to generate a vaccine — use of a related virus causing a mild disease in humans, which protects against the dangerous virus. A cowpox instead of a smallpox virus is such an "attenuated" successful vaccine. The virus is considered extinct. Surprisingly there are occasionally still some pox outbreaks — viruses never disappear completely.

There are only a few laboratories left, in the USA and in Russia, where smallpox viruses are kept under safe conditions. How safe are they against abuse by a bioterrorist attack? This possibility seemed to become real after the anthrax bioterrorism in the USA in 2001, which caused some deaths and resulted in a hectic search for pox-vaccine leftovers, which had been stored for decades. They had been produced on animal skins and by no means corresponded to any acceptable safety standards, but they were then prepared as vaccine again after fivefold dilution to increase the number of doses! In my diagnostics department in Zurich we quickly developed a pox virus test. The necessary information about the viral sequence was freely available to everybody through the Internet! We even practiced how to react in the case of a pox virus alarm, how to prepare samples for a highly sensitive diagnostics method. The fear was a bioterrorist attack against the participants at the annual World Economic Forum at Davos. We dressed up in special safety coats including masks, which many people have seen in the movie "Outbreak" starring Dustin Hoffman. We practiced with a low pressure chamber, under the strictest safety conditions, which would prevent viruses from escaping to the outside — however, the chamber collapsed owing to someone opening the wrong valve by mistake, leaving us with a pile of wreckage. We were lucky, and never experienced an alarm. The fear was slowly forgotten worldwide!

Twenty years earlier I had witnessed the last pox alarm in Berlin. A patient was kept in a quarantine building with policemen sitting outside on high chairs to keep the exits in view and prevent the patient from escaping. Meanwhile, next door at the Robert Koch Institute, the only experienced technician who still knew how to test for poxviruses was busy opening chicken eggs to inoculate the virus and to incubate it for propagation — the only test available in those days. Pox alarm is shown in the movie "The Virus Empire — Silent Killers". This very informative movie was produced by experts for teaching purposes, and I always showed it at the end of my virology teaching series at the University in Berlin; the series ended on Sundays — and was still attractive enough for the students to come. In the second part of the movie a fictive poxvirus alarm in Berlin is demonstrated.

Another virus, frightening us to this very day, is the influenza virus, the Spanish Flu, called H1N1. About one hundred years ago during World War I influenza viruses killed at least 20 million and possibly up to a hundred million people. The virus was only recently isolated — in 2005, from soldiers and an Inuit woman buried in the permafrost in the Alaskan tundra — and reactivated in the laboratory. The recovered virus was even able to infect animals. This was technically very demanding and in addition also extremely frightening. Rightly, the public media complained about it. Scientists wanted to know why this particular influenza virus was so deadly, especially for young men. There were only few special virus sequences which may have increased the virus' affinity for lung cells and increased pathogenicity. (*Other changes are in the polymerase or the nucleoprotein or the hemagglutinin, so there is still some controversy about which of the changes are deadly.*) Major reasons for the pandemic were the war, hunger, humidity, cold temperatures, wounds, lack of hygiene, overcrowded shelters and field hospitals — all factors that cannot be deduced from the viral sequence, of course. All these factors contributed to the catastrophe. We also have to blame ourselves for it.

In 2009 the swine flu started in Mexico, caused also by influenza A H1N1 but distinct from the Spanish Flu. The World Health Organization (WHO) graded the outbreak as worldwide risk and declared it as an epidemic. This was due to a miscalculation. The ratio of people who died relative to the hypothetically infected people was wrong, because nobody knew the real infection rate in a country such as Mexico, where people do not go to see a doctor just with fever. The death rate was 5%, and not 50%, of infected people, no more than in a normal seasonal epidemic, so the alarm was a false one. However, this was a real pandemic because of the high number of countries affected. Safety measures had already been taken quickly and a vaccine production initiated, but it came too late for the Western world; the wave of infections was already flattening off. People in the Southern Hemisphere did not even want the vaccine free of charge. Nobody there took the swine flu seriously. I got infected in China, possibly in an internet café in Shanghai. I was rather sick back home and cancelled my flight from Berlin to Zurich out of fear that I could infect someone and thereby make newspaper headlines as a professor of virology who spreads the virus. I indeed had the swine flu, as verified by my own diagnostics department.

The influenza viruses responsible for the bird flu only became dangerous through manipulation in the laboratory by scientists. Researchers produced from a bird-only virus to a virus infectious for humans. The necessary mutations were even introduced into the viral sequence twice, in two independent laboratories in the USA and Holland. Why do scientists perform such risky experiments? This question was only raised when the scientists were naïve enough to publish their studies. Only then did the research funding organizations begin to ring the alarm bells. A mandatory break, a moratorium for 6 months, was imposed on the studies and their publication. This lasted longer than ordained, and the ban on publication was then softened: details had to be omitted, so that not everybody could repeat the experiment and convert a relatively harmless virus into a dangerous one.

A moratorium with self-restrictions had happened once before, with limitations on the use of recombinant DNA technologies — the construction of new genes by combinations of gene fragments — at the Asilomar conference in 1975, as well as restrictions on gene therapy of human diseases using viruses against cancer. Even today, viruses are still not allowed to replicate when applied for therapeutic purposes in order to prevent the possibility that replicating viruses might infect the germ cells of a patient, which in turn would open up the possibility that the virus could then be transmitted to the next generation. This restriction is strictly fulfilled and accepted. As a consequence, gene therapy is safe — but it is also ineffective for exactly the same reason. It would be much more effective if virus replication were allowed. Other approaches are being pursued now.

The prohibition of the influenza viruses studies can be summarized: "no dual use", which means, that publications are not allowed to serve two potential purposes, scientific ones *and* also bioterrorist interests or other abuses. The results on the manipulated influenza viruses were published by omitting technical details, and they were not unimportant, because they showed that four mutations out of 13,500 nucleotides were sufficient for the virus to become "humanized", to be transmitted from people to people. This is always the main threat. Surprisingly, certain influenza strains already carry three of the four mutations — so we are only one mutation away from a dangerous virus in the wild. The danger is real, and therefore a worldwide surveillance system has been installed, the Sentinella survey of local influenza outbreaks. From this study the annual influenza strains are predicted as a basis for vaccine production for the coming winter. Vaccines are still often produced in chicken eggs, one egg per dose, which requires billions of "special pathogen-free" (SPF) eggs. There are essentially only two drugs against influenza, very few compared with HIV: Tamiflu and Relenza. Tamiflu became a blockbuster and was sold to panicking governments around the world. It is now stockpiled in many storerooms and waiting to be aliquoted when needed. There is a rather strange law according to which only the day of aliquoting is the basis for calculating the expiry date of the drug. In Scandinavia resistant viruses have already shown up, and in Japan Tamiflu seems to lead to increased suicide rates among young people. Influenza should not be underestimated. I caught it, not just a cold by some rhinovirus, and was so ill and semi-conscious that I did not even remember that I had Tamiflu in my refrigerator for exactly this possibility. It is only effective if taken early after infection, for simple reasons, because then the virus load is still small. By the way, paper handkerchiefs should not be thrown into a paper basket next to a desk, but into a bin with a lid, and even an irreplaceable secretary should stay home instead of spreading the virus at work.

There is a broad virus-monitoring system, the Global Viral forecasting initiative (globalviral.org). Google also participates in forecasting, in a surprising way: it is assumed that users of the Internet google "influenza" more frequently if it is spreading, "Google Flu Trends" has predicted the arrival of waves of influenza reliably for more than 100 cities in the US, weeks in advance. Very clever!

Ebola viruses had theoretically for a long time been considered a potential danger. However, in 24 outbreaks during 1976 and 2013 it never became epidemic, with about 1500 fatal cases in total. The outbreaks stayed local and small in West Africa. Yet they were frightening enough for people to flee; this sometimes even included health-care workers, who are at greatest risk because almost every second patient dies of hemorrhages, internal bleedings. People were thought to get so terribly sick, that spreading would be impossible. That changed in 2015. Mobility had increased and markets, schools and other crowds contributed more than before. The epidemic spread through three countries — Guinea, Sierra Leone, and Liberia — with 11,000 deaths reported out of almost 30,000 cases. There is a lack of hospitals, and the families are used to care for the patients, thus exposing themselves to the disease. The funeral traditions, with highly infectious corpses and body fluids were always blamed as a major source of contamination, but this may have changed in the meantime by educating people. There is no therapy except supply of fluid by infusions, which however requires sterile needles. During quarantine, which may last up to 30 days for the infection to manifest itself, people were afraid of attracting the disease there. Those who recover are resistant and could help the others. Even their blood was tested for the presence of potentially protective antibodies. Vaccines are now under investigation that were available in research institutions but too expensive to develop — except under new pressure. The virus is spread by bats and bushmeat, whereby the carriers do not have the disease. Carrier animals comprise also dogs, pigs and perhaps rodents, which can be found at many locations worldwide and may be infectious. Endogenization of Ebola sequences has been shown to be a characteristic property of healthy carriers and potential transmitters. Endogenization means the presence of viral sequences in the genome of the animals (see below).

Unexpected was the observation that Ebola viruses can hide in reservoirs like the human brain and caused severe encephalitis several months after recovery of a patient. This was observed for the first time with a nurse who was thought to have survived Ebola but came down 6 months later with viruses in her brain. Viruses can also last for four months in semen of men who have recovered. The newly developed vaccine may become important in the future.

Germany has a certain tradition with Ebola, since it is a close relative of the Marburg disease virus. The Behring company — close to Marburg, Germany — experienced some transmissions of the virus to animal keepers by imported apes in the 1960s. About 30 people became infected and one-third of them died. Then, 40 years later, the television host Gunther Jauch invited survivors and scientific specialists to take part in a talk show. First of all we had to learn to applaud loudly, and long enough, when the host entered the scene. Then there is always the discussion about bioterrorism with this virus; however, since nobody has a protection or cure yet, this would be dangerous for the terrorists themselves and not attractive for bioterrorism. Based on the tradition of Marburg there is now a safety laboratory at the highest Biosafety Level (BSL4) for Marburg and Giessen Universities. Also, the Robert Koch Institute in Berlin opened a new high-containment laboratory in 2015. There are only about half a dozen such laboratories in the whole of Europe. So Germany has a big chance to contribute to knowledge about newly emerging dangerous pathogens and diseases. It is known that viruses rarely disappear completely, therefore we may have to be aware of viral "come-backs".

SARS (severe acute respiratory syndrome) is a disease which also requires laboratories working at the highest safety level. In Hong Kong the coronavirus, which causes SARS, has escaped more than once from a research institute. Most surprising was the press release in 2014 that thirty containers with SARS samples had disappeared from the liquid nitrogen tank, the storage place for the samples in the safety laboratory in Paris. They were not declared as such for safety reasons, so that nobody would know about the dangerous contents. I guess what happened is that someone simply needed space in the tank and cleaned it up. Tanks are notorious for always being full and often ancient samples reappear, years after they have been put there. Sometimes this opens up new research fields, as in the case of a retrovirus with a then new oncogene Jun. Even in the refrigerator of the seminar room which I once cleaned up by myself, I discovered a pot of margarine which had survived there unnoticed for seven years in spite of daily use of the place. Probably the samples from Paris ended up safely in the steam sterilizer according to the rules — but were lost for research purposes. Even in my safety laboratories in Zurich we discovered a hole in the wall — which was not really dangerous because of low pressure in the room and sterile cabinets — but still, what an embarrassing surprise.

One late night a nurse came to my office in Zurich, her arms full of sample tubes. A pilot from the Philippines, potentially sick with SARS, was stationed in the hospital. This rang an alarm bell. First I called a colleague at the Institute of Tropical Medicine in Hamburg for advice. I had to wait, until he returned around midnight from a TV show on SARS. He then dictated a list of reagents, some of which had just that week become available in Berlin, primers for a sensitive laboratory test (a polymerase chain reaction, PCR, which I shall explain below). I thought that that would simply take too long for a high-risk patient. I suggested flying to Hamburg with the samples and analyzing them there. The transport was made possible, even without my accompanying the samples, by World Courier. About ten containers had to be packed around each other like a Russian doll. The bicycle courier for the airport was already waiting, when one container was left. I stealthily let it disappear in the pocket of my lab-coat, so as not to cause any further delay. At night the Institute from Hamburg gave the green light and informed me that the result was negative, which was also a big relief for the health-care workers at the hospital. A SARS patient arrived in Frankfurt, in Germany, and was successfully quarantined and treated — and recovered without causing a local outbreak as in Canada or Singapore. There paramilitary actions — such as daily reports of the body temperature of employees — were taken successfully. The World Health Organization was very alert and helpful in managing the dangerous outbreaks all over the world by issuing frequent reports and daily advice. Again, bats were the reservoirs of the virus, as in the case of Ebola. Bats live in colonies at high population densities and transmit viruses without becoming sick themselves — like resistant survivors — those that did become ill succumbed. Some people say that bats may have different immune systems, especially a much higher interferon level, therefore getting sick more rarely. In the meantime, therapies against SARS have been developed: inhibitors of its protease. A similar approach was also successful against HIV. Meanwhile in the United Arab Emirates, a new SARS outbreak has occurred, killing a sheik and his son and, later, others. It is most probably transmitted by camels. The virus was isolated at the Erasmus Medical Centre in Rotterdam. The location of isolation is normally used to coin the name of an isolate, so in this case it was first called EMC-virus, but is now known as Middle East respiratory syndrome, MERS-coronavirus. Infection rates are increasing. A short episode was experienced in Korea in 2015 in a hospital — and was spread through family members, because they take care of patients, while nurses as such do not exist. The virus was rapidly brought under control. Please note SARS was controlled by very old-fashioned hygiene measurements and paramilitary actions, not by a fancy vaccine or therapies.

One day my co-worker Alex came into my office: "I am so sick I am dying." I remembered pictures of measles from virology textbooks (I am not a physician) and diagnosed measles. Alex had just come back from the Ukraine, from his grandmother's birthday, where the Internet indicated a measles outbreak. Political uncertainties had prevented vaccination of a whole generation of people. Everybody in the Institute in Zurich was tested for measles antibodies and was vaccinated immediately if necessary; that is not too late. Alex did not go to the hospital as he was told to, but took the tram and infected a child. Fortunately, both of them recovered. Even a doctor for children recently infected the children in his practice office with measles because he had never been vaccinated. This raised a debate about mandatory vaccination of such a profession. 1.7 million people die every year of measles. Many parents decline the vaccination out of fear of side effects (autism was claimed to be such a side effect, but falsely). Measles is especially dangerous for adults, with encephalitis as a potential complication. Measles has written history, and depopulated islands; it contributed to the decline of the Mayas, influenced the outcome of the battles of Charlemagne, and accelerated the end of the Roman Empire. Measles can also affect animals. Some people may remember the extinction of numerous seals in the North Sea.

The fear of measles vaccination teaches us a lesson — vaccination is based on trust and education and correct information. The interest in vaccination plummeted when the information spread that the American Secret service ran a fake vaccination campaign by taking blood samples in a rural area of Pakistan to uncover the hiding place of Osama bin Laden. This increased refusal rates for vaccines for years.

Noroviruses are not very dangerous, but they are extremely contagious. A few virus particles are sufficient to cause an infection. The television cruise ship "Deutschland" was the first ship to be hit by noroviruses in the 1990s.

It is based in Neustadt, on the Baltic, the city of my childhood. The newspapers speculated about the cause, the yellow and black quarantine flag was raised and nobody was allowed to leave or enter the ship. The head of my diagnostics department in Zurich, whom I called, immediately gave the correct answer: noroviruses. They were contaminating the drinking water that had been taken on board. This is a threat since then it could happen on any cruise ship and perhaps cause deaths among the elderly or weak passengers. Every crew now has to learn how to cope with this — isolate the infected passengers completely! A hospital close to Zurich had to be locked for several days. And a city nightline train was isolated at night in Frankfurt because of infected school kids. I should probably avoid the six-bunk couchette cabins in this train in future, though in fact I am still using them.

And what about ticks? If the doctor does not know what to say, he may have to be informed whether you were bitten by a tick. You should save the bug, if you can, for diagnostic purposes. It may have transmitted bacteria, *Borrelia burgdorferi*, causing Lyme disease — or a virus, the "tick-borne encephalitis" virus (TBEV). The bacteria can be treated by antibiotics, but TBEV cannot be treated at all. This can be diagnosed with the bugs. A bite is sometimes but not always noticeable by a moving red halo on the skin. At Princeton everybody knew and talked about this risk. One day, the famous John Hopfield — known for the Hopfield artificial neural network — pretended to look crazy and pointed out that the deer in his garden very often have ticks and could make him look like that. He took antibiotics just to be on the safe side after a bite. One needs to remember this if one cannot explain the cause of a serious headache. By the way: for a summer holiday in Austria, vaccination against ticks is mandatory and has to be done well in advance. Their chief virologist is a tick specialist — he must know!

New viruses are often not new, but they are easier to diagnose or show up at unexpected locations. One day the crows fell down from the sky over New York City. Within only a few years a new virus had conquered the whole of the USA, the West Nile virus (WNV). As indicated by its name, it came not from New York but probably by airplane from Israel. Mosquitoes get infected by birds and can induce encephalitis in humans. Air travel is the fastest way for viruses to spread. Within 24 hours viruses can get around the world in an airplane, often lurking in the air-conditioning system.

The public used to complain if health authorities do not take precautions or actions fast enough. We all have to learn that there are newly emerging viruses, which take us by surprise. The Zika virus is one of them I had never heard of it until early in 2016. It was for decades a local virus of monkeys in Uganda, the "Zika" forest, and then spread to over 32 countries in 2016, so that the WHO announced a "global health emergency". The virus may cause a microcephalus in newborns in Brasilia, and brain infections and may be sexually transmitted. Again the poorest are hit the hardest. Two types of mosquitos (*Aedes aegypti, which also transmits malaria and is called the yellow fever mosquito, and Aedes albopictus, the tiger mosquito*) can transmit the virus, which is similar to dengue fever virus and yellow fever virus. The mosquitos need to be eliminated by insecticides or infected with bacteria so that the virus is blocked, or made sterile, so that no new females produce progeny insects — all very difficult to do. The tiger mosquitos have reached south European countries. They can resist cold temperatures and multiply in water puddles. This is a new outbreak — but will not be the last one.

HIV as an example

Early one morning, about thirty years ago, during my summer vacation in the Provence in France, I tried to contact my co-worker Jutta in Berlin. "Jutta, they are reporting accidents in a newspaper and the death of a technician working with HIV in US laboratories. Stop working with HIV, throw everything directly into the steam autoclave and kill all the virus samples immediately — this may be life-threatening." She did not want to! "Can't I finish the experiment first? I will be careful!" This was the typical reaction of a scientist. I remember, when we looked at a virus preparation of HIV in a centrifuge tube, it was turbid, from all the viruses that we had grown in the laboratory. We were playing with our lives without knowing it. I once gave a virus sample to the stewardess during an overseas flight to put into the refrigerator, next to food for passengers — it was in a capped tube and there was no danger.... Still! Jutta even ran an HIV test in the laboratory for a co-worker to determine his HIV status. What would have happened if he had been positive? Being HIV-positive was a death sentence in those days.

Infectious diseases are, worldwide, the most frequent cause of death. In Germany 60,000 people, and worldwide 80 million people, die of

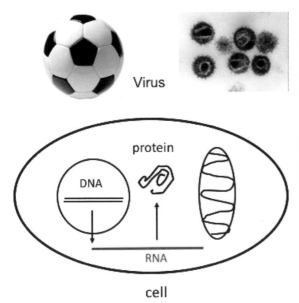

Virus

protein

DNA

RNA

cell

Top: Football and retroviruses are icosahedrons. Retroviruses can cause AIDS by immune suppression but once built the human placenta, so that we do not need to lay eggs.

Bottom: Cell with nucleus and mitochondrium (former bacterium), DNA is the genome, RNA is the messenger for protein synthesis.

infectious diseases every year. HIV contributes 2 million cases of death per year. A similar number of people become newly infected. By comparison, 8.2 million people die of cancer worldwide each year. At the end of the 1970s infectious diseases were thought to be under control by antibiotics. This optimism vanished with the appearance of HIV/AIDS and turned into fear and pessimism. Even panic arose, as reported during times of the terrible Plague of Justinian in the Eastern Roman Empire around 541. For HIV everything was unknown: spreading, infection routes, diagnosis, and the course of the disease. It was a lucky coincidence that just at that time novel technologies for understanding diseases were becoming available: molecular biology and gene technology. The age of molecular cloning, recombination of DNA fragments of genes, the highly sensitive polymerase chain reaction (PCR) for detecting very small amounts of genetic material by amplification in the test-tube became available. All that was ready to go into research and ended up in one of the most incredible success stories of medicine, such as had never been experienced before. The only element still missing today is a vaccine. This is the conclusion after 35 years of intensive research. The virus was sequenced in almost no time; one could grow and amplify virus production in the laboratory, which enabled researchers and pharmaceutical companies to screen for drugs and to develop reliable diagnostic tests and methods for screening blood supply. In the meantime more than 30 drugs have been approved for anti-HIV therapy. Recently, a rapid diagnostic test has become available to test the HIV status of a potential sexual partner before entering a male sauna; it is fast, albeit not entirely reliable. Self-testing leads to fewer infections. Today nobody should die of AIDS in the Western world, with health insurance to cover the expenses of therapies — which still amount to about 20,000 dollars per year. The challenge is to support the poor, even in rich countries. The approach of suppressing multi-drug-resistant HIV infections by three drugs, "HAART" (highly active anti-retroviral therapy), the "triple therapy" has become a concept for other infectious diseases and also for cancer — if there are enough drugs available. HIV infection then allows the patient to lead an almost normal life, with a life expectancy similar to that of an uninfected person, i.e., about 75 years in the Western world. In the Third World the survival time is about 11 years after diagnosis, often with only a double therapy (ART, anti-retroviral therapy) and late beginning of treatment. In the Western world a normal family life can be led, with uninfected children being born. "Therapy as prevention" is the motto, indicating that no infection should occur if a patient adheres strictly to the therapeutic regimen. Pre-exposure prophylaxis (PrEP) and post-exposure prophylaxis (PEP) were launched in 2012 and have enjoyed great success, leading to an "almost-healing". Depending on the drugs the "morning-after-pill" has to be taken within hours or a day. Thus, healthy people can stay healthy, while therapy of infected people is at the same time a prophylaxis for the uninfected partners. This may be the best we shall have for quite some time, until a vaccine (perhaps) becomes available one day. Complete removal of the virus from a person's body, a cure, is not yet possible but is a new ambitious goal. The virus changes and persists in reservoirs where it is inaccessible to therapies. However, the viral load (VL) is reduced by the triple therapy from a billion particles to about 20 per milliliter of blood. This is the limit of detection in diagnostics based on the highly sensitive PCR gene amplification method mentioned above.

This dramatic reduction of the viral load raised the question as to whether someone treated successfully would be infectious at all. That would mean that precautions during sexual intercourse would no longer be necessary — so ran the argument. Interestingly it was Switzerland where this was discussed first, where people are more safety-oriented than in any other country. The proposal set off a worldwide debate, but was finally accepted. Consequently, one would no longer have to inform a partner about one`s status, one's own HIV infection. But adherence and the patient's compliance are strict prerequisites — and severe limiting factors.

35 years of HIV research were celebrated in 2013 with a symposium at the Institut Pasteur in Paris. There the Nobel price laureate for the discovery of HIV, Françoise Barré-Sinoussi invited researchers from around the world. A noted non-participant at the meeting was Luc Montagnier, head of the department where the discovery was made, who shared the Nobel prize (2008). It was interesting to learn that Montagnier has meanwhile completely abandoned research and evidence-based therapeutics against HIV infections and turned his interest towards "alternative medicine" with natural plant extracts. This is not generally accepted.

The first electron micrograph of the virus was shown again, a historical slide. Already in the initial picture one could see that the virus changes its internal structures owing to the action of the viral protease: the virus matures after it has left the cell. This step is prevented by protease inhibitors — one can *see* the effect of the drug. The immature virus is no longer infectious. This electron-microscope picture was of historical importance, because Robert C. Gallo, at that time at the National Institute of Health in Bethesda, published it as his own picture. Françoise showed me the two pictures on her pinboard. Protests from the scientific community finally resulted in the declaration that there had been a mistake, pictures had been confused, the "wrong" one was published in error. Some co-workers then emigrated to other countries. The journalists, however, did not want to let go. Gallo had received the virus from Luc Montagnier, who generously gave samples of it to other researchers. Out of a mixture of viruses prepared in Gallo's laboratory a viral candidate grew up, which however, turned out to be Montgnier's isolate. Virologists normally do not mix virus isolates — on the contrary, they isolate individual clones — however, mixing and masking is what had happened. The French and the US governments finally reached a compromise according to which both sides would share the patent royalties. After all, Gallo had identified a very important growth factor (called Interleukin IL-2), which allowed propagation of the virus in cell culture, an important contribution to the development of a diagnostic test. Only HIV has such a fast mutation rate that no two virus isolates in the world are identical. A few point mutations, which were different in the "Gallo virus isolate", were the basis for a defense by the lawyers, but did not prove an independent isolate. For ten years Gallo was under attack. This cost him the Nobel prize in spite of three excellent papers in the Journal *Science* shortly after the discovery of the virus. Furthermore, Gallo discovered the human T-cell leukemia virus HTLV-1, which is endemic in Japan and only distantly related to HIV. A friend of Gallo`s described him as a football player: you lose — ok — try again. So Gallo came to the Paris celebration. The show must go on. All the participants in his annual meeting at the Institute of Human Virology (IHV) in Baltimore celebrated an "almost-Nobel-prize-party" with a Ms Kennedy as event manager. The participants paid.

Has enough been achieved with the developments in therapy? Interestingly, new drugs seem still to be required, even though there is not a single virus or disease against which we have so many different pills. Most successful turned out to be the rather late developed integrase inhibitors. A new drug in 2015 prevents the release of the virus from host cells — not replication, as prevented by most other drugs. A very unusual approach is that of persuading the virus to leave its hidden reservoir that no medicine can reach. The battle cry is "purge the virus", or "shock and kill" the virus, first to produce more viruses and afterwards treat them all. Making the infection worse before making it better — that sounds a bit risky.

The pills have dropped in price significantly in Africa, thanks to the commitment of Bill Clinton, down to several hundred dollars per year. However, this has the side effect that the pharmaceutical industry is not enthusiastic about developing new drugs. The pills in Africa, heavily subsidized, were smuggled to other countries, thus undermining their pharmaceutical markets. Today the pills have two colors, light and dark blue, depending on their origin, so that one can trace them back to the source. Are they equally good? I do not know.

There are still problems with resistance of the virus in patients undergoing therapy. The virus contains 10,000 nucleotides, and these mutate — about 10 mutations per replication round in about 24 hours. The reverse transcriptase, the enzyme for virus replication transcribing the viral RNA into double-stranded DNA, is highly error-prone. Therefore, many mutants accumulate and thus evade the treatment. There is always a swarm of different viruses in the patient, described as quasispecies. All the members of this swarm are related, but not identical. Some viruses are mutants, escaping from one therapy and therefore requiring another treatment, often a different combination of drugs. If one replaces one treatment regimen by another one, the original virus can grow up rather quickly, within weeks, which was surprising to scientists when it was first noticed. A triple therapy is used in the Western world, a quadruple therapy in a single pill is already forthcoming, and an injection shot that should last for three months is in preparation and initial testing. The volume to be injected is still rather large (4 milliliters) and hurts a bit.

What is the situation in the world? Is it as good as the results from research suggest? 37 million people are HIV-positive, 21 million are diagnosed as infected, 2 million new infections are estimated to happen per year (figures for 2013). UNAIDS, an organization of the UNO focusing on AIDS, regularly publishes these numbers. 1.5 million people died of AIDS in 2013, one-tenth as many as 10 years earlier. 15 million people with HIV were receiving a therapy, 22 million patients are still waiting. Treatment reducing the viral load, VL to an undetectably low level means that the person is no longer infectious and does not spread the virus. In the US this is achieved in 30% of the people, in Switzerland and the U.K. about 60%, and in Russia 11%. Almost a quarter of the newly infected people are females between 15 and 24 years of age. In Africa they have no chance to improve their living standards, and young females become prostitutes of "sugar daddies". But the epidemic will not end if this continues. A new effort has been initiated to keep the girls at school by "cash incentives", giving them money to prevent prostitution and allowing them to buy desired phones.

Similarly, male circumcision is promoted by cash payments ($12.50), which reduce the infection rates by about twofold. I remember when circumcision was first proposed, 30 years ago — it sounded like an emergency solution and a declaration of bankruptcy of all research then — and the audience burst out laughing. It is still performed as in Biblical times.

"Save the children" is the goal, save the next generation, after almost a whole generation has already been lost in Africa. No newborn child should become infected. In the Western world it is regarded as medical misconduct if an infection happens during delivery. Mother-to-child transmission (MTCT) is preventable if pregnant women can be reached by therapies and treated during childbirth. Reaching them all is still a problem. We can hardly remember the model proposed by George W. Bush, ABC: Abstinence, be faithful, (use) condoms. It is the biggest challenge for public health workers to translate knowing into doing — this is difficult enough to achieve with smoking or eating, let alone with sexual intercourse.

Worst of all — many infected people in the USA do not know that they are infected. The number amounts to 200,000 out of 1.3 million. One-half of new infections can be attributed to this. Unfortunately, newly infected people are especially highly infectious. A convincing new strategy is to treat as early as possible — but you need to know! Another serious problem is the lack of compliance. About half of the people who receive therapies do not take them regularly and remain infectious. Education is needed, as it has a strong protective effect against HIV infections.

Up to 2020 the hope is described by a rule of thumb, "90:90:90", meaning: 90% of people infected should be diagnosed, 90% of these should be provided with antiretroviral therapy, and 90% should be below detectable viral load, VL leading to about 72% successfully treated. Then no infection of partners would occur and the pandemic could come to a halt. This has already been put into practice successfully in some countries (Botswana) in Africa in 2016. The goal is that HIV RNA should become undetectable, thanks to treatment. This ambition is to target the viral load, VL and make HIV undetectable in all people living with AIDS. Could then AIDS be overcome by the year 2030? That is a prediction. It should be possible in principle — it is mainly a matter of supplying antiviral therapies, learning, teaching, improving sanitary conditions and health-care systems — and that means money.

The Berlin patient and the Mississippi baby — cure HIV?

A unique therapeutic effect to cure a patient was achieved at the Charité Hospital in Berlin. A patient with HIV and a lymphoma was due to receive a bone-marrow transplant against the lymphoma. For that, a special donor was selected, one who was naturally resistant against HIV. The hope was that the "Berlin patient" would become resistant too and be cured — and this was indeed what happened. About 15% of Europeans are resistant to HIV infection because of a natural mutation in the surface receptor that the virus needs to enter the cell. (*The receptor is a chemokine receptor CCR5 and the mutation is called CCR5delta32 because the molecule is shortened by 32 amino acids.*) Some research was performed with survivors of the Black Death, to see whether they survived because of exactly this mutation in the Middle Ages. Graves were opened to sequence some of those survivors — however, the speculation appears probably to have been unfounded.

A bone-marrow transplantation and replacement of HIV-resistant lymphocytes was successful. So the lucky survivor Tom Brown was "functionally cured" since ten years — that is the description for someone whose viral load, VL is undetectable but may lead to a later remission. Even better would be a "sterilizing" cure without remission. Eight repeats of the same procedure failed subsequently. Repeats for AIDS patients cannot be easily performed because it requires almost a complete removal of the patient's immune system — a risky procedure. Thus, this case is at best a model for novel therapeutic approaches. They are indeed pursued by either blocking or knocking out the receptor CCR5 (by silencer RNA), so that the virus cannot enter the cell. People can dispense with the receptor and live normally without it.

Another case of near-cure was the Mississippi baby. It was treated directly after birth (within 30 hours), which can be done because the therapies are not as toxic as they were. The baby was treated for 19 months with ART. Nineteen other babies were treated in the same way, by this so-called post-treatment control (PTC), which controls the virus. The control is, however, not complete, as the virus came back after four years in the Mississippi case. But another child has now been "functionally cured" for more than 12 years, so we may hope that the residual virus will stay under control for the person's entire life. (*At least five logs (100,000-fold) reduction of the viral load VL down to about 50 viral copies/ml is required.*)

New approaches aim to cut out the integrated HIV with molecular scissors, which will easily be more possible with a brand new technology with molecular scissors described below (CRISPR/Cas9). This would require *ex vivo* gene therapy of selected cells, perhaps viral vectors for treatment and reinfusion, and a second therapy to fight against "minimal residual disease", left over cells, because no therapy works 100 percent. Who can pay for this type of research and later treatment — where? In Africa? Yet even a whole new journal *Cure HIV* was founded. A big hype. Another novel approach is to activate the virus in the patient's reservoirs (by a drug Vorinostat) to make it accessible for therapies. This is being tested in a study called DARE — a meaningful acronym, because it sounds a bit risky first to increase the number of viruses in order then to kill them.

Furthermore, the so-called "Elite Controllers" are being investigated to find out what mechanism can control the virus. They have special features in their immune system (HLA-B57), which may be exploited; this possibility is being explored in a study called VISCONTI (Virology and Immunology Study on CONtrollers after Treatment Interruption). A major simple message is: treat early!

No vaccine against HIV?

What is the result after 30 years of vaccine development? There is no vaccine and after so many wrong promises nobody dares to make any more predictions. When HIV was discovered, molecular biologists assumed that a vaccine could be produced immediately. The model was the hepatitis B virus (HBV), where a surface molecule induces antibodies that neutralize the virus, counteract its infectivity and prevent its replication. By analogy to HBV, the envelope sequence of HIV was selected as the basis for vaccine development, and the sequence was known within a few months. The DNA of the surface protein of HBV was then used to produce a protein in yeast or in bacteria by recombinant DNA technology — the first successful vaccine ever to be produced by this novel technology. However, this did not work out for HIV, even though it has some similarities with the hepatitis virus, HBV. The reason is that HBV is genetically more stable, and it carries an almost complete double-stranded DNA inside the virus particle, which does not change easily in contrast to the single-stranded RNA genome of HIV, which is so variable that all efforts to "capture" it have so far failed. The prediction of virologists was thus far off the mark that all the people involved are still ashamed of it, since the differences between hepatitis B virus, HBV and HIV were even then known to all researchers. The vaccines against the surface protein were at best capable of neutralizing the virus used in the test-tube, but not the viruses in real life, which do not consist of one species but of many different ones, the quasispecies, and furthermore can rapidly change.

An unintended "vaccination" took place in Australia when patients with a blood-clotting deficiency received supplementation with a blood product called Factor VIII. However, the product they received was contaminated with a virus that was defective and lacked an accessory gene of HIV, the pathogenicity (originally "negative") factor Nef. This was a slowly replicating virus and would have been an ideal vaccine, as known from previous vaccines, e.g. the vaccine against poliomyelitis, an attenuated (slowly replicating) virus. That is one of the best designs for a vaccine. However, after a dozen years some recipients of the contaminated Factor VIII fell ill: the virus had reverted to a faster-replicating virus, replenishing the defective gene and causing the disease — to the great surprise of the virologists.

The surface protein has been analyzed down to its last possible detail, but so far without success. The virus needs this surface molecule to find its host cell. Therefore, surface molecules are still in the focus of attention. Today a new approach is being taken, exploiting the antibodies of long-term survivors, termed long-term non-progressors (LTNP) or elite controllers (EC). Out of the whole antibody reservoir of such an infected but resistant person, the antibodies that bind to the surface of HIV and neutralize it, to make it non-infectious, are selected in the laboratory. 200,000 antibodies were screened in a *tour de force* to select a few candidates, which are now under development for producing a novel vaccine.

An early large scale clinical trial, RV144, was performed in Thailand with 16,000 volunteers, and evaluated more than once, with always better interpretations, depending on the statistics between 2009 and 2015. The vaccine is a prime-boost combination of a canarypox virus vector (ALVAC) and gp120 viral surface proteins (*from clades B and AE, as monomers not the trimers, it was not known initially that the coat protein naturally trimerizes, so this vaccine was not optimal*). It reduced the risk of infection by 31%. The Bill and Melinda Gates Foundation is sponsoring a variation of this vaccine in Africa with 5400 volunteers (clade C) and predicts — or hopes for — success by 2030.

New technologies identify "broadly neutralizing antibodies" (bnAbs) with a new strategy. If one vaccine fails, maybe several of them will succeed — by sequential immunizations, though. The immune response could then be guided stepwise to develop special classes of antibodies to fight HIV — several sequential vaccines shooting at a moving target? Difficult. Adeno-associated virus (AAV) expresses the bnAbs in an ongoing clinical trial in the U.K. A vaccine "Ad-Env" is tested by Dan H. Barouch at Harvard with an adenovirus modified to express several HIV proteins (Gag, Pol, and Env) and a subsequent boost with the viral surface Env protein alone.

Then there is a "cheating approach" on the way. Instead of using the real viral surface protein one synthesizes a similar one as antigen, more immunogenic than the native viral surface Env, in the hope that antibodies will also recognize the native one better. This approach is known in animal studies (we also used it successfully), it may work but it is new for HIV. Now a "replicating" vaccine virus is going to be tested: remember, replicating viruses are the most efficient ones as vaccines — if they do not cause diseases. In this case the virus is a modified herpes virus CMV, so no intact HIV can arise from the vaccine again. Let us hope and wait — 15 years? 30 years?

"Naked DNA"

One day in the 1990s I received an unexpected invitation to Malvern, Pennsylvania, USA. I had earlier been an adviser there, for several years, for the company Centocor for the application of cancer genes in the diagnosis or therapy of cancer. The project now was a vaccine against HIV, and a spinoff of Centocor by the name Apollon was founded. I was supposed to direct the development of that vaccine and received an impressive business card as CSO, Chief Scientific Officer, for a very small team! I was on leave from the Max Planck Institute in Berlin for one week per month; my time at the computer was not recorded. The science was exciting: the goal was a naked DNA vaccine, a novelty then, whereby the muscle of a vaccinee is injected with DNA to produce parts of the virus, some proteins, simulating infection. This should lead to antibodies in the recipient as vaccine. The DNA was a complex combination of a variety of viral genes, amplifying genes, some of which originated from totally different other viruses or even bacteria. That is the playground of virologists but also the art — the secret lies in the selection, cutting and recombining genes to intensify certain properties and to avoid others, as modules of an artificial virus. That was the challenge. No intact pathogenic virus is allowed to emerge from the process. And in clinical tests, or the customs at the US border, it should be clear that the carrier has not been infected but only vaccinated.

While we were producing large amounts of the DNA, suddenly all samples, text-tubes and reagents were found to be contaminated with the DNA. There was a leak somewhere — until we could repair it in the safety laboratory, we were all already vaccinated through the air and our nose and lungs. The DNA was safe! It was further modified to a "dummy" so that it could not be stolen or imitated by competitors — a normal precaution in industry, but less so in a research laboratory — and new to me.

The DNA was then tested in some HIV-infected patients in Zurich. This was the first DNA-based vaccine in Europe. "Adverse events" did not occur in the patients within several years. (*The most important thing was that the volunteers did not develop autoimmune responses.*)

However, unexpectedly, the local immunologists at the University showed "severe adverse" behavior — maybe because we were invading their "territory"? I tried to co-operate before we started injecting the patients, but failed. I had initiated the project before I came to Zurich — still something to remember for others if one wants to avoid trouble.

After the injections we noticed that one patient had been infected with a "wrong" HIV strain, which could not be recognized by the vaccine. After a first shock this was not a real problem because a Phase I Clinical Trial in patients is normally only a safety test. Of course we would have liked to see some efficacy, some protection, even though that is not the purpose of such a trial. The DNA construct is now in use in the U.S. military, but in combination with other vaccines. Especially a sequential approach is attractive, "prime-boost", in this case a DNA-prime — and virus boost. The virus is a modified Adenovirus with HIV landmarks, because virus particles lead to best immune responses. The DNA vaccine has apparently a training effect on immune cells, making them into long-lived memory cells. We observed that as an unexpected effect. The virus for the boost is sometimes constructed to contain rare triplets, designated as "codon deoptimization"; this slows down the virus production. Such attenuation, which is frequently intended for vaccines, results in the viruses being unable to cause diseases. Real but slowed-down viruses have been used as the basic principle for vaccines in the past. To administer DNA, injection-needles are no longer used; pistols are now applied to shoot the DNA into the muscles. I have such a pistol in a safety box with a red velvet lining, delivered by the producer, making it look very valuable — however, we did not find this type of injection very effective. Better pistols may have to be designed. This type of DNA vaccination is used worldwide in many variants: against influenza, respiratory-syncytial or Ebola and the newly emerging Zika viruses. Yet DNA alone is too inefficient, so combinations with other vaccines are required. In contrast to viruses, cancer cells often do not expose foreign proteins on their surface, so that one cannot train the immune system easily to recognize them as targets for antibody production. One therefore tries recently to exploit not antibody production but rather the stimulation of cell-mediated immunity, by growing lymphocytes with DNA coding for cytokines. We therefore injected such an (interleukin-12-producing) DNA into malignant melanoma patients in a clinical trial — with some effect (see below). Help for self-help is the idea.

Microbicides as "condoms" for women

Anti-HIV drugs which failed as oral therapies have recently been developed into drugs for local application in the vagina as protection against sexual transmission of HIV. Worldwide, women want to be responsible for their protection and do not wish to leave that to their male partners. It turned out that the vagina, as a target of microbicides to inactivate microbes and HIV, was surprisingly poorly characterized. Two microbicide trials by the Bill and Melinda Gates Foundation failed at late stages of development and had to be terminated prematurely, as the microbicides were found to enhance, instead of preventing, infections with HIV. In spite of its competence and reputation, the Gates Foundation failed — which must indicate that our state of knowledge is inadequate.

Together with a dozen European scientists I once organized a Global-Exchange EMBO course on HIV/AIDS in Stellenbosch, South Africa, in 2010 with support from the European Molecular Biology Organization (EMBO) and my grant. EMBO started this as one of the first steps to go global outside of Europe. South Africa is the country with the highest HIV prevalence worldwide. 60 young Africans participated, some of them hesitated to apply and did not dare to inform others about it — HIV was a stigmatized topic. The former President of South Africa, Thabo Mbeki, proclaimed for many years that HIV infections could be avoided by taking a shower. This must have killed people.

The students liked the congress bags with the EMBO logo and almost fought for them. We did not have enough, because half the bags disappeared at the customs and never showed up again. The course took place in a safe location at Stellenbosch in a building sponsored by the Swedish Wallenberg Foundation. The students barely dared to travel back and forth to and from Cape Town, and never after sunset. They were afraid to pass by the local townships.

In a township close to Cape Town a million people lived in primitive shelters built of cartons, car tires or corrugated iron. We estimated that for these many people there were about 50 dry toilets, blue cabins aligned next to each other almost outside the village. Instead of visiting the Dutch vineyards we attended a church service in a township. It was all about AIDS — the sermons, the prayers, the decorations and the discussions afterwards, people approaching us foreigners: "We need help" — men think they can get rid of HIV by having contact with a virgin — what a fatal error! On the whole, sadly, I saw little output for my investment of time and effort into the course. Bigger organizations are now increasingly successful.

All microbicide studies have so far failed. Even the results of the most promising one from South Africa, the CAPRISA-trial, which was presented during the EMBO-meeting and at an international AIDS Congress in Rome, where it caused a big splash — turned out to be not directly reproducible. Another microbicide study, PrEP-VOICE, comprising 10,000 women was no success. More than 30% did not take the drug as prescribed — lack of compliance was the explanation — which is often lack of education and perhaps their living standard, with poor housing. A dozen microbicides have failed by now. To prevent unreliable intake of pills by the volunteers, a systemic long-lasting injected medicine is an option, which is under investigation with an integrase inhibitor. Ashley Haase from the University of Minnesota, Minneapolis showed that HIV does not infect the cells lining the vagina, but that the virus bypasses them through cellular junctions. A short-cut taken by the virus — so that cellular compounds are ineffective, as the virus stays outside the cells. Our compound would fit.

Thus the trials were not designed properly because of lack of knowledge. Researchers have to get their "homework" done, was the conclusion. The virus has to be killed before entering the cells. Two clinical trials with microbicides are ongoing in 2016 ((*CONRAD 128 and MTN-030/IPM 041*) and a vaginal ring was tested with dual functions, against pregnancy and HIV infection with moderate results.

Driving HIV into "suicide"

Together with my colleagues we designed an approach in Zurich, to kill HIV before it infects a cell. Application in the vagina as microbicide seemed the most urgent application to us. "Driving HIV into suicide" was the headline of a commentary in the journal *Nature Biotechnology* where we published our results (2007). The University of Zurich printed postcards with this slogan, which one could pick up in the streetcars during the jubilee of the University. However, this did not help us to put the approach into practice. It is just too difficult and too expensive. The suicide of the virus is based on "molecular scissors" called RNase H, causing cleavage of the RNA in RNA–DNA hybrids. This process normally takes place inside cells after the viral RNA has been copied to DNA and is then useless. The scissors are present inside the virus particle, ready to go to work after the virus has entered the cell. Its activation inside the particles destroys the RNA before it is copied, so that the virus cannot replicate — a "suicide" or a dead end for the viral life cycle. (*The scissors RNase H is one of four retroviral enzymes, the only one not yet targeted by an inhibitory drug because there are too many similar enzymes inside the cell. We activate this enzyme: treat the virus before it enters a cell and offer a piece of DNA which leads to a hybrid and activates the viral scissors, which kills the virus. We described the effect correctly as "silencer DNA" — but got a furious response from a reviewer, who said that only "silencer RNA" was acceptable — we gave in.*)

A co-worker of mine introduced the DNA hairpin into the vagina of mice, using a gel as carrier. Only five years later we noticed that the same substance was approved by the U.S. agencies as a "wellness factor" during sexual intercourse and a lubricant for medical instruments in the trachea etc. The Bill and Melinda Gates foundation accepted our compound for testing in a standardized procedure — to make data directly comparable. Their procedure was, however, designed against HIV inside cells. This did not apply to our compound and therefore it failed, what I foresaw but could not prevent. The hairpin destroyed HIV in a special immune-deficient SCID mouse model successfully, it completely prevented infection and it also reduced significantly the size of tumors in mice. A tumor virus mouse model is a surrogate for HIV, which does not infect mice — which is a big problem for research, so for years only precious monkeys were used. New mouse models are being developed by "knocking-in" new properties. We submitted our results on tumor reduction and prevention as a manuscript to *Nature Biotechnology*. The reviewers requested a higher number of mice for better statistics within four weeks. This presented me with problems: the first author became pregnant and did not want to work in the animal house, another had left, other team members refused point-blank to take part in animal studies, and the only technician willing to help had no license. Out of 50 co-workers nobody could help. So I cancelled all appointments in the institute and did it myself — to my own surprise I had not forgotten, during the many years at the desk, how to handle and inject mice.

Meanwhile, we tried to fulfil the request made by the Bill and Melinda Gates Foundation that HIV microbicides should be long-lasting. We tested the DNA hairpin-loop after long storage and also in the presence of semen — a critical parameter for a useful microbicide. At the Heinrich Pette Institute for Experimental Virology in Hamburg we placed a little container at the men's room asking for anonymous "donation" of semen. That worked out perfectly — but when we submitted the manuscript we were asked immediately before they even processed it for the written consent of the donors, including their names and in addition the written approval by the Ethics Committee at the Hospital — this took weeks. Such requests are totally new. (Even the donor of a feces sample for feces transfer, described below, was required retrospectively to submit written consent.) Thus: No body fluid can be taken from anybody without written consent — no tears or a drop of urine — because genes in the fluid may indeed be sequenced and reveal a lot of personal information.

What next? A bottleneck is the high-quality production of the DNA hairpin, according to Good Manufacturing Practice (GMP). This is almost impossible to pay for. I went to Russia, twice to China, once to Africa. HIV is still a taboo subject, and the outrageously expensive GMP production is required (even if I find it in some cases unnecessary) — no short-cut is allowed in the Third World, and one can even risk going to jail. During my visit to China the head of the admission committee was sentenced to death and shot for insufficiently strict obedience to the Chinese rules, which Western companies often try to circumvent by going to China. His organs may have saved some lives.

The origin and future of HIV

It is surprising, that one of the most recent new results on HIV refer to its origin. Some time ago a committee analyzed the possibility that HIV was caused by a contaminated polio vaccine. It was "most likely not" caused by an HIV contamination of the polio vaccine produced in monkey cells.

An extensive sequencing analysis of hundreds of samples from patients in Kinshasa all the way back to the 1920s was performed only recently.

One strain HIV-1 (M is the major strain) can be traced back to 1900 and another one (O) to 1920 in Congo, which increased in 1960 with unexpected speed in Kinshasa. Were hospitals treating hepatitis B, HBV and sexually transmitted diseases short of sterile needles? Social and political changes and in the 1920s, the train system to the north and south-east of the then Congo State may have supported spreading of HIV. AIDS is the result of four independent viral cross-species transmissions from non-human primates to humans, two times from chimpanzees (M and N) and two times from gorillas (O and P) as judged from more than 6000 monkey feces samples collected in the wild. Subgroup O infection was observed in 1964 in a Norwegian sailor thought to be the "first" case. The O and P strains clustered in the Cameroon, 1000 km away from Kinshasa. The Congo River to Kinshasa served for transport of goods and people — and may have helped spread of the virus.

Only HIV-1 group M out of the four went global and became pandemic. It was transported by Haitian refugees who returned home from Kinshasa to Haiti, then the virus spread to San Francisco and to the rest of the whole world. Other strains clustered in Africa (group C), and Russia (group A), which are due to singular "founder effects", transmission from a single person causing local spreads. HIV-2 can be traced back to 1940 and further to 1985 and has undergone 9 independent primate-to-human infections, but never did spread. HIV is thus clearly caused by contact with primates, — this is worrying because 80% of the people there still consume bushmeat — 20% of which is HIV-contaminated today in the Democratic Republic of Congo, DRC. Individual new infections have already been monitored, continuing the alphabet of the various groups. We do not know what made only one of the strains, M, pandemic! A specialist on this topic, Martine Peeters from Montpellier, answered: many parameters just coincided and caused the pandemic.

What about the future of HIV? HIV/AIDS is not primarily a problem for research any more. It is the best-studied virus of all. "Ending HIV/AIDS" was the motto by Toni Fauci, Head of Infectious Diseases at the NIH already in 2012. A colleague stated "I'd rather get HIV than diabetes" (not me — to be honest). A major goal is to identify persons who are newly infected, so that they can be treated early, when they are most infectious. "Therapy as prevention" is an important achievement and will help until one day (perhaps) a vaccine arrives.

A surprising effect of HIV is the appearance of late-stage cancer. Three cancers are at the center of attention: lymphomas, cervical cancers and Kaposi sarcomas (KS). These cancers arise by synergism between HIV and other viruses such as the papilloma and herpes viruses. HIV itself is not an oncogenic virus. It has a relative, the human T-cell leukemia virus, HTLV-1, which is oncogenic and endemic in Japan, where it can cause fulminant leukemias. (*It can be prevented by avoiding breastfeeding.*) Retroviruses were the basis for cancer research and still belong close to one another. Their "teaching effect" for human cancer researchers has been enormous and will be discussed below in the context of oncogenes. Today, there is a new research institute founded under the direction of Harold Varmus, who won the Nobel Prize for the analysis of the first retroviral oncogene Src. Why HIV patients develop cancer, in spite of therapies and at least partial restoration of their immune systems, is one of the questions. Are there general defects, which could help us to understand the effect of the immune system on cancer development? The incidence of cancer increases with age, even in the absence of an immune-suppressing virus such as HIV. It is faster in HIV patients in spite of therapies — why?

In Russia I participated in a congress on HIV, and the two hepatitis viruses HCV and HBV (2013), supported by the U.S. health organization NIH with sessions about diagnostics and therapy in addition to science. Bureaucracy, regulatory affairs in respect of HIV/AIDS and tuberculosis, *inter alia* in Russian prisons. Tuberculosis is often a secret surrogate description for HIV/AIDS, which it is not socially acceptable to mention. In spite of a simultaneous translation system and announcements of English talks, there was no translator. When I finally found one, he refused to translate the session. This made my trip redundant. I wanted to understand the system and find the people in charge of a putative microbicide. No way.

Instead, I interviewed students at the reception and at some industry exhibitions who knew some English. Nobody visited them — no free pencils or other little presents? No interest? Not affordable anyway? The prices of therapeutics were skyrocketing. Western standards. One person had been in Kiel and spoke German, another spoke English. No, we do not know anything about HIV/AIDS. Condoms were once discussed several years ago, but people were too ashamed to buy them; homosexuality is a forbidden topic, as of course were drugs. No, my 14-year-old daughter

3 Retroviruses and immortality

Reverse transcriptase — a personal retrospective

A journalist asked Howard Temin, on the occasion of his Nobel Prize celebration in Stockholm in 1975, what was he awarded the Prize for. "I cannot explain that — you would get lost in 30 seconds" was his answer, "but why don't you write 'stop smoking' instead? That is much more important!" He was an anti-smoking activist — he even brought up the topic during his acceptance speech in front of the King of Sweden — but, sadly, he died of lung cancer without ever having smoked.

Temin did not even try to explain his discovery to laymen. I will try anyway. I witnessed it. He discovered the reverse transcriptase, RT, the enzyme required for replication of retroviruses such as HIV. The enzyme has a much more important role than mere virus replication, as Temin foresaw: it plays a part in the composition of our genome. There relatives of retroviruses amplify our genes, change the genetic composition of our genomes but never leave a cell. Today, there are even speculations that it was the reverse transcriptase by which DNA evolved from RNA during evolution. When I started research on the reverse transcriptase, RT was not known, now 45 years later it is recognized as one of the most abundant — and possibly most important — proteins in biology.

How did I witness the discovery of reverse transcriptase? "Replication of retroviruses" was the title of my doctorate, my PhD thesis. I remember well some of the details. At the end of the 1960s I came back from Berkeley, USA, where I had had the courage to quit physics and jump across into the then relatively unknown field of molecular biology, at a time when student riots were raging across the campus. Back in Germany I looked for a PhD project.

PRO-RNA ⊂▥⊃▥⊂▥⊃ viroid or ribozyme

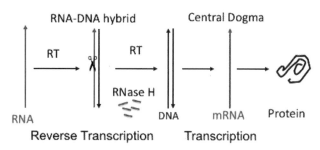

Top: Primordial "PRO-RNA", first biomolecule, with hairpin-looped structure, perhaps our oldest "ancestor", related to viroid, ribozyme, circRNA, and piRNA today.

Bottom: Replication of retroviruses and "central dogma": flow of information in biology from RNA or DNA to proteins, with reverse transcriptase (RT) and the molecular scissors ribonuclease H (RNase H), which removes the RNA in RNA-DNA hybrid, messenger RNA (mRNA) and proteins.

Gunther Stent, a famous geneticist, who had emigrated from Berlin to Berkeley, showed me in his office in Berkeley a desk-top model of the Max Planck Institute in Tübingen and suggested that I might go there. That is what I did, finding three topics within a day, and I decided to accept the one that seemed the most interesting, and the most likely to yield rapid results. The magnificent view of the mountain range, the *Schwäbische Alb*, had a substantial non-scientific influence upon my decision. I introduced molecular biology, which I had learnt in the U.S., to a virology institute. However, first I had to cross a difficult barrier. I had to learn how to isolate viruses from living chickens that I had infected beforehand with the avian myeloblastosis virus (AMV) causing a blood disease. This involved drawing blood directly from the chicken's heart with a hypodermic needle — a method that would not be permitted today. How this virus replicates was the subject of my project. At first, work stagnated for months. Then, during one of the famous Monday evening lectures at the Max Planck Institute in Tübingen, I heard from Friedrich Bonhoeffer (who had just come back from a Gordon Conference) that Howard Temin had discovered a reverse transcriptase. Some people laughed; "he keeps saying this" was their comment. However, my project was not progressing. So I decided to test for the reverse transcriptase. On the following day, at 8 o'clock in the morning, I went next door to meet Heinz Schaller, a DNA replication specialist, to get building blocks for DNA (including some radioactively labeled ones) instead for RNA. Using these was the only difference to my previous studies — yet it was a very important one. I had lots of virus from heart-puncturing of chickens. In the evening Schaller and colleagues assembled at the radioactivity detector to look at the result: an incredible amount of radioactively labeled DNA had been produced, and the counter hit to the top end of its measurement scale. With the right idea, one can perform a Nobel-prize-winning experiment in an afternoon. But, unfortunately, I did not have the "right" idea! I looked for the wrong enzyme, an RNA polymerase instead of a DNA polymerase. In a few weeks all the DNA building blocks (nucleotides) were sold out worldwide. The non-ionic detergent used to disrupt the virus was obtained from gasoline stations — that information quickly spread by word of mouth, so that the detergent also quickly became unavailable. The whole world seemed to be studying the reverse transcriptase, RT!

Temin himself had to overcome the skepticism of his own co-workers, who considered the idea of a DNA intermediate for an RNA virus to be a highly unlikely one. But on the basis of these experiments Temin pursued his vision much further, and predicted the integration of this DNA intermediate, which he designated DNA provirus, into the host organism's DNA. This RNA virus can "hide", in the form of a DNA provirus, in the host's DNA, and this trick helps it to survive. This led to the names "retrovirus" and "reverse transcriptase". These endogenous viruses and related structures populate our genomes in large quantities, almost 50 per cent — thanks to the RT. But the reverse transcriptase, RT can even do a lot more than this — as we shall see.

How can one get DNA from RNA? By more than one step? Yes, by help of molecular scissors, a nuclease, an enzyme called ribonuclease H or shorter RNase H. (*In more detail: the virus enters a cell, then the viral RNA is first copied by the reverse transcriptase into DNA via an RNA-DNA hybrid. By doing so the RNA has fulfilled its function and has to be removed by the RNase H from the hybrid to allow a double-stranded DNA to be produced.*)

At lunch at the Max Planck Institute, MPI in Tübingen I had talked to a student who analyzed calf thymus glands, Werner Buesen. He used to go to slaughterhouses to pick up the thymus tissue, and he isolated from it an RNase H. However, he had no idea what the enzyme might be good for — in fact, nobody did. It occurred to me only months later that a scissors-type enzyme of this kind could be used by the virus to get rid of the RNA, so I tested this idea. I was given some precious radioactively labeled RNA by my colleague and, sure enough, I could prove the existence of the RNase H activity when the RNA was dissolved. (*Later I found that the RNase H is fused to the reverse transcriptase, in a single molecule, the two are linked and move in tandem. The RT copies the RNA and a few nucleotides behind the RNase H cleans up, getting rid of the RNA — very efficient.*) The discovery of such a cleavage enzyme as the RNase H in a retrovirus earned me a paper in *Nature*, a PhD, invitations to give talks in the USA and surprisingly numerous co-authors. All of them claimed they had contributed to the paper or ideas. Good ideas have many authors, as I learnt later. I also learnt about frustration, because from that time on I failed almost completely to keep up with the explosion of papers that appeared on this topic. Everyone else seemed always to be faster than I was.

There is one more aspect to it. People often ask me how do scientists get good ideas or make discoveries? Here is one example — have *lunch* with other scientists, eat — ask, listen and learn! No fast food, though!

The then famous virologist Peter Duesberg, who was an excellent scientist and teacher at Berkeley when I was there (and known not least for his provocative pictures of pin-up girls on his tissue-culture incubators) could not find the RNA-removal enzyme RNase H in mouse viruses. I had looked at chicken viruses. They do not exist in mouse viruses, he claimed, so in his view my observation was a singular exception with a bird virus. Fortunately for me he published this opinion, and nothing better could have happened to me as a young student than the opportunity to prove a world-famous scientist wrong. We became lifetime friends, he often came to Berlin, where he visited his mother and we met even for big podium discussions about the danger of the most important retrovirus HIV — but we were almost never of the same opinion.

I was lucky in having had such huge amounts of mouse viruses as well, thanks to "Eveline" cells, a permanent cell line established by a technician with this name — enough virus to solve the scientific controversy and for experiments that led to several papers. However, Eveline payed a high price — the deterioration of her joints from repeated pipetting.

Later, after a move to the Robert Koch Institute in Berlin, I noticed that the purified reverse transcriptase RT consisted of two subunits not one as everybody expected. One could see the two bands (*by staining the protein after separating them by size in an electric field*). Once again in Cold Spring Harbor Laboratory I heard during the symposium several famous speakers describing the second band as a co-factor of cellular origin. Other polymerases were known for having this, a sigma factor. However, their results were more speculations and based on an analogy, not on experiments. I summoned all my courage and approached Jim Watson, then Director of the Laboratory, telling him that I had seen increase of the lower band at the expense of the upper one. The lower one is a degradation product — not a host cell factor. Adding a bit of protease (which promotes the degradation) would accelerate the process. "Do you have slides? You give a talk tomorrow." I did. Three months later he called me at the Robert Koch Institute in Berlin. I still wonder how he found my telephone number. "Didn't you describe the cleavage of the reverse transcriptase and its breakdown to two subunits? Now, all of a sudden, four papers on it have been submitted for publication in the Symposium volume." Not a word about a foreign sigma factor from the cell any more! Not a word about my result either! All this was published in the Symposium book that appeared in 1975 and included my contribution but also all the "corrected" other ones. Since then Watson has never forgotten my name. "Hi, Karin," he still says at the age of more than 85 years, and invites me to his table for the lobster dinner. Watson wrote a book with the title *Avoid Boring People*, in which he describes the secrets of success: "never be the smartest at the table", sit next to people who are smarter than you are — so there I was, doing precisely the right thing.

As a physicist and PhD student, I did not know how to isolate such an enzyme as the RT. I rented a room at the St. Joseph's Home in Zurich (with a Bible on my bedside table). I was allowed to watch an enzyme being isolated in the institute of Charles Weissmann at the University, by watching his technician and shivering violently in the cold-room, where the isolation has to be done — otherwise the enzyme would lose its activity. Weissmann was isolating the Q-beta replicase from bacteria (which will be mentioned again in the last chapter of this book). I then went home to the Max Planck Institute in Tübingen to isolate the replication enzyme RT from chicken viruses. I had also learnt a typical Weissmann lesson: calculate input and output, know the numbers, test their balance — after that, I never forgot to quantify whatever I did. Together with Weissmann we came up with several models on virus replication with an RT and an RNase H, which we published together — and all of our models were terribly wrong! The replication of retroviruses turned out to be much more complicated than we had envisaged, Nature`s result by trial and error. Years later, the Boehringer Company contacted me and ordered massive amounts of the RT enzyme to sell it: every laboratory wanted to use it as tool to make DNA from RNA. The trouble with the two subunits turned out to be very important but solved automatically during storage of the enzyme — just as it happened in my freezer. I was lucky to bring research funds back to the Institute in Berlin.

Early in the 1980s, about 15 years later, HIV was discovered, the most important retrovirus, and we started analyzing the replication enzymes, the RT/RNase H of HIV. Today I am trying to "kill" HIV by *activating* (rather than inhibiting) the RNase H scissors prematurely in virus particles outside the cell, in the hope that this may ultimately lead to a therapy and prevent virus production and spread of the disease. When we started our HIV work, we wanted to prepare monoclonal antibodies, which had just been developed and which are useful for purification and characterization of various substances because of their high specificity. The first step requires immunization of mice with the purified RT. A little sign "HIV-RT" on the mouse-cage caused a panic at the Max Planck Institute in Berlin. An inspector who had flown in from Munich to solve the case did not dare to shake hands with me — so afraid was he of the virus. The scandal escalated when the cage was moved, after cleaning, to a slightly different place in the same animal room. This was interpreted — absolutely unjustifiably — as a removal as "hiding evidence". To make up for this trouble I was allowed later to prepare the reverse transcriptase for a trip into orbit around the Earth, for an attempt at crystallization under low-gravity conditions, together with other (ribosomal) proteins from the Institute. Disappointingly, this did not work out — crystallization turned out, much later, only to be possible when the RT-enzyme is locked into a certain conformation by a drug, in my case it was too flexible! Crystals are needed for drug design for therapies.

Later in my research I more than once encountered fear of our work. This happened with influenza, SARS and cancer viruses, and it affected not only colleagues but also their co-workers and often their spouses. This suspicion was not unjustified, because there have indeed been several accidents, such as the escape of SARS (thee times!) from a high-security containment laboratory in China; moreover, some laboratory frogs (*Xenopus laevis*) once escaped and became a plague in California.

The RT is a specialty of retroviruses, and it came as a surprise, because nobody expected RNA to be transcribed into DNA for replication. This was the reversal of the "Central Dogma" of molecular biology, which posits that the path of biosynthesis is from DNA to RNA to proteins, in two successive processes termed respectively "transcription" and "translation". Temin and, independently, David Baltimore showed that the reverse was also true: DNA can be biosynthesized from RNA. This reversal of transcription was consequently called "reverse" transcription leading to the name "reverse transcriptase", RT, for the enzyme. (*Interestingly, the RT can perform two reactions: RNA to DNA and DNA to DNA, and between these two the original RNA, after it has been copied, has to be removed by the RNase H. The final result is double-stranded DNA.*)

The discovery of the reverse flow of information came as a great surprise. As a consequence the DNA can then be integrated into the host genomic DNA and be transmitted as long as the cell lives and divides as a cellular gene. Retroviruses normally do not lyse or destroy their host cells but "bud" out of the cells' membranes. If we believe that RNA came first and that DNA came later during evolution, then "reverse" transcriptase is in fact the wrong name: the step from RNA to DNA is not a reverse one but a straight-ahead one. A strictly correct name would therefore be "real transcriptase"! Indeed, this would allow the same abbreviation to be used — however, nobody is interested in such a change.

Now that, figuratively speaking, every possible genome has been sequenced, the latest surprise has been that there are numerous reverse transcriptases around. They are present in many organisms, in all eukaryotes (animals and plants), and also in archaea, in bacteria, in spliceosomes, in retrotransposons, in a strange chimeric multi-satellite msDNA, and in human and bacterial immune systems. In bacteria alone there are more than 1000 different kinds of reverse transcriptases. What are they all there for? In mammalian cells we know about the retrotransposons, which code for reverse transcriptases necessary for the "copy-and-paste" mechanism of cellular DNA described below (retrotransposons are reminiscent of simplified retroviruses). How unexpected this was can be described by an anecdote. In 1978 one of the co-discoverers of the reverse transcriptase David Baltimore got up in a meeting when someone described the existence of the reverse transcriptase in flies: "To my knowledge flies do not have retroviruses." Only much later now we know the answer — also flies have reverse transcriptases not from retroviruses but from their relatives, the retrotransposons, which are precursors or truncated "crippled" retroviruses and widely distributed including flies, they are extremely abundant. The retroviruses as special case happened to be discovered first!

Thus, we find reverse transcriptases not only linked to retroviruses. Since phages are also viruses mainly containing DNA, one could imagine that they also once had RNA and only later DNA genomes and in between retrovirus-like properties, just by evolution. Then one would expect to find "retrophages" too. I made this word up. But, to my knowledge, only a single "retrophage" harboring a reverse transcriptase exists — in spite of an intense search that I also joined in with (see below).

We now know that the reverse transcriptase is a key enzyme in biology. Perhaps it even "invented" the DNA in the first place; the RT enzyme was certainly essential in building up genomes, including our own, during evolution, where it is recognizable more than 100 million years back. Perhaps there initially existed a simpler, more primitive, precursor RT. I stress the importance of the reverse transcriptase not (only) because as a specialist for this enzyme my view is biased — but there is even novel evidence for its outrageous importance: it is one of the most abundant proteins worldwide and even the top one in an ocean sampling project, being 13.5 percent of all proteins there in the plankton — 45 years later. How come? Retroviruses, retrotransposons, copy-and-paste mechanisms belong together and built the genomes in biological systems as drivers of evolution. The RT is the top number one molecule — together with the RNase H following in the next chapter.

The RNase H — molecular scissors

A paper that I once published in a relatively low-quality journal is cited most frequently. This is normally not the case. Citation frequency is an important quality criterion — which however, obviously, can go wrong. It came about like this: I had discovered the scissors, the ribonuclease H or RNase H in animal retroviruses and then also in HIV. (*It is essential for replication of retroviruses on their way from RNA to double-stranded DNA, the RNA in a hybrid RNA-DNA has to be removed, cut away by the ribonuclease*

H (RNase H) with hybrid-specificity.) In the 1990s I got a phone call and was asked "Did you prove that this enzyme is indeed necessary for viral replication? Is it a validated target?" The call came from a company that was interested in finding an inhibitor against the RNase H. A company will only start with drug design against an essential viral component, a "validated target", because drug development is such a big and costly effort. "Yes, of course" was my first reaction, "Otherwise a double-stranded DNA would not be possible — the RNA has to be removed once it has been copied into single-stranded DNA." That was, however a premature answer. "Is there a formal proof of it?" No! It had never occurred to me before that there was none. For that purpose, one would have to mutate the RNase H (that means destroy its activity) and show that the virus cannot replicate any more. That would be a "loss of function" mutation. I could not perform that experiment, since it required an infectious DNA copy (clone) of HIV, directed mutagenesis and testing of the infectious virus mutant; this could only be done in a high-containment laboratory at the highest biosafety level 4. The only group capable of that was at Wellcome Co. in London. I called them and we set up a co-operation. For mutagenesis we looked at sequences that had been described in the bacterium *E. coli* and were known. Therefore, it was easy to find the important sites for mutation, because they are conserved. This worked out fine. To my own surprise it never occurred to me in those days to wonder what an RNase H was good for in bacteria. How could there be a relationship between a bacterium and a human retrovirus? At that time, I was not so used to thinking of evolutionary implications as I am today! (*The common theme is that all DNA replication systems need a starter — a piece of RNA to hold on to, which is later removed by an RNase H*).

I flew to London, we wrote and submitted the paper to a journal for publication. It came back rejected — we tried again, and it came back again. It was repeatedly rejected because every scientific reviewer was already convinced — as I initially was too — that the RNase H scissors was essential, so that nobody felt any need to see proof of this any more. It was accepted as a fact! Finally, the paper was published in the *Journal of General Virology*, not a prestigious organ in those days. Today that paper is considered to be of fundamental importance and is among the most frequently quoted papers that I have ever published, in spite of it not having appeared in a top-level journal — something to think about!

The RNase H is very abundant, which is an explanation of why there are no inhibitors for it: too many enzymes would be affected, and that could in turn cause side effects. Even integrases belong to this family. They also cleave as molecular scissors but can also heal the cut. The scissors really "cut" molecular bonds. They all contain a triad of three conserved non-contiguous aminoacids DDE to fix a magnesium, out of 100 which are not conserved. I once tried to find a related protein by sequence comparison. I did not exploit the DDE and failed, only the crystal structure finally proved the relationship — and a very general conclusion is: the sequence does not matter, but the folding, the structure does — as in arts and design "form follows function".

In Japan I was given a farewell present before my obligatory retirement from the University of Zurich: a model made of epoxy resin, with the crystal structure of the RNase H lasered into it on the basis of the sequence published in *Nature*. Robert Crouch from the National Institute of health in Bethesda in the USA had it made for me, and the beauty of this present touched my heart. Today you can buy such blocks, with two hearts, on market stalls as a present for a close friend — even without the coordinates published in *Nature*.

RNase H and embryos

Recently RNases H without retroviruses have also been detected in humans. Andrew Jackson and his colleagues in Edinburgh, Scotland analyzed a genetic disease in humans, mental retardation, the Aicardi-Goutières Syndrome (AGS). Much to their surprise, they identified an RNase H as the culprit. The enzyme was not yet known in humans. It consists of three subunits, RNase H2A, B and C, and only one of these is active. Then there is another RNase H1 in mitochondria. This raises the suspicion that it is of bacterial origin, since bacteria are the precursors of our mitochondria. The cellular RNase H removes individual RNA nucleotides in our DNA genomes if, by accident, some get incorporated there. Our genome contains hundreds of such misincorporations, so this seems to be a frequent accident. The RNase H scissors purifies our genome of these wrong nucleotides, preventing genetic instabilities. Failure leads to diseases. One such disease is AGS. Now specialists for genomic stabilities are refreshing the slightly old-fashioned biannual RNase H meetings.

Why is one type of RNase H enough for HIV, when humans need three? There is a principle behind this: the higher an organism, the more complex it is. The cellular RNase H in mammals has additional functions to fulfil, which are distributed on several proteins. If one fails this is fatal, "embryonic lethal".

It is frequently observed that multiprotein complexes in mammals have a part to play during embryogenesis. HIV does not have embryos, and therefore a simple RNase H is sufficient. But the RNase H with the name PIWI is extremely important in silencing especially in germcells to prevent jumping of genes (moving retrotransposons). Thus PIWI (together with some RNA) is the guardian of our genome, required for semen fertility.

The RNase H can diversify its functions by combining various building blocks in a modular way. Fusion to many different domains lead to new functions. All the enzyme by itself can do is cleave, but where and when and what it cleaves is dictated by the fusion partner. About a dozen such fusion partners are known. Only one of them is the reverse transcriptase. It pulls the RNase H as a nuclease, cleavage enzyme, along the RNA, and when it stops or stumbles, the pause is used by the RNase H to perform a cut (*its catalytic activity is slower than that of the reverse transcriptase*). It is surprising how many RNases H there are with unexpected functions. I studied the viral RNase H for decades, probably very narrow-mindedly. Many of the other functions were only recognized later as a result of the new sequencing technologies. I am surprised myself that RNase H is the most frequent and ancient structure in the protein world, even topping the RT, as shown in recent phylogenetic analyses based on gene sequencing determined by Gustavo Caetano-Anolles from Urbana, Illinois. Wherever there are nucleic acids, also scissors are needed, to trim or to "silence" the nucleic acids. They have many different names, such as Drosha, Argonaut, Cas9, and PIWI. The RNase H and reverse transcriptase are among the most important components for the development and integrity of genomes — both ours and those of other species. I am not saying that because they are the focus of my research. It's really true!

Telomerase and eternal life

Can we live for ever? The oldest dream of humans is eternal longevity. Does it exist? Yes, it does. There are indeed cells that can live forever —

a fact that I find astounding. Unfortunately, these cells are tumor cells. All the cells of our body are short-lived, even more short-lived than our body as a whole. All cells need to be constantly replenished by stem cells throughout our life-time. Not so with cancer cells. In almost every laboratory one can find the "HeLa" cell line, which is more than 60 years old. HeLa cells are the workhorse of many cancer researchers. They originated from a cervix carcinoma of a woman whom we always called Helene Lange, but whose actual name was Henrietta Lacks, an Afro-American who died, after bearing five children, at the age of 31 from cervix carcinoma. The cells were brought into culture in 1951 and grown up, a procedure that had never been successful before. Probably, this was a very aggressively growing tumor. Now, all of a sudden, Henrietta's descendants are starting to complain that they did not know about these cells of their ancestor. A television broadcast and a new book with the title *The Immortal Life of Henrietta Lacks* has appeared (2010). The genome of the cells has been sequenced, and the relatives worry about possible disclosure of their genetic traits, perhaps even defects, that they may also be carrying. From now on, results with this cell-line need to be reported to funding agencies. However, it seems rather unlikely to me that the cells — after an estimated 6000 passages under non-standardized conditions — are the same everywhere. There will be many chromosomal changes. I find it surprising that tumor cells can live "for ever" — why not just a bit longer than the normal cells of the patient, why for ever? (Under laboratory conditions, of course, but still!) How do tumor cells live for ever? Can one learn from them for longevity without cancer? Which other cells live for ever — I am asked this question during interviews. Our germ cells are passed from generation to the next one — immortal? And stem cells — how long could they live? They need a niche, thus age and longevity also depend on the environment. The beetle-type little animal tardigrade lived long, stayed viable for 40 years in a deep freezer at −20°C.

A hallmark of tumor cells is the telomerase, a specialized reverse transcriptase. It is named "TERT", Telomerase Reverse Transcriptase. It is located at the ends of all chromosomes, the telomeres, and is active there during embryonic life, elongating the ends to generate a buffer zone. The elongation consists of very boring repeats, seven nucleotides, over and over again. After birth the ends become shorter with every cell division. This is because the replication machinery of the DNA needs some space like a "seat" to turn around. This piece is lost, it is a molecular clock, during each round of cell division the ends become shorter and shorter until important genetic information is reached — then it becomes dangerous! At this point, the ends cluster and stick together, and the cell will die. This was already discovered by Barbara McClintock, who will be mentioned again later. Could one use the length of the telomeres as a predictor of life expectancy? No, that is not possible, because different cells have different molecular clocks, ticking at different speeds, so that the telomeres are all of different lengths. Could one then at least restore the deleted ends, to stop cell death and thereby yield a longer life? Indeed, that could be a goal — however, a long-lived cell of that kind would be a tumor cell. HeLa cells are such tumor cells. It is a general feature of tumor cells that they can divide for ever. In tumor cells, the telomerase is active again, leading to elongation of the ends just as in the embryo. Archimedes asked for a location outside of the earth so that he could lift it. Here, the location is not outside the chromosome, but by the extended ends.

Could one reverse the argument and ask whether there exists an inhibitor that is capable of stopping tumor cells from growing, by blocking the telomerase and thus acting as a universal anti-cancer drug? Yes, indeed: in 90% of all tumors the telomerase is active and is a potential target for therapy. This approach is under development, among other places at the Pasteur Institute in Paris and Geron Corporation in the US.

One can also activate a telomerase to extend the life of certain cells. That would be an anti-aging therapy, for eternal youth — maybe for skin without wrinkles! Biotech companies are interested in creams, which are advertised on television as "activating telomerases". How many people understand that? It just sounds scientific especially with the advertisers wearing a white lab-coat, intended to motivate people to buy the product almost like a medicine — but the promise is not true.

Telomerases in cancer research were long misunderstood, because most studies on human cancer are performed with mice as animal model, yet mice get tumors without any contribution from telomerases. That raises the question of whether mice are good tumor models. But they are difficult to replace.

Elizabeth Blackburn, from San Francisco, and her former student Carol Greider received the Nobel Prize for the discovery of the telomerase, together with Jack W. Szostak. In a single experiment Szostak showed that the protected telomeric ends are a general phenomenon (*he showed in a small single cell living species, tetrahymena, later also in yeast, mice, and frogs*). He founded recently an "Intellectual Powerhouse" for researchers to apply evolutionary approaches, to "out-evolve Nature" applying evolutionary technologies. The approach of using evolutionary biotechnology for the improvement of chemicals was originally developed by Manfred Eigen. Szostak also tries to fill lipid droplets to learn what makes drops divide — a model including oscillating waves for cell division.

It is not so frequent that a Nobel Prize is given "twice", for two such closely related molecules as TERT and RT. Perhaps the similarities were not noticed initially. Recently, evidence has begun to accumulate that telomerases can transcribe RNA into RNA and can repair defective DNA — not obvious functions for a telomerase.

We used the telomeric structures, so called pseudo-knots, to make the knots tighter and to block telomerases, as an anti-cancer approach. In mouse studies we were able, with the help of Elizabeth Blackburn, to stop the growth of tumors (malignant melanoma). Tom Cech also analyzes the telomerase. I once rather stupidly asked him whether the telomerase has an RNase H — I am still embarrassed by that question, as I should have known: The answer is no — no RNA primer needs to be removed by an RNase H. DNA is elongated and the RNA stays fixed to be copied for the growth of the DNA chain. Seven RNA nucleotides (TTAAGGG) are copied up to a thousandfold elongating the telomere. Tom Cech also explained to me that he thinks the RT is older than the telomerase: RTs are all over the place and telomerases are more specialized and came later during evolution. Or the other way round?

Interestingly, insects have developed another procedure to protect the ends of their chromosomes, they use *transposons* as short mobile DNA fragments that migrate and bind to ends of the chromosomes. That resembles hair elongation at the hairdresser's without hair growth. However, even the transposons carry an enzyme, transposase, which is related to the telomerase and the reverse transcriptase, one is fixed at the telomeres while the other one is mobile. Are telomerases degenerated transposons? To protect the ends of chromosomes various mechanisms have evolved — all related to retroviruses and their reverse transcriptase.

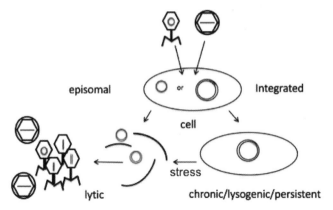

Life cycle of phage or virus inside cell, integration or episomal (separate) state of the DNA, environment leads to either chronic, lysogenic or persistent state (*right*) or stress-induced lytic state with release of progeny (*left*).

Viruses as cellular nuclei?

Eukaryotes contain nuclei that gave them their name. By analogy, bacteria, which do not have a nucleus, are termed prokaryotes.

Bacteria and cells melted by a symbiosis. The best-known examples of this are the mitochondria, which are endogenized and degenerated bacteria and can never leave the cell any more. They gave 90% of their genes away to the nucleus of the host cell. Some genes also got lost. The mitochondria kept only 300 genes from their former 3,000 genes. They became highly specialized, the powerhouse of the cell, supplying it with energy. Everything else that they need is supplied by the cell.

We notice a general principle here: symbiosis with loss of genes for the benefit of specialization. Something similar occurred with the cyanobacteria, which were taken up by plants and today as chloroplasts have specialized in photosynthesis. This happened only twice on our planet — surprisingly rarely, especially if a cell has so much to gain from it.

Perhaps this happened a third time after all. Who supplied the nuclei to the eukaryotes? There are virus candidates such as the poxviruses. Even today a retrovirus can integrate as a DNA provirus into a poxviral DNA. Integration takes place inside the nuclei with every retrovirus infection. Also herpes viruses are sometimes discussed as possible candidates for nuclei. Another candidate would be a pseudo-retrovirus such as the hepatitis B virus. Its DNA does not normally integrate into that of the host, but it could allow the integration of retroviral or other DNAs. Such DNA could lead to the integration of more retroviruses. The integration of DNA proviruses from retroviruses into our genome at some time in the past can be detected today. Up to 50% of our genomes are related to integrated retroviruses. They have built up our genomes — also the nucleus?

So far we have not discussed where the first cells came from. They are referred to as LUCA, the last universal common (or cellular) ancestors. Was a virus a precursor of LUCA? Did the viruses not only supply the nuclei but, possibly, the whole cell?

The giant viruses, which will be discussed below in detail, strengthen the suspicion that viruses could be precursors of cells. If LUCA was a virus then it should be named LUCAV, V for virus, as proposed by Eugene Koonin from the NIH, Bethesda, USA. He does not mean a pre–virus, not a single viral sequence but a mixture of many different virus sequences, reminiscent of the quasispecies of Manfred Eigen. Yet Eigen does not believe that viruses got there first and were important at the beginning of evolution. Are the bacteria blown-up viruses? Newest analyses support this notion, because there is such a broad spectrum from tiny viruses to giant viruses; the latter are even bigger than some bacterial cells. As well as Eugene Koonin from Bethesda, Luis P. Villarreal from California has long been a proponent of the virus-first concept, evolution from viruses to cells. The very first cells and viruses resembled each other, a little bag made out of lipid membranes and filled with increasing numbers of biomolecules, the first biomolecules of life. The transition is not sharp, it is more gradual, so that one does not need to be dogmatic about whether the first simple structure was a virus or a cell.

Viruses for detecting viruses — the PCR

Reverse transcriptase is essential not only for replication of retroviruses and for the "copy-and-paste" events that take place in our DNA genomes, but also in daily life in the laboratory. The reverse transcriptase RT is important for the "polymerase chain reaction" (PCR), an amplification system. This amplification resembles a simplified virus replication in the test-tube where genes are amplified exponentially. This technique revolutionized the field of diagnostics, because it is by far the most sensitive method to detect nucleic acids from microorganisms, viruses, bacteria, tumor cells, any tissue samples, fossils, mummies, hairs from extinguished bisons, etc. Even mutations — single genetic changes — can be detected. Many routine visits to a doctor's practice include this reaction for diagnostics or for monitoring the status of a disease or response to therapies, because the reaction is also extremely fast. Microgram amounts can be produced, enough for staining methods that allow one to "see" the amplified DNA by eye. Then amplified DNA undergoes further analysis (sequencing). There is a variation: If the starting material is RNA, then an RT is included to make DNA — just as for retrovirus replication. The amount of RNA reflects the activity of a gene, which is informative for research and medicine. Aberrant expression may be indicative of a disease. Therefore, it is important that the reaction can be carried out quantiatively in real time, and indeed "qRT-PCR" is a method of analysis in very frequent use. "Expression profiling" is the name for this analysis, performed to determine the activity of all the genes of a cell. Its sensitivity is enormous, and for example in the context of HIV the viral load can be detected down to a level of sensitivity of 20 viral copies per milliliter. This is what newspapers describe as "cure"; it is no real cure, but the amount of viruses is below the level of detection. Technical tricks make the reactions faster and more specific, and thermostable enzymes for the copying reaction allow fully automatic amplification. The enzyme originates from a 60-degree hot geyser fountain in the Yellowstone National Park but then from many hot springs too. A researcher trying to catch heat-resistant bacteria from the rim of a similar hot spring to isolate the thermostable RT was reported to have slipped and lost his leg — a sad story if it is true. The name of the enzyme is Taq-polymerase, with Taq standing for the heat-loving bacterium *Thermus aquaticus* (*not an archaeum*).

The method is so sensitive that it can amplify amounts of nucleic acids so low that they may be irrelevant — or may not relate to diseases — yet! Thus the ultra-high sensitivity could be counterproductive, and blind faith in it could be wrong or even dangerous. Indeed, we do not know whether small amounts of pathogens or even cancer cells can be removed by our immune system and would never make us ill — but could be detected by PCR and worry us.

However, it did help us once to find a leak in a safety cabinet for dangerous HIV DNA fragments that had escaped from test-tubes, and since we were developing a DNA-based HIV vaccine, we were all vaccinated by

4 Viruses and Cancer

The Tasmanian devil

An important note to begin with: cancer is not an infectious disease. This is very important for those who deal with cancer patients — so do not worry.

Is it really true? Always? Are there any exceptions? Unfortunately, there seem to be some: Tasmanian devils are among the very few exceptions. These animals have been carefully investigated to find out why their tumors are infectious. They bite each other wildly and they look frightening — like devils! Through bites into each other's faces they cause facial cancer in 90% of their victims. These animals are about to become extinct. Surprisingly, no contagious agent such as a virus has so far been detected. The reason why they transmit cancer is that they are already sick to begin with. They have a compromised immune system, so that cells transmitted by bites from one animal to the next one are not recognized as foreign, but are "tolerated" by the bitten animal's immune system. Recent genome sequencing analyses showed that the devils are all highly interrelated. Cells from other animals are normally recognized as foreign by the immune system, and are rejected. Not here. Therefore, the tumor cells can be transmitted and lead to new tumors. The furious biting behavior leads to bloody wounds so that the tumor cells get directly into the bloodstream. The tumor cells reach all regions of the body immediately, like metastases. This result is a big relief for all cancer researchers, because it means that the Tasmanian devils represent a really special case. So, everything fits again: cancer is not transmissible. Not normally.

Similar to the devils are human patients who are immune-suppressed for organ transplantation. In a patient, a foreign organ is normally not "self" but rejected due to individual cell surface markers, the Major Histocompatibility Complex (MHC) molecules, which are not compatible. The patient needs to be immune-suppressed, to allow acceptance and survival of the transplant. This is already present in the Tasmanian devils.

Researchers use mice without an immune system as animal model for studying cancer cells. The mice have a genetic defect resulting in loss of T cells, components of immune systems, and also of body hair, which gave them the nickname "nude". In contrast to normal mice, nude mice can be injected with tumor cells, and these are then not rejected but grow into tumors, even though they may originate from other distant species such as humans. We, like almost every cancer researcher, used these mice routinely. Some other special laboratory mice are the SCID mice, which allow testing of foreign tumor cells: SCID stands for Severe Combined Immune Deficiency — a disease that can also occur in humans (somewhat similar to AIDS). Such patients have to live under sterile conditions, under special tents, to avoid the risk of infection. Such mice are difficult to breed too. So far I have not considered using Tasmanian devils as animal models for studying human cancer cells — they are just too furious and, indeed, diabolical. Only two other naturally occurring animals with transmissible tumors are known, one in dogs (Canine Transmissible Venereal Tumor, CTVT), which is transmitted sexually and affects their sexual organs, and one in hamsters — both of these may resemble the genetic disease of the devils.

Retroviral oncogenes

Especially the retroviruses have taught researchers how viruses can contribute to tumors. Such retroviruses are therefore also designated as tumor viruses. They carry cancer genes or oncogenes. Consequently, many retrovirologists think of themselves as cancer researchers — as indeed I do. Many of these viruses are laboratory strains and rarely occur in humans, and, if so, then only in combinations with other factors, which then all together cause cancer — often after decades. However, they can cause cancer in animals, mice, cats, cows, even in plants, and also in monkeys.

About 100 viral oncogenes are known. They all are characterized principally by their ability to stimulate cellular growth. Those that do not induce growth advantages, or even suppress growth, do not manifest themselves so easily and are often overlooked. Tumor viruses can lead to small piles of cells in the culture dish, "mini-tumors" designated as foci.

They can be big enough to be seen by the naked eye. Researchers picked the biggest foci and thereby selected the viruses with the strongest growth-promoting oncogenes. In animals, viruses with the "best" oncogenes caused the biggest tumors. These viruses helped us researchers to select for the most dangerous cancer genes, now — 40 years or so later — often used by industry as the best targets for anti-cancer therapies which indeed led to today's most efficient drugs. How test tube oncogenes play a role in human cancers requires explanations, please wait.

Normal cells stop growing as soon as they touch their neighboring cells. This phenomenon is well known to everybody who works in the kitchen, where a cut in a finger will be healed by growing cells, which stop dividing as soon as they touch their neighbor cells; this phenomenon is called contact inhibition. This stop signal is missing in tumor cells, which keep growing on top of each other — an important aspect of tumor cell growth.

There was a long search to understand where the oncogenes of tumor viruses came from: from the inside or the outside of a cell. To answer that, researchers first had to learn about gene exchange between viruses and cells, Horizontal Gene Transfer (HGT). This is a process whereby genes are transferred in both directions: genes from the inside of the cell get out by means of viruses and the viruses can transfer them back into the inside of new cells by infection. The newly acquired viral genes can be deposited by integration into the cell's genome. Thus, gene transfer proceeds in both directions. The genes picked up by the viruses are random cellular genes — initially. When we, the experimenters, selected for fast cellular or tumor growth, we ended up with viruses that have special genes, ones that confer upon the cells the greatest advantages in growth — a hallmark of cancer. An additional phenomenon plays a role. The high mutation frequency of viruses contributed further to the most effective growth-promoting carcinogenic genes. The mutations arise during virus replication through the reverse transcriptase, which is highly error-prone, "sloppy". During each replication, about 10 mutations occur within 10,000 nucleotides, the size of a retrovirus. This high error rate makes the RNA-containing viruses so inventive. Mutagenesis is part of the survival strategy of viruses. Mutations have increased the oncogenicity of the oncogenes. Cells with DNA genomes such as our own cells have developed proofreading mechanisms that largely prevent mutations. That is good for us, because otherwise we would have a higher chance of getting cancer. The cellular genes are known by the abbreviation c-onc (or proto-onc) and after mutations by the virus they are called viral oncogenes v-onc (or viral-onc) — which are the "better", because stronger, oncogenes. Thus, oncogenes originate from the inside of the cell, but the virus modifies them.

A huge sequencing project was performed by the Sanger Sequencing Centre in Cambridge in the UK, where tumor cells were screened for cancer genes. Humans have 20,000 genes. How many of those are cancer genes? The researchers detected hundreds of them, but interestingly the most efficient ones were often the viral oncogenes which the researchers selected and isolated. Thus it was the viruses, and not the sequencing machines, that helped us to find the oncogenes. Furthermore, they were first found with tumor viruses in the laboratory, and later in human cancer — independent of viruses. About a dozen of them are today targets for rather promising drugs against human cancers.

The viruses taught us what causes cancer. Nowadays we know of some 100 oncogenes from retroviruses. How do they cause cancer? It may be worth looking at the discovery of the first oncogene — the "sarcoma" or "src" gene; this is part of the Rous Sarcoma Virus RSV, which causes tumors in chicken. Here comes the story of its discovery.

The sarcoma saga

Peyton Rous, the discoverer of the first tumor virus, gave up. Just like Galileo Galilei, who lied before the Inquisition to save his life and abandoned his research, Rous terminated his cancer research under pressure from his colleagues. They could not reproduce his results, and instead of solving the problem Rous simply stopped bothering! About 100 years ago he had shown that cancer could be transmitted, like an infectious disease, in chickens. A farmer brought his chicken with a big tumor to him at the Rockefeller University in New York City. The farmer was afraid other chickens might become infected, too. Rous was an experienced pathologist and diagnosed a sarcoma. He isolated the tumor, homogenized the cells, filtered them and injected the flow-through into healthy chickens. A new tumor arose. This fulfilled the "First Postulate" of Robert Koch, which Koch had coined around 1880, according to which an isolated agent can only be regarded as the cause of a disease if it is able to cause the same disease from which it was first isolated. Rous talked about a "filterable agent"; not a virus. He published this result in 1911. Similar results were obtained at the same time for leukemias and published in 1908 by two Danish scientists, Vilhelm Ellermann and Oluf Bang. The problem was that nobody could reproduce Rous's results. At that time nobody knew about genetic resistance against viruses and cancer. His competitors had used different chickens that were resistant against the tumor agent. Simple today. That was bad luck for Rous, and he missed the point.

I did too, 80 years later, not knowing about Rous's problems. What a nuisance! I tried to grow a leukemia virus in chickens to isolate large quantities for research for the isolation of the reverse transcriptase, which I had succeeded in doing for months. However, after the move of my laboratory from the Max Planck Institute in Tübingen to the Robert Koch Institute in Berlin nothing worked any more. The chickens simply did not get ill, and the virus did not replicate. In despair, I ordered chickens from the original supplier, a farm at a castle close to Tübingen — even though this meant a complicated transport of thousands of day-old chicks by Pan Am flights to Berlin, because (West) Berlin was in those days an isolated political entity in the middle of the former East Germany. With chicks from the old supplier, the infection worked again. The chickens became sick with the virus again (*the name is avian myeloblastosis virus, AMV*) and produce enormous — visible — amounts of virus. Many years later HIV research began. Again, it was first recognized with some surprise that there are resistant humans: 15% of Europeans do not get infected, because of genetic resistance, as they have a defective viral entry receptor (*CCR5delta32*).

It did not occur to Rous and his colleagues that they had used the "wrong" chickens. In fact, this knowledge might not even have helped Rous, because with such exceptions there would have been no confidence in his idea. He gave up! He did not understand the importance of his observation of a transmissible tumor in animals! Later on, he nevertheless became the "father of tumor viruses", and the first tumor virus discovered was named in his honor the Rous sarcoma virus (RSV). In 1966 he received the Nobel Prize, 55 years after his discovery when he was 87 years old. He had been nominated for the Prize 40 years earlier, but without success. He established a method for growing viruses in the chorioallantoic membrane of chicken eggs (close to the little air bubble) — and this old-fashioned method is still in use today for producing millions of doses of influenza virus vaccine — hard to believe. We used this technique even for testing a putative pox alarm case in the 1970s at the Robert Koch Institute in Berlin, using a little mill to open the shell, insert the suspicious sample, close the shell again and incubate the egg to allow the virus to grow — they grow like crazy in such nourishing milieu!

The viral v-Src protein is multifunctional. Several developmental steps of tumor formation are performed by this single multifunctional molecule. v-Src is shorter at its terminus than the analogous cellular protein c-Src — the truncation amounts to seven out of 536 amino acids. This sounds like a very small change, but viruses always find simple solutions, and the small deletion has huge consequences: loss of binding to the inner side of the cell surface, causing uncontrolled cell growth and cancer. Even the loss of a single amino acid at the very end has dramatic consequences: metastatic behavior of the cell. We detected this and found it so surprising and hard to believe that we analyzed it. We attributed this to the loss of binding to a tumor suppressor, which is a dramatic consequence, leading to loss of contact of cell-cell interaction, allowing the cells to run away to new locations and causing metastases. The next surprise was that this tiny loss of a single amino acid resulted in changes in about a hundred other cell functions. (*We showed this by "expression profiling" analysis of the metastatic cell. The procedure is described in the -omics chapter.*)

The v-Src protein is further intensified by the viral promoters (*the long terminal repeats LTRs*). These cause overexpression of the oncogene — a dangerous dose effect. Promoters normally direct the virus replication inside cells, bypassing cellular functions, the so-called egoistic behavior of viruses. These viral promoters are the strongest known in the whole of biology; they up-regulate the viral genes, thereby heavily overdosing the oncogene "promoting" cancer. They are the favorite tools for gene expression in all molecular biology laboratories.

The Src protein became a model oncogene, investigated by John Mike Bishop and his then postdoctoral assistant Harold Varmus in San Francisco in the 1970s in their effort to understand cancer. They tried to find the origin of Src, and they discovered it everywhere — all the way from flies to elephants, in normal cells completely independent of any cancer. That was enigmatic, and it seemed to suggest a systematic error. They withdrew their paper, which had been accepted for publication and was already making its round in our laboratory at the Max Planck Institute for Virology in Tübingen.

The trouble was that the technologies did not allow one to distinguish between a normal src gene from an oncogene: the slightly shorter oncogenic version looked the same. They solved the problem and received the Nobel Prize in 1989 for their work. I was present at their celebration in the Cold Spring Harbor Laboratory. Mike with a tuxedo and a manuscript — which he normally never needed for a talk. Varmus I remember riding his bicycle to meetings throughout Europe and arriving dripping wet — with more energy than most other people seem to have! They both visited my laboratory in the Max Planck Institute in Berlin and asked about the wild pigs in the Grunewald, mentioned in a travel guide — nobody had ever asked me about them before. Then there was the contribution by Peter Vogt from Los Angeles, who supplied the Rous Sarcoma Virus. Also the Berliner Peter Duesberg together with Steve Martin, both today at Berkeley, showed in an important experiment that a genetically modified inactivated Src does not cause tumors. That was the final hard proof. Peter Duesberg is the devil's advocate, who always stresses the opposite of what the research community believes (reminding me of Mephistopheles in Goethe's *Faust*). He has been right more than once but he was terribly wrong when he rejected the idea that HIV is the cause of AIDS. I recollect him telling this to hundreds of students during a podium discussion at the Humboldt University in Berlin — a view that I regarded as dangerous and misleading. We spent a lot of time sparring, never in agreement, but always friends.

Viral oncogenes without viruses — a paradox?

Retroviruses can cause cancer even without an oncogene. The integration of a DNA provirus into the cellular genome is highly dangerous for the cell. Wherever integration takes place, it occurs at random and can lead to "genotoxic" effects, damaging our genes, the DNA. These are among the most important factors that can lead to cancer: lifestyle and environmental factors such as cigarette-smoking, alcohol, poisons, car exhaust, contaminations, radioactivity and (though less so today) smoke from chimneys. Many chimney-sweepers succumbed to cancer of the testicles — that is how environmental carcinogens were first noticed. More recently some stomach bacteria have been identified as potentially carcinogenic. We do not know all the factors yet. Also unknown are the effect of the time points of exposure, the dose dependence, or combinations of factors.

Tumor viruses did teach us how cancer can arise through oncogenes. It is paradoxical that the best-studied DNA tumor viruses are animal viruses that do not normally cause diseases in humans: simian virus 40 (SV40) and polyoma viruses (PY), both are not normally associated with human cancers, yet they were both critically important in helping us to understand the molecular mechanisms of malignant transformation. The simple reason for this was that good cell-culture and animal systems were available. We have also learnt from these viruses that only in rare cases do they cause tumors, in so-called non-permissive cells, in which the viruses do not replicate but rather integrate their genomes into the host cell. In permissive cells the viruses replicate and lyse the cells, releasing the progeny virus particles without cancer. Integration in non-permissive cells is dangerous and can promote cancer in animal cells. Integration is also the hallmark of oncogenicity of retroviruses, even in the absence of an oncogene, then mediated by their strong promoters, the LTRs, which normally guarantee virus expression but can reach further to adjacent genes and muddle up their expression. Overexpressed or deregulated genes lead to a wrong dose or to activation at a wrong point in time. This phenomenon is termed "insertional mutagenesis" or "downstream promotion". Only recently it was discovered that our genome is full of "solo-LTRs" viral promoters, which are minimal leftovers from retroviral insertions that took place up to 100 million years ago! We have about 0.5 million of such solo-LTRs, the rest of the viruses is lost. That is a bit frightening, and researchers are presently asking whether they can still cause cancer. Our cells can counteract and "silence" their activity, especially during embryogenesis, as a safety measure for the next generation. Is that enough?

But now comes the real paradox: When human cancer cells were analyzed, many of the oncogenes known to researchers from studying the retroviral oncogenes (v-oncs), were detected in human cancers — without any sign of the tumor viruses. This sounds strange, but the explanation is simple: tumor cells normally grow faster than normal cells, no matter by what stimuli — irrespective of whether the initiators are chemical carcinogens, cigarette smoke, genetic defects, lifestyle, viral oncogenes or viral promoters. Thus, any mutated cellular gene can lead to increased cellular growth — a hallmark of cancer. What counts is the result, not the cause (not the virus!).

Viruses and cancer

Viruses can help to cause cancer — but they never act alone.

Members of six distinct families of viruses are capable of contributing to cancer in humans. About 15–20% of human cancers involve viruses as "cofactors" — a word that needs to be stressed.

The viruses comprise: a rare retrovirus, the human T-cell leukemia virus (HTLV-1, distantly related to HIV); the hepatitis B virus (HBV, a pararetrovirus), and hepatitis C virus (HCV, an RNA virus), both of which cause liver cancer; two human papillomaviruses (HPV) leading to cervical and other genital cancers; herpes viruses such as Epstein–Barr virus (EBV) contribute to two tumor types: Burkitt's lymphoma (BL) in Africa and the nasopharyngeal carcinoma of the nose in China (*EBV also contributes to mononucleosis, not a cancer*); and Kaposi's sarcoma-associated herpesvirus (also called human herpesvirus 8, HHV-8).

HIV is not a cancer virus, but it cooperates with other viruses and factors, only indirectly increasing the risk of cancer. The only known human retrovirus causing cancer is HTLV-1. It is endemic in Japan and can lead to a fulminant disease, adult T-cell leukemia (ATL) and in Africa to a tropical spastic paraparesis (TSP). The virus in Japan is transmitted in mother milk and can be easily avoided by educating the mothers to avoid breast-feeding. The virus has some kind of an exceptional oncogene, Tax, which is not related to a cellular gene as has been described for other oncogenes above, yet it behaves in a similar way: it is a transcriptional activator and activates cellular genes, leading to a potent chemical loop consisting of a growth factor and its receptor. This combination drives cellular growth. Tax is thought to play a role in seminomas, a non-malignant tumor in young men, which often is recognized in preventive medical check-ups of young military recruits. Other leukemia retroviruses are known in animals, such as sheep, cows, or mice. A mouse mammary tumor virus with hormone dependency has been extensively studied, but has never been found in human breast tumors. Efforts to prove its existence went so far that Sol Spiegelman, in New York, published scales with counts per *ten* minutes (instead of the conventional counts per minute), to make the data look more impressive, perhaps hoping that nobody would notice. We all did! And he was wrong! No cancer virus — as far as we know — causes human breast cancer.

Goats, sheep and horses can fall ill from retroviruses (*caprine arthritis encephalitis virus* and *equine infectious anemia virus*) but do not normally die of them. Human equivalents of these animal viruses have never yet been found. Surprising.

Important human cancer viruses are the hepatitis viruses HBV and HCV, both of which cause liver cancer. The genome of HBV consists of a double-stranded DNA with some single-stranded areas looking unfinished, and it replicates through a reverse transcriptase as do retroviruses, which led to the name pararetrovirus. The main difference between the two is what gets packaged into virus particles: retroviruses package RNA and pararetroviruses package DNA. Then the replication cycles are very similar. HBV does not normally integrate into the host genome. If it does so against the rule, then it becomes genotoxic and very oncogenic. HBV is supported by a chronic liver inflammation — sometimes enhanced by lifestyle and alcoholism. Then the multifunctional X protein, a kind of oncogene, contributes to cancer. The cancer caused by HBV is hepatocellular carcinoma (HCC), which is most frequent in certain areas of the world through contributions by other factors, such as aflatoxins in China from the fungus *Aspergillus flavus*, a contaminant of food. After World War II we scratched fungi off the marmalade or cut off the end slice of mouldy bread. That was not enough. Born from a shortage of food, it should not be repeated, because the dangerous mycelium is not removed in that way: everything needs to be discarded. A vaccine against HBV was the first successful one against two problems, against a virus and cancer. It was, furthermore, the first recombinant vaccine, produced by means of recombinant DNA techniques in yeast (DNA pieces are combined) and without any danger to the recipient. This admirable success was achieved by the legendary Maurice Hilleman from the company Merck — who also developed a dozen other antiviral vaccines. He saved millions of lives with the hepatits HBV vaccine — without getting the Nobel Prize. The vaccine against HBV was the model for HIV, but has so far still failed. HIV is highly variable, HBV is not.

"How many of you have been vaccinated against HBV?" was my question 20 years ago when teaching medical students in Zurich. Only a few had been, out of a hundred. My niece went to work at a hospital in England and I packed into her luggage the syringe for the necessary third vaccination against HBV. Nobody cared then. Now it is no more an accepted occupational disease if a health-care worker gets infected. The vaccine is now obligatory for all of medical employees. It should be given to everybody, with a subsequent test to verify the immune response or repeat the injection — especially before traveling to distant places.

Then comes the hepatitis C virus, HCV — a relatively new disease; it is caused by an RNA virus and can become chronic, leading to hepatocellular cancer, HCC. It was long searched for as a "non-A-non-B-hepatitis" virus. Finally, it was identified with the help of public research but then blocked by a US company from further analysis — a terrible story, which delayed research significantly. The route of transmission is not completely clear. Many do not know how they could have acquired an infection with hepatitis C virus — by blood contact, by use of needles, or even in some clinical setting. About 40 years ago contaminated needles unintentionally caused the infection of millions of Egyptians, who received three injections with a drug against schistosomiasis (also named bilharzia). This is a parasitic infection spread by urine through contaminated water, such as the river Nile in Egypt. The parasites were reduced successfully, but HCV infections occurred instead through non-sterile needles. Nobody knew about HCV yet, and people got infected by reuse of contaminated needles. Even though reusing needles is now known to be dangerous, it is still often a matter of cost. Now liver cancer is on the rise, and Egypt has a high degree of expertise in liver transplantation. Such needle contamination also happened in the former East Germany, when 3000 females were injected against Rhesus factor incompatibility, a characteristic protein of red blood cells. This genetic disposition can prevent having a second child — as had been the case with Goethe and his wife. And nobody wants to be reminded of the football "miracle of Berne" in 1954: several team members went down with hepatitis — not yet known to be due to HCV. Only dozens of years later was this understood, when several tumors arose. They were attributed to vitamin injections with a contaminated needle passed around among the players.

Two pioneers of HCV research are Charles M. Rice from the Rockefeller University, New York, and Ralf Bartenschlager from the University of Heidelberg, who developed a simplified replication cycle of HCV in the testtube by using a "replicon", a truncated virus, which allowed drug screening under low stringency safety conditions. Recently, new drugs have become available killing HCV in three months with complete "clearance" of the virus from the body of a patient — a dream still out of reach for HIV. This is a big success story, even though one therapy costs €54,000. This should be available to everyone affected — but there are restrictions by health insurance supporting only late stages. For many years an Interferon therapy was the only — but much less successful — option. Fortunately, the Egyptian patients are receiving the drug at a price of 900 US dollars per cycle. But why is HCV still spreading in Egypt is unclear. 15% of the population are infected. Is it due to the custom of being shaved at a barber's shop — and getting infected there? Foot hygiene by podologists is not popular in Egypt (!), but in Western countries where there may be some risk involved as well. About 170 million people are infected worldwide by HCV. Rice and Bartenschlager both received the Robert Koch Prize in 2015. They insist that we still need a vaccine to eradicate the virus fully. Perhaps this will have to be developed by academia, because the companies are currently satisfied with the lucrative drugs!

It may be worth mentioning that chronic inflammation is a cause of cancer even in the absence of other factors. We have learnt that, in animals, a piece of metal if fixed underneath the skin can lead to cancer. Similarly, permanent scraping by a dental prothesis can lead to cancer. And we all are aware of the cyclist Lance Armstrong's cancer — which he may have acquired from his saddle, because of riding his bicycle for too many hours — or years! The tumor was probably caused by constant abrasion.

Papillomaviruses belong to a group named "Papovaviruses": the papilloma, polyoma and SV40 viruses. They look like prototypical viruses, consisting of icosahedral structures with double-stranded DNA. The human papillomavirus, HPV, attracted a lot of attention when Harald zur Hausen from Heidelberg received the Nobel Prize together with the two HIV researchers Françoise Barré-Sinoussi and Luc Montagnier in 2008. Now mothers of young daughters are confronted with the question of whether girls should be vaccinated against the cancer-specific types of human es before they become sexually active, because the vaccine against the viruses only works prophylactically, even though the virus may cause cervical cancer only decades later during their lifetime; this is sometimes detectable as genital warts, but only a few develop cancer. Other factors such as hygiene, genital infections and even smoking can contribute to it. Only two out of 178 known human papillomavirus-types cause cancer, numbered HPV-16 and HPV-18. Also, young men should be vaccinated, according to zur

Hausen. HPV contributes to cancers of the penis, anus, mouth and throat. It is involved in 5% of all tumors worldwide. We do not yet know how well the vaccines will work, because it takes decades for cancer to develop until we see a protective effect. Now young men are advised to get a vaccination. (*The oncogenic variants of HPV, the HPV-16 and-18 express two oncoproteins E6 and E7, early transcription factors, which abolish tumor suppressors (p53 and Rb) and act as growth stimulators.*)

Papovaviruses do not normally integrate, but if they do then they are genotoxic and dangerous. Polyoma (Py) viruses can become threatening in immune-suppressed patients after transplantations. (*These especially so-called BK or JC viruses named after patients, can lead to tumors, while SV40 is normally active only in animals.*) Normally these viruses do not cause diseases: 75% of people are infected, but only kidney transplantations and associated immune suppression are a risk to activate them. Only 1% of infected cells integrate the viral DNA into their genomes, almost like an accident. This then contributes to cancer development. The simian virus SV40 once contaminated a poliovirus vaccine which was produced many years ago in cells from monkeys (simians), African green monkey kidney cells. It was a major concern whether this vaccine could cause cancer as a late consequence. This vaccine was given to millions of Russians in the former Eastern Bloc. For decades they were monitored for potential tumor development. This catastrophe did not occur.

Poliovirus production for vaccination was also considered as a possible source of SIV, a monkey relative of HIV, with potential consequences for the spreading of HIV. A committee concluded that this was unlikely.

Cancer causes about 10 million deaths per year worldwide, and this figure is increasing, especially in poor countries. In about 15 to 20% of cases, viruses are involved. To identify new viruses as causes of cancer, old-fashioned epidemiology is still extremely informative. Zur Hausen finds candidates for new viruses, that cause colon cancer, in steaks — especially in Argentina, where beef is eaten very frequently. Steaks there are very thick and often not sufficiently well heated to kill any viruses further inside. Another epidemiological study showed that Norwegians, who drink less milk because of frequent lactose intolerance, have fewer breast tumors. Are there cancer-causing agents in milk? Too early to say. Both possibilities are being discussed by zur Hausen — who was right, with the human es, HPV once before!

Cancer caused by microorganisms such as bacteria has recently attracted attention. The bacterium *Helicobacter pylori* may lead to cancer in the stomach or a bit further up in the oesophagus. Also, a sexually transmitted bacterium, which replicates inside cells, *Chlamydia trachomatis*, contributes to ovarian carcinoma, as described by the Max Planck Institute for Infection Biology in Berlin. It has the dangerous genotoxic effect and also blocks gene repair and prevents cell death — all features promoting cancer. Antibiotics have also been discussed as indirect cancer promoters. If patients need antibiotics while undergoing chemotherapy — that may be a serious disadvantage reducing the effectiveness of the chemotherapy, because bacteria in the gut are required for an effective cancer therapy and should not be destroyed by antibiotics. "How many patients may I possibly have killed without knowing it?" was the question raised by a famous oncologist during a recent meeting.

So one virus causes only one disease? No, that is wrong. A virus and many co-factors can cause many different diseases, a fact that has often caused confusion or misunderstanding. The herpes virus, Epstein-Barr virus (EBV) is a virus with many manifestations. It causes Burkitt's lymphoma (BL) with swollen lymph nodes often in children in Africa and a nasopharyngeal cancer in China. Textbooks describe a single virus plus different cofactors as cause. Indeed, new sequencing techniques have made it possible to determine genetic differences in the virus in both cases and in a third disease as well: mononucleosis (a glandular fever) or "kissing disease", often occurring in the USA among college students, which is not a cancer but also involves the same virus, EBV. Genetic variants have now been identified by sequencing the three viral genomes. The viruses differ. Various cofactors also the Myc protein contribute to the growth of these tumor cells.

Another human herpes virus is the HHV-8 or Kaposi-Sarcoma-Associated herpes virus (KSHV). It was a hallmark of HIV during early days in San Francisco, because young men were infected whereas the disease was formerly only known in older men. HHV-8 leads to angiogenesis (blood-vessel growth), so that anti-angiogenesis drugs were developed, such as the anti-vascular endothelial growth factor (VEGF). This could also help against other tumors, because new blood vessels are necessary for supply of nutrients for growing tumors. The drug is used for treatment of ovarian cancer, but also against macular degeneration, which leads to blindness. This 100,000-dollar therapy is controversial and often has only transitory effects.

HIV does not cause cancer, yet cancer is often a late-stage consequence: HIV is effective in concert with other viruses, e.g. HIV together with EBV leads to lymphomas, HIV with KSHV to Kaposi sarcomas, and HIV with HPV to cervical carcinomas. Several viruses cooperate late during HIV infections. Important is the role of the immune system. Even with anti-retroviral therapy, the immune system is reduced and not good enough to prevent cancers. Is that also true of aging immune systems? Indeed, cancer rates increase with age. What gets lost during aging? Why does cancer increase with age, with or without HIV infection? To find out more about this, a new Institute for Cancer Research was founded under the direction of the Nobel Prize winner Harold Varmus, who was one of the discoverers of the oncogene protein Src.

Strange fatalities

Many years ago my mother sent me a note, a newspaper cutting, which I still have today, about unexpected mortalities among rather young scientists who worked with a tumor virus in Paris. I worked as a scientist at the Max Planck Institute for Molecular Genetics in Berlin on oncogenes of viruses (*more than a handful, such as Myc, Myb, Mil/Raf, Src, Ets, and Erb-B2*). In an institute in Paris several scientists had passed away. "Could this happen to you?" was the question of a worried mother — rightly so. A committee in Paris reached the conclusion that all tumors were of different types, so that a common factor was unlikely. A causal relationship with research on animal tumor virus SV40 was not proven, because all tumors were different. The tumor antigen (*T-antigen of SV40*) was excluded as a cause — wrongly so? This is 40 years ago now. The role of cofactors is complex. One person may smoke, another works with radioactivity or poisons such as ethidium bromide, routinely used in laboratories, or has genetic mutations in his genome — in combination with the T-antigen all this may lead to *different* tumors. I know of a number of cases, in which I suspect the T-antigen to have contributed significantly to tumors and premature deaths of colleagues. I always warned my students to be careful with SV40 and the T-antigen — even though the virus rarely or never causes cancer in humans. However, in the laboratory the doses are extremely high compared with that in Nature. I always worry and consider chemicals in the laboratory as potential carcinogens and causes of cancer. Just as an example, take mouth-pipetting (which today nobody can imagine any more). A colleague of mine developed a tumor in his mouth. He had worked for years with the tumor virus SV40 with mouth-pipetting. Was there a connection? These are case reports and no solid proofs for the cause. Rubber bulbs, for some reason always red, were developed to replace mouth-pipetting. Today we have the automatic pipettes, so that mouth-pipetting is now banned.

The researcher Hans Aronson took a sip of diphtheria toxins through a mouth pipette at a research institute in Berlin, and passed away. His wife was devastated when he died, and she founded the Aronson Prize — full of grief. She had diamond balls on her shoes, I was told, but she could never get over her husband`s death and starved to death herself. Her niece survived the pogroms in Germany and was present when I received the Aronson Prize for Cancer Research in 1987. I bought a small pipe-organ with two keyboards and a full pedal-board, because the award was sponsored explicitly for a private hobby and not for research. For decades I had absolutely no time for practicing — exactly as a colleague predicted when I started working at the University of Zurich — but now, after retirement, I can try again!

The T-antigen of SV40 had a follow-up. It causes tumors in animals and was therefore selected by many researchers wishing to understand cancer better. A single gene — that was not expected to be too difficult to find its function in cancer. President Nixon declared in 1971 the "War on cancer" in a National Cancer Act, based on the success of the space program. He thought simply that technical progress would solve the cancer problem, just like flying to the moon. This initiated the research on the T-antigen. But Nixon was wrong. In research it is still not clear, even today, how many different functions such a protein can fulfil — hundreds. It is almost omnipotent. Worst of all, I think it is dangerous in the laboratory.

Retroviruses as teachers of cancer research

Retroviruses have shown researchers the direction in which to look for oncogenes. This initiated, about 50 years ago the field of molecular cancer research. My group was part of the crowd. With enormous luck we fished out seven oncogene proteins at a single stroke — normally a success limited to fairy tales. That is what happened: I was organizing in Berlin a laboratory training course under the auspices of EMBO, the Molecular Biology Organization on viruses and oncogenes for 30 students from all around Europe and the U.S. A guest from the U.S., Mette Strand, introduced a technology that was brand new to the students — and to us — the production of monoclonal antibodies. We learnt this technique by teaching it — also an interesting variation how to make progress.

This technology is standard now. One can even order tailor-made monoclonal antibodies from university service centers free of charge. The principle is the fusion of two cells: one is an immortal tumor cell-line derived from a tumor called multiple myeloma or plasmacytoma, and the other is the antibody-producing spleen cell from immunized mice. The fused cells, called hybridomas, have to undergo selection, picking individual cells, and propagation. This finally gives rise to a single clone of identical cells, all producing one highly specific antibody against one specific antigen (epitope). That explains the word "monoclonal". They can be adapted to humans away from the mouse and are then the basis for various highly specific cancer therapies. The name of such therapeutics ends with "…mab" for monoclonal antibody, and they are among the best cancer therapies available today.

A co-worker in Berlin was so enthusiastic about the new technology that she continued producing monoclonal antibodies, almost as favors for colleagues, against the most unusual antigens such as the variant surface proteins of trypanosomes, the cause of sleeping disease. One "mab" we tried to design against an oncoprotein — but failed. The "mab" recognized the wrong part of the protein, not the oncogenic part (*which was too highly conserved and not immunogenic*). Yet this part was universal to dozens of other oncogenes, so we had a fishing rod to isolate them all. We spent a hot summer day and night characterizing those numerous oncoproteins and could not keep up with handling all the information. This is a rare event in science and it never happened again to my co-workers or myself.

We fished oncoproteins named Myc, Myb or Erb-B2 and a novel kinase Mil/Raf. For that we organized a control antibody — and added the name of the supplier as co-author of our *Nature* paper in 1984. That he had taken this antibody out of someone else's freezer I only learnt years later, when the colleague who had produced it initially asked me how I had got it and why I had not thanked him for it, while we were standing in Cold Spring Harbor waiting for an airport limousine. The paper was out and I could not include his name any more or take the other name out. The new kinase was not immediately accepted: was it a cell contamination? No, the "mab" we had developed allowed such efficient purification that we were right. (*We proved this further by mutating the kinase so that it lost its function. This is always the final proof.*)

The newly identified Raf kinase works in a cascade, whereby one kinase delivers a phosphate to the next and activates it. This is like a relay race, with four runners designated Ras, Raf, MEK, and ERK. ERK is the final sprinter, crossing the nuclear membrane barrier and delivering the phosphate to Myc (or others), which then turns on or off other cellular genes and programs as transcription factor. The pathway is universal and can, depending on the first stimulus, regulate phenomena such as growth, aging, hormone production, stress response, smell or even death. We may compare it to a TV antenna, which can transmit any program to the TV set in the living room where we chose the channel for a specific "program".

Four consecutive "runners" allow amplification, better regulation, interruption, and safer transmission than one. As everywhere in biology, the result depends on the input, the context, additional factors or cross-talk. However, a central switch is the Raf kinase. Raf is a target of Ras. It looks like a closed shell, but it can be activated to a more oncogenic, partially opened shell (*c-Raf turns to B-Raf*). The oncogenic version is active in many cancers, in half of all malignant melanomas, in ovarian, colon, oesophagus, and in breast tumors. Only recently did industry succeed in producing inhibitors against Raf, about half a dozen of them — 30 years after our discovery of this kinase. Raf inhibitors can be applied. The survival time of patients is only a few months on average — rather disappointing. A triple therapy should do better. HIV did teach us a lesson about multiple therapies — also needed against cancer. Recently, I met a former co-worker from my Institute, now a professor of dermatology, who successfully treated melanomas with Raf inhibitors. "I did not know you discovered it! We worked with it but you never told us." True. 30 years is a long time.

Another clinically important oncogene protein is Ras (Rat sarcoma), with a single point mutation in codon 12. Inhibitors against Ras still do not exist today, in spite of gigantic efforts on the part of industry and of its prominent role in 80% of human cancers. However, Ras has so many targets which would lead to side effects — but a new progam has been initiated with lots of financial support.

Also breast cancer has an activated oncogene as target for therapies. HER-2/neu or Erb- B2 is the receptor tyrosine kinase, which is overproduced and oncogenic in 25% of breast tumors. Only in those cases is the expensive drug Herceptin applied. Personalized medicine is already practiced with this drug, simply because the expensive drug should not be given if it is not expected to be effective. This oncogene was originally discovered in the avian erythroblastosis virus (AEV) which has today almost been forgotten. This chicken virus has indeed nothing to do with breast cancer — which is still an enigma to me.

The oncoprotein Raf can induce two opposite effects: to activate growth, but also to stop it. We called that the "Raf paradox". We attributed this unexpected Janus-like behavior to another kinase (Akt) and other cellular factors. This result was accepted in the journal *Science*. But then something stagnated, a second paper from a former co-worker, who went to New York, was apparently also submitted to *Science* causing trouble and delay to ours. About their negotiations with the Journal we only heard rumors, and no real facts. I never dared to negotiate with a Journal. The information about our data must have crossed the ocean through my Swiss co-worker. He liked to talk about his wonderful results long beforehand, also to his friend when they shared a room at a meeting in the US, and then the telephone did not rest. It was a detective story with a happy ending — because both papers finally ended up "back to back" in *Science*, the second one including us also. I fought for it! And what is the take-home lesson? *Talk* about your results — you may end up with *two Science* papers — if you are lucky!

Another lesson refers to the patients: The "Raf paradox" has consequences for a therapy, inhibition of growth is intended but what about inhibition of growth arrest? The tumor will grow — opposite of the desired effect. Recently authors described the dangerous side effects of new drugs in patients, as we had predicted and shown in cell culture. They did not mention or remember anything about our previous publications. This has become a new trick: ignore the literature and hope that the reviewers will not notice. This will then upgrade your discovery and priority. And indeed, very often the reviewers do not notice. I — to my own surprise — got carried away during a meeting, complaining to the speaker in front of a scientific audience about our "Raf paradox" predicting his trouble. "We used to be friends", I started my attack on a colleague whom I had known for 40 years and whom I had first informed in a personal discussion, that he seemed to have "forgotten" our two *Science* papers. "What exactly do you want?" asked the chairman. "Correct citation!" was my answer. The chairman and the speaker were upset, but other participants applauded and later brought me chocolate and congratulations — even the next day — for having been so "courageous". No, I was only furious! And a bit ashamed. On the following day several editors initiated discussions on this well-known "scientific misconduct". There are literature-search machines for correct quotations. The pharma industry seems to use them — at least they send me all the information about Raf-reagents for research purposes — which I now do not need any more after retirement.

The Myc protein and reactor accidents

Another class of oncogene proteins comprises transcription factors, which regulate cell programs at the level of DNA transcription. A very prominent one is the Myc protein, a universal player in gene regulation.

All our genes are present in every cell, but only a special subset of them is activated. This is how the cells of our body, of which we have 200 types, are specialized. Activation can go wrong: oncogenic transcription factors of tumor viruses can turn on wrong programs and cause cancer. An important example is the Myc protein.

To study it, we had to grow the virus in quail eggs (nobody could ever explain to me why we needed to use eggs from quails and not chickens!). The eggs tended to disappear in the animal house — I was told that they are even better aphrodisiacs than ground-up African rhinoceros' horns! Once we hatched hundreds of eggs, producing little quails to bring them to a huge birdcage, a volière on the Peacock Island of Berlin to clean up the ground. We took an enormous box, full of little birds, across the lake by ferry to the island. After this successful experiment we were given a number of big peacock eggs by the delighted farmer, the only person allowed to inhabit the island. We tried to hatch them. This required a modification of our chicken-egg incubator to fit the huge size of peacock eggs, which the workshop of the Institute immediately made

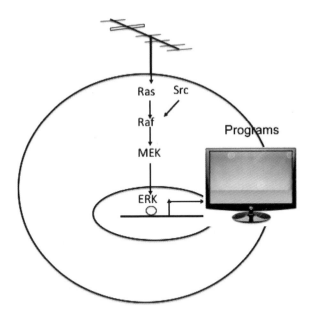

Signal transduction in the cell via Src or the Ras, Raf, MEK, and ERK kinase cascade for gene regulation and choice of programs in normal cells, but uncontrolled in tumor cells.

possible — sharing our enthusiasm with such an "applied" experiment — but sadly we failed: either the eggs were not fertilized or they had dried out. No new peacocks — what a sad story. So we continued learning about Myc from quail cells.

The normal Myc protein, the cellular c-Myc regulates normally about 15% of all human genes by binding to an enhancer. Myc itself is overexpressed in many tumor cells. Deregulation or untimely expression of Myc protein and its target genes can be tumor-promoting. Myc is one of many final targets of signaling pathways such as the Ras-Raf-MEK-ERK pathway regulating cell functions. There are two tissue-specific versions: L-Myc special for lung and N-Myc for neuroblastoma, a brain tumor where up to 10,000 fold overexpression correlates with very aggressive tumor growth. What makes tumor proteins dangerous is a wrong dose and a wrong time point. This happens frequently after nuclear reactor accidents which lead to broken chromosomes and often result in blood tumors, leukemias and lymphomas. Blood cells repair broken chromosomes with their recombinases, also the wrong breaks leading to blood tumors. An illegitimate combination of chromosome 8 and 14 leads to overproduced Myc protein and leukemias as observed after the disaster at Chernobyl. It is abbreviated by t for translocation, t(8,14), Another chromosomal translocation t(9,22) leads to the chronic myeloid leukemia, CML. Thereby the so-called "Philadelphia" chromosome arises, in which two growth-inhibiting proteins become truncated and fused to a highly growth promoting oncogenic protein Bcr-Abl, which lost both inhibitor functions — also a consequence of nuclear power failures.

In this case, luckily, an inhibitor was developed, Gleevec, an "orphan drug" (with fewer than 10,000 new cases per year worldwide) which a Swiss company started and abandoned. However, one of their co-workers, Nick Lyden, set up a spin-off company (Kinetix Pharmaceuticals) in the US, which resulted in a blockbuster, because Gleevec surprisingly turned out to be active not only against the leukemia CML but also against stomach cancer and other tumors. The parent company then stepped back in! Now drug resistances are on the rise, we need more drugs!

Myc was suspected to be a primary oncogene in a lymphoma, Burkitt's lymphoma (BL), but it is only a driving cofactor there. Our Myc results on its DNA binding ability for gene regulation brought me to a lymphoma congress about 30 years ago to Israel, organized by George Klein from the Karolinska Institute in Stockholm. As a German and guest of a refugee from a concentration camp it was quite a privilege for me to be invited. I was recognized for our results on Myc by an award that ended in an unfortunate way, because the prize committee did not agree to include my co-workers. When I shared the prize money anyway, this did not go down well with my co-workers, because money conveys no prestige. Moreover, the director of a research institute in Seattle, who happened to be sitting next to me in a taxi on the way to a congress, was annoyed, claiming that his co-workers had had the results first. We published our results in *Nature* first without knowing theirs. However, the frustrated colleague was influential and cut me off from the "Myc-scene" for my lifetime! What did a Director of the Max Planck Institute in Berlin tell me? You will always have trouble with important papers — but never with boring ones. That was true for the rest of my life in research. By the way, most papers are boring!

Even without a myc gene the integrated retrovirus can cause cancer with its promoter by "downstream promotion", activation of unknown cellular genes, whereby the cancer in chickens we observed is similar to the myc-induced one, except with myc the tumor arises very quickly, without it takes much longer. This promoter insertion happened once with a gene-therapy which then caused cancer instead of curing it. The gene therapy was quite successful in ten children, but two of them developed leukemia, because the viral promoter turned "on" a cellular gene (*a transcription factor LMO2*) and induced leukemia — which, however, was then also controlled successfully. In theory this danger was always predicted, but that it would happen so frequently in two out of ten patients was shocking!

An international lymphoma study included, among many other substances, our monoclonal antibodies, "mabs" against the Myc protein to test it for diagnostics within the "Kiel classification" of lymphomas. This resulted in my being invited to take up an appointment as professor at the University of Kiel. However, the (female) minister of research, while commenting on my nice green-blue handbag (favorite colors of people from the seaside), never found the money for the disposal of infectious waste — which would have contaminated the Baltic. Don't get confused: I did not make it to Kiel, and did not end up in my great-grandfather's house. This was intended to be the Institute. The street name "Moelling Street" has nothing to do with me, but is in honor of my great-grandfather, who built the harbor, the North-East Sea Canal — one of the busiest waterways today — and railroads for train connections. He would have deserved a nicer street, though.

Myc was indirectly honored 2012 when the Nobel Prize was awarded for stem-cell research and tissue regeneration. Myc is one of four factors that can convert an adult cell back into a multipotent, actively dividing stem cell. However, that lies in dangerous proximity to tumor cells, which are characterized by deregulated growth, as will be mentioned later. Myc can make cells "fat" — a new property correlating cell growth and increased cell volume. Could that lead to high body mass? Nobody knows yet.

Are there drugs against Myc? As an adviser, some decades ago, I voted against a drug screen: A Dutch colleague invited me to Philadelphia to meet colleagues at the company Centocor (now Johnson & Johnson), waving the flight tickets. He was Jo Hilgers — known as the "Flying Dutchman", which indeed he proved to be. I had to evaluate oncogene proteins as drug targets. My advice was "no drugs" — blocking these proteins, which have so many important roles in normal cells, seemed to me to be too risky to treat. I was right concerning Myc: it is too important in growth, including stem cells (as we now know) to be targeted by inhibitors. Only recently, discussions have started in which L-Myc is being considered as a short-term target for compounds to treat lung cancer. A short, transient therapy, a small therapeutic "window", may be possible without too many side effects. This concept is new and now commonly encountered. However, I was totally wrong with my advice concerning the Raf kinase. I could not imagine that one out of a thousand kinases, which were already known, could be inhibited specifically, so I did not propose it as a drug target. Nobody thought of patenting either Myc or the Raf kinase as cancer targets, and no-one started an inhibitor screen. Now, more than 30 years later, there are potent Raf kinase inhibitors available, especially against the oncogenic form, which is already preactivated for cancer. (*The oncogenic B-Raf together with the normal c-Raf are important in cancer, because they hetero-dimerize and inhibitors react unpredictably.*) The new drugs are now the stars for cancer treatment and are already being used in clinical settings. Some other kinase inhibitors are very supportive in combination therapies. Inhibitors of Myc are not yet available.

All patents would have expired today anyway, and — just to make it clear — I am not getting rich, as some people suspect, and indeed I should not, because my research was paid for with public money. And my input has been completely forgotten about anyway — but I am glad that the (almost) lifetime that I spent on cancer research contributed a bit to the war against cancer!

Tumor suppressors and car crashes

What events cause tumors? Think of a tumor as a car accident; then there are two possibilities: too much speed, or brake failure. In both cases the car is driving too fast — symbolizing too much growth of a cell, and an accident, a tumor. Loss of the brakes corresponds to loss of a "tumor suppressor". A normal cell has two sets of chromosomes and thus two times the same brake. If both fail cancer arises fast — what happens if only one of the two brakes fail? Not a slow tumor but a late tumor arises.

This question was answered by Alfred Knudson from the Fox Chase Cancer Center near Philadelphia, analyzing an eye tumor, retinoblastoma. He detected this tumor either in very young children or in elderly people.

He was wondering about these extremes. If the gene, which he called the retinoblastoma gene, Rb, was defective twice, in both chromosomes, then the tumor developed in young children. Those born with only one defect, not two, developed the tumor much later in life and acquired additional defects during their lifetime. Thus, tumor development was either dominant or recessive, fast or late. He received the Lasker Award for this "two-hit hypothesis". The Lasker Award is a copy of the Greek goddess of victory, Nike, with two wings but no head, which is displayed in the Louvre in the stair hall because it is too big for the exhibition rooms. A copy also dominates the main building of the University of Zurich. Is this statue meant to stimulate the students striving at one of the biggest awards in science — one step away from the Nobel Prize? That would be a big incentive — if the students knew it, which I doubt. They may know the bright blue Nike by Yves Klein, or perhaps not that one either.

Thus only the disposition for cancer is inheritable, but one defective chromosome can still allow a healthy phenotype and remain unnoticed. Half the dose of the Rb protein is apparently enough to suppress a tumor for decades. Only additional mistakes will increase the progression of cancer. There is something to note — a factor of two in biology can be extremely important, as shown here, but it is often neglected as too small and irrelevant! This is terribly wrong — a factor of two can really matter!

There is another tumor suppressor protein, p53. It has a structure reminiscent of a hand with fingers, which touches and detects defects in the DNA, strand-breaks or mismatching mutations. Then p53 activates an alarm program, and stops cellular growth to gain time for carrying out the repair.

Alternatively, p53 induces cell death if the situation is hopeless. If both of these safety measures fail, the cell will become a tumor cell.

Tumor suppressor genes are the favorites of gene therapy against cancer. In spite of the multifactorial development of cancer cells, the phenotype can be reversed by a single therapeutic with an intact p53, supplied by an artificial virus. This is surprising, and it must indicate that p53 is a key player in the development of cancer. It must override dozens of defects to "normalize" a cell. Therefore, p53 is a favorite gene for replacement by gene therapy. A physician in the city of Canton (today Guangzhou) in China offers this therapy already, on the Internet. I would recommend waiting, though, until the approach is offered by scientists or medical doctors with approval by the regulatory authorities. Several groups — including one of the main discoverers of p53, Sir David Lane from the UK — are busy working on it in Singapore.

It is worth mentioning that many viral oncogenes owe their very high oncogenicity mainly to their ability to bind and remove tumor-suppressor proteins; thus, they have dual functions, they are oncogenes but also anti-tumor suppressor genes at the same time. This can be interpreted as a viral survival strategy, because a viral oncogene removing a growth inhibitor guarantees better survival of the host cell, and high viral progeny, by a mutualistic mechanism. Viruses can go even further and prevent apoptosis, a programmed cell death, for their own profit. A living cell is better for the virus than a dead cell. Therefore, inhibitors of apoptosis can be found in many viruses, adenoviruses, poxviruses, herpes viruses and the rather rare African swine-fever virus, in which the anti-apoptotic effect of viruses was discovered. An anti-apoptotic effect by a virus can lead to cancer, in which cells never die.

The various stages during development of human tumors were analyzed by Bert Vogelstein from Baltimore. He used colon cancer as a model. This resulted in frightening numbers: cancer development takes 30 years! The various stages are characterized by diverse activated oncogenes and loss of tumor-suppressor genes, about 54 and 71 respectively. Up to eight genes can function as "drivers" for cancer development, whereby drivers are single cells with mutations leading to proliferation rates higher than those of the neighboring cells. Thus, the cellular population changes constantly, making therapies complicated. Moreover, each person develops his own individual tumor, which requires individual therapies — personalized medicine is needed.

During the first seven years of a tumor development, hyperproliferation of cells comes first, characteristic of an early stage through benign polyps, so-called adenomas. Also, changes in cell-cell contact by adhesion molecules such as catenins will take place. Thereafter, small polyps grow up and become bigger, the survival kinase Akt is activated, tumor suppressor genes such as p53 or a cell-specific gene "deleted in colon cancer" (DDC) are lost, and then oncogenes such as mutated Ras and Raf kinase drive cancer progression. This process is estimated by Vogelstein to take 17 years. In the subsequent two years, tumor progression includes advanced carcinomas and, as late-stage activation of metastasis, genes such as Met or mn23 for metastasis, when the tumor penetrates into the bloodstream and disseminates individual tumor cells which have lost their cell–cell contacts. In summary, this process adds up to about 30 years. One defect in a cell leading to a growth advantage will increase the chance of the next mistake. When should one start a therapy? When do we want a diagnosis?

Metastases — and how cells learn to run

Long before tumor-suppressor genes raised general interest, a tumor suppressor was discovered decades ago by Elisabeth Gateff. She received the Meyenburg Prize for Cancer Research for it in Heidelberg in 1988, and otherwise her work would have been forgotten. She named the gene "discs-large". If it is missing, this leads to embryonic defects in flies and abnormal development of wings, legs and antennae. In my laboratory we ended up by accident with a project on this type of tumor-suppressor protein, analyzing a "PDZ" protein that binds to the oncoprotein Src. First we planned to inhibit the PDZ protein, aiming at preventing tumor growth. However, the D in the PDZ domain name should have warned us, because it stands for discs-large, indicating the protein is a tumor suppressor protein (similarly P and Z). An inhibitor against a tumor suppressor would activate cancer, not inhibit it. Meanwhile we were awarded a significant grant from a governmental program for Research and Technology, the BMBF, to intensify collaboration between academic research and companies, with Evotec in Hamburg for drug development. We debated how to avoid losing the grant, and rephrased the question, how PDZ domains inhibit tumor growth — perhaps they might be stimulated by drugs? If this protein is missing in newborns, they can suffer from cleft lip syndrome or spina bifida, where cell contacts do not close. Another name for the protein describes the phenomenon meaningfully as "canoe". There are about 600 such proteins, which keep cells in contact and prevent growth. Herpes viruses exploit this contact, to creep from cell to cell ("herpes" means "creeping"). Many people have experienced a cold sore on their lip caused by a herpes virus, which creeps, migrating by stress-activation from distant ganglia to the lips.

I almost lost this €1.8 million German grant, because at the time I was professor at a Swiss, not German institute. "Share it" I was advised unofficially. I went around on the "open door" day of one of the biggest hospitals in Berlin, offering many colleagues on the spot the chance of sharing this grant with me, to make the project at least partly a German one. Finally, the Dean himself accepted it and, very generously, did not even request his share — and even allowed me to use his letterhead, that is the Berlin "Charité" hospital for further correspondence. ("I trust you"!!) He wanted the scientific co-operation, which indeed worked out very successfully.

We published our results on the tumor suppressors which organize membrane receptors, in *Nature Biotechnology* in 1999, with countless co-authors. One of them left Zurich for his home country Australia and added some more names from there to the author list. When I complained about that, the head of his institute expressed his interest in even more additional co-authors and considered that to be an "honor" for his university. We included them all, more than half a dozen. I do not know any of them — even today. Before this issue was resolved, I received a letter from a lawyer in Hawaii. His client also claimed co-authorship — a former summer student, who had been waiting for his visa for the US in my laboratory. For years I had not heard anything from him. How did he know about the preparation of our paper? That remained an enigma for me, and I preferred not to find out the answer. I just included his name.

PDZ-tumor suppressor proteins bind to the cancer protein Src. If one single amino acid is missing at its end, the PDZ suppressor effect is gone, cells lose contacts, form foot-like structures, pseudopodia, and use them for movement, running away and become metastases. Together with specialists from the sequencing group at the Max Planck Institute for Molecular Genetics in Berlin we tried to find out what the difference was between metastatic and non-metastatic cells by expression profiling, that is, analysis of the whole set of transcripts or "transcriptome". We detected 600 genes to be differently expressed, a figure that we could hardly believe. How we did this will be described next.

-Ome and -Omics

The student of bioinformatics at the Max Planck Institute sits in front of his computer, without noticing people entering or leaving the room, half-empty cookie boxes on the table, and piles of plastic water bottles under his desk. He is analyzing the transcriptomes of normal versus metastatic cells. Everything ending in –ome comprises all the components within a given sample, all mRNAs (transcriptome), all proteins (proteome), all metabolites (metabolome), all kinases (kinome), etc. "Do them all" is the idea. Especially components that are completely unknown individually are analyzed in bulk, not individually but all together. (*A friend, who is not a scientist, did not like this section, and suggested that I delete it. You can skip it. The principle is in any case already clear. But perhaps some of the technical expressions are worth remembering if an elderly boss wants to understand the conversation of his young students!*)

To continue: First, there is a computer with the beautiful name Illumina, which does it all, sequencing all RNA after transcription into DNA — a high-throughput machine, which generates and stores the sequences. Such a machine is by now available at every Institute in a university. Known genes can be identified easily, unknown genes need to be sequenced in total, leading to billions of sequences, called "next-generation sequencing" (NGS), meaning a newer version than the previous ones. Repeated sequencing is performed to reduce statistical errors, called "coverage" — so 10 repeats correspond to tenfold coverage. All genes of a sample are analyzed by "genomics". The DNA is fragmented by a "shotgun" method and the pieces are fitted back together into "Contigs" (contiguous stretches). A new vocabulary has to be learnt here. Then the sequences are aligned by comparison with a "reference genome" to identify them. This is a sample or model genome which scientists have agreed upon. The official human reference genome is not derived from a single person but from a virtual human. It was composed of DNA sequences of more than 20 "donors" and is constantly being refined.

Two "normal" genomes differ by more than 3 million nucleotides — SNPs, small nucleotide polymorphisms (pronounced "snipps"), point mutations. This number is enormous — are there really so many "normal" differences? Almost 0.1% of each genome are different — but what does that mean? Probably it does not mean anything special, at least not immediately a disease. There is plenty of work to be done, to find which of the differences are indicative of a disease. Those genes will be one day determined by personalized medicine — not a trivial task. Which genes are they? We asked about the differences between the transcriptomes, the levels of expression of genes that differ between normal and metastatic cells, taking the Src protein as an example. One important difference is visible by microscope. While normal cells can grow in 3-dimensional cultures as acini, ball-type structures, the metastatic protein makes them leaky (*by the action of some digesting proteases*).

The sequence analysis indicated that 600 transcripts were differentially expressed: in the metastatic cells: some transcripts were 10-fold, 100-fold, even a million fold higher — but some also only twofold. A twofold difference is often ignored as irrelevant, but I think incorrectly so. We have two chromosomes; if one gene is missing, then the cell is already halfway to a tumor. A computer program allows one to identify these 600 genes, by using the Gene Ontology (GO) database or similar ones. After excluding from the list all the genes that also occur in the normal control cells, we were still left with 435 genes potentially typical of metastases. Some genes have an influence on migration, adherence, or apoptosis and 17 are transcription factors — which in turn can turn "on" many more genes. How complicated!! Are some genes more relevant than others? Are there master genes? Many people are searching for them — and so are we.

Another –ome is the metagenome, the sum of all the genomes of an organism. A very fashionable metagenome is the microbiome, the totality of all the microorganisms that one can find associated with a human being: in the gut, on the skin, in the birth canal. Equally interesting are samples from the soil, sewage, lakes, oceans, feces of living animals or fossils, the lungs of patients with cystic fibrosis; even the smog of Beijing is being sequenced (it contains microorganisms), and the river Ganges is being analyzed to find out why the holy water is so holy (it contains some phages, which destroy some of the bacteria)! This requires hundreds of bioinformatics specialists and better and better computers. The next topic could be the metabolome — all components of metabolism. And then? Kinome all kinases? — There are already 1000 known. I researched only one of them — and therefore this raises the question: what do we need more, all kinases as a whole or the analysis of a single kinase, of one special case? The answer is that information from both is required.

Freeman Dyson, one of the best living mathematicians and physicists, who missed the Nobel Prize but helped others to obtain it, whom I got to know during my stay at the Institute for Advanced Study (IAS) at Princeton, wrote an article with the title "Birds and Frogs", in which he discussed the different views of a project, and of science in general, from close with jumps or from distant perspectives. Both are required. But researchers normally prefer one or the other. Dyson thinks of himself as the "frog type".

(British understatement?) He became best known for his book entitled *The Origins of Life* — why plural? Why more than one? That we shall discuss later.

And what about the virome, the world of all viruses? They are especially difficult to find and to characterize. In our genomes, which are full of viruses, they are integrated at different locations, and so far only the ones in conserved regions of the genome are thought to be "of interest". Only what is conserved in our genomes is considered important. Virus integration sites are normally not conserved. Most viruses are therefore missed. Some viruses, however, were detected in our genomes, and it came as a big surprise to learn how many there are — there are millions of "fossil" retroviruses (see below). In total, 3000 types of virus are presently known, with about 150 of them sometimes causing diseases. In sewage 50,000 types of phages have been catalogued. Remember: there are in total 10^{33} viruses and phages on our planet! How can one possibly analyze them all? There are just too many. So one will have to focus again on the disease-causing viruses, and revert to the "frog perspective" — in direct contradiction to the intention of this book, in which I wanted to direct the reader's attention to all the many viruses that do *not* make us sick. Yet, there may be just too many.

A discussion on how to cope with all the sequences — even when using the Cloud — ended up with a surprise for me: don't save your sequencing data, throw them out! If you need them later, just sequence again. Repeated sequencing is faster than storing and finding them in the database — wow!

Cancer — completely different?

Our genome consists of 98% "empty space", and only 1.5–2% of it codes for proteins: "Empty" means non-protein-coding sequences originating from introns or other information or simply "junk". Some of the "empty space" comprises defective endogenous retroviruses. The introns are the regions which increase our human genetic complexity by the splicing mechanism, which allows a combination of various exons excluding the introns. (*The intron sequences are deleted by splicing at the RNA level by loops or lariat-type exclusions.*) This allows a fantastic increase in the number of new combinations for new proteins. Then there are more DNA sequences expressing regulatory short microRNAs, "miRs", 20–30 nucleotides long, and the opposite, long non-coding RNAs, "lncRNA" (pronounced "link" RNA), hundreds of nucleotides in length. They bind to related RNAs and regulate them. This has a role in leukemias and currently constitutes areas of active research. Carlo Croce, an oncologist in Ohio, USA, searched for chromosomal abnormalities and cancer *proteins* in leukemias and instead discovered miR-15 and miR-16. Carlo Croce was a frequent visitor and speaker at the cancer meetings of the Mildred Scheel Foundation. M. Scheel, the wife of a former German president, was a medical doctor and very committed to funding cancer research in Germany — and passed away herself of colon cancer, but her foundation has survived for decades by now. I had the privilege to get a special grant from her personally, the highest ever awarded by her fund (€1.3 million). But nobody could help her. (And I could not help my partners against their deadly cancers.)

Carlo Croce started a network with cancer researchers and was a bit secretive about how he detected the cancer-specific micro RNAs, miRs — yet now, only a few years later, they are in everybody's hands worldwide. The search for all kinds of non-coding, that means "not for protein-coding" RNA, ncRNA, has exploded. Thousands of them have been detected in cancer. Companies offer chips for their detection in cancer specimens. Various combinations of them characterize tumors. In connection with cancer: do the miRs replace oncogenes? Is then everything wrong that was mentioned above? In some ways yes, in diagnostics, because miRs are easier to detect: chips for finding cancer-specific miRs have become available. miRs can now be determined in blood, saliva and feces, and thus from non-invasive sampling for diagnostics. However, some of them feed into the oncogene-specific pathways described above. Two striking examples are the Myc protein and the tumor suppressor p53. Myc protein acts through the miR-17-92 and activates many other genes — including oncogenes such as Raf — and no tumor arises if the miR is deleted — suggesting a key function and a possibility for future therapeutics targeting miRs. The tumor suppressor p53 acts through miR-34 and so on!!

miRs are detected in almost any biological system that one can think of: trypanosomes, bacteria, plants, mammals — everywhere. Fast kit-based tests are available for human diseases all the way to "happy chickens", where miR is a biomarker for welfare and stress during egg-laying in production systems such as chicken batteries, and useful for inspectors of animal health. This is what the suppliers recommend to apparently well-educated farmers!

Also, in viruses the small regulatory RNAs are present. They were discovered in HBV and also in HIV, where they were heatedly debated. The small miRs can regulate transition from early to late genes in viruses. (*Antisense RNA can also do that, though.*)

Furthermore, small RNAs cause epigenetic effects also in relation to cancer. Epigenetic changes are not encoded in the DNA, and they arise not through mutations, but by small RNAs. They lead to chemical modifications of the DNA in response to environmental factors (*by the attachment of methyl groups to nucleotides, or by modification of the chromatin by acetylation and packaging of the DNA.*)

The epigenetic effects are normally only transient, and not passed on to the next generation. Environmental factors are due to lifestyle, nutrition, habits such as smoking or drinking, stress, or diseases. They can lead to epigenetic changes, to special methylation patterns. Such changes can sometimes be detected by the naked eye and by the color of mice: in a special cancer-indicator mouse, the Agouti mouse, the color of its fur is indicative of environmental effects by carcinogens (see below). Black fur is healthy, while yellow fur indicates carcinogenic food!

Diagnostics of cancer can no longer be based on the search for mutations or on overexpression of oncogenes, but epigenetic changes, which require new diagnostics of methylation patterns (*by bisulfite conversion*), make up a rapidly moving research field.

Then there are the long non-coding lncRNAs, which can be very long and regulate gene expression in normal or cancer cells.

Last, but not least, Otto Warburg needs to be mentioned. He wanted to solve cancer — and was terribly afraid of dying from it. He was attacked severely in Berlin shortly before he died in 1970 because of his then unorthodox theory of cancer. He hypothesized that tumor cells, instead of respiring, ferment. Even the German redtop newspaper *Bild-Zeitung* did not hesitate to publish mockeries of him, as I remember quite well — embarrassing enough, because he was certainly one of the greatest biochemists that there ever have been in Germany. Tumor cells, according to Warburg produce energy by anaerobic chemical pathways, breakdown of sugars without oxygen. We all know of such metabolism from the muscle-ache that is caused by insufficient oxygen supply during sport or exercise. Cancer cells thus have a sort of "muscle-ache". They need glycolysis inhibitors, which are now under development. Warburg received the Nobel Prize in 1931 for solving the problem of fermentation. I witnessed some kind of his resurrection in 2011 in Princeton at the Institute for Advanced Study (IAS) during a symposium on Warburg's cancer hypothesis — including photos of Warburg's — in three successive talks acknowledging his concept of cancer. Hypoxic conditions in tumor cells are engendered by cancer-causing mutations and metabolic changes. Warburg himself was so terrified of the possibility of dying of cancer that he produced his own milk and wheat directly next to his Institute in Berlin–Dahlem. Today the Max Planck Institute for Molecular Genetics is located on the former fields. Only his technician was allowed to process the food. Was he afraid of poisons? After all, there are anecdotes about fights with his family, which led to a death notice in a major British newspaper — while Warburg was healthy and well. He sued the newspaper successfully for compensation: as he was such a public figure, the newspaper should have confirmed his death before reporting it. In honor of Warburg, the independent research groups in Berlin are named the Otto-Warburg-Laboratories. As head of such a research group, receiving superb support, I published my best papers on oncogenes — however, I failed to pursue the idea that tumor cells do not need oxygen, and should really have investigated it as obligation to the Institution. I simply missed it, embarrassingly enough.

23andMe — will I get breast cancer?

Will I get cancer? This question obsessed not only Otto Warburg's mind but is relevant to almost everybody. Is there an answer to it? Let us try:

A colleague at the IAS in Princeton told me about the company "23andMe" and encouraged me to get my genetic analysis. 23 refers to the number of chromosome pairs in people. However, he then started to get worried, and indicated that in the event of bad news I could contact another company and hope they might produce a better result! Indeed, more recently the companies have to fulfill quality control (QC) standards, to ensure that their results are comparable. There was a ban on 23andMe by the FDA for a short period because of QC, but no longer. Two diagnostics laboratories have to reach the same conclusions, independently of one another. I had to establish such a QC procedure, on viruses, in my diagnostics department — with endless amounts of paperwork, protocols, instructions, training of personal, and signatures. The documents were piled meters high — quite a challenge.

What do I learn from such a personalized medicine and how would I cope with possible negative results? This is what I wanted to find out. My gene analysis cost $200 in 2011. By far not the whole genome was analyzed but only certain relevant regions, which can be screened by ready-made chips. One million selected base pairs are tested for 200 genetically determined diseases and 100 predispositions for certain diseases. One has to supply five milliliters of saliva into a given tube; collecting this was a little effort, because no sweets and no toothpaste were allowed. A thick cap secreted some preservatives after closing. Even the postage was already included. One has to agree, however, to the anonymous use of the genetic results for research purposes and statistical analysis. That is why the analysis is so cheap. No anamnesis is required — everything is deduced from the genes. The head of 23andMe is a relative of the inventor of Google. A friend from New York was unable to participate in such an analysis, which is forbidden in certain States of the USA, including in New York. Why so? Is the analysis not FDA-approved, are there too many uncertainties, is there a risk of suicides after bad news? We used my address in Princeton. Four weeks later I received the results by e-mail. A stepwise interrogation began, on whether I was really prepared to read about my chance of getting Alzheimer's disease and breast cancer and I was counselled not to panic if I continued reading. I was informed how many breast-cancer genes are known and which of them were tested, including the references to the scientific literature. Regarding my chance of getting breast cancer I did not get a clear answer, only statistics. I learnt, however, that my Erb-B2 gene is not mutated — a hallmark of 25% of breast cancers. That was good news.

I was also informed which parameters had not been considered, such as "lifestyle". If more information became available in the future by better analysis, I would be informed. I have been on their mailing list for the last five years, with continuous questions and information. I have learnt that I am a descendent of Marie Antoinette and of the daring and frightening Vikings from Scandinavia. I am indeed of French Huguenot origin and was born in Northern Germany. Jewish genes they did not find: "bad for you" was the comment by Eric Kandel, when we talked about it, followed by his unforgettable loud laugh, which reminded me of the very last scene in the movie about his life, *In Search of Memory*, referring to both his life and his research topic. He is one of the brightest memory researchers, using an animal model with the beautiful name *aplysia* with particularly thick nerve strands (a slug, some kind of a jellyfish).

I have too high eye pressure, too high cholesterol and cannot tolerate certain drugs. I will not develop emphysema, as my father did (he died of it), my risk of getting diabetes is low. In reality everything was rather vague and the results expressed as probabilities of the various risks. So I do not really know anything for sure. But a good consequence could be to go to some preventive analyses more often. Freeman Dyson had his whole family tested, about 60 people. Perhaps one will discover the gene for his genius, and find out who inherited it? By now 750,000 customers from 50 countries have been tested, the more the better.

"Who wants to know this?" was a frequent question of friends and relatives, even medical doctors, when I told them about my genome analysis. One would be scared and health insurance could abuse the information — so the suspicion. Is that true? One could be happy to learn which kind of therapeutics may be good or bad for one, and whether one is eligible for them — on the basis of genetic facts! This is already practiced by health insurance: A breast-cancer patient will only get the very expensive drug Herceptin as therapy if a mutation in the receptor Erb-B2 is identified, which is the case in only 25% of the women patients. (*Erb-B2 is the same as HER-2.*) Individualized medicine will come anyway. The sequencing apparatus needed for these diagnostics will be a desktop version, and a prototype, called MinION and about the size of two fingers, made by Oxford Nanopore Technologies, was already on the desk of a colleague at the ETH in Zurich. (*The sequencing principle is completely new and based on the nucleic acid passing through a flow cell, a nanopore, influencing the conduction of an electric current by each passing nucleotide, sensitive enough for its specification and sequencing, a technical miracle!*) At a price of only a few hundred dollars this will soon be on every doctor's desk.

Craig Venter established the sequencing of whole genomes in a very efficient and rapid approach and was one of the first to analyze his own whole genome. In the light of his results he tried to reduce his cholesterol level by changing his nutrition. There would have been cheaper ways to find out! J.D. Watson, also one of the first to have his genome sequenced, mentioned his cancer risk — which, however, did not worry him at the age of above 80 years. Other genes potentially concerning his family he did not discuss. My ophthalmologist laughed at me when I gave him my data sheet from 23andMe with some of the mutated indicator genes suggesting a glaucoma. He ignored all of them, and simply asked: Did your grandmother or anyone in your family have glaucoma? That was all he cared about.

Viruses and prostate cancer?

Do viruses cause prostate cancer? That would be a sensation if it were true. Indeed, it provided one — but the opposite of what was expected: it set off a false alarm, as in fact no viruses are involved in prostate cancer. The purported viruses were a laboratory contamination. Searching for viruses as a cause of a disease is a permanent fight against contaminations. Laboratories that can identify viruses and do research on them also often have their laboratory space — the refrigerators, deep freezers and sometimes even the air in the laboratory — full of viruses, which get around in aerosols that arise during pipetting or when centrifuges are used. It took as long as six years recently to find out whether retroviruses cause prostate cancer or not. Why did it take so long, with so many new methods available? This can teach us a lesson.

The virus under suspicion was a xenotropic mouse retrovirus, XMRV. It can infect not only mice but also other species, as indicated by the word xenotropic. This virus was detected in samples taken from some prostate cancer patients. With the highly sensitive polymerase chain reaction, PCR (described earlier), the viral sequences were even sometimes found in normal people and were absent in samples from Europe — but instead of becoming suspicious, researchers simply postulated that there was an epidemic raging — in the US only. There, blood donations were restricted, because retroviruses such as HIV can be transmitted by blood. Some people even started taking anti-HIV drugs, thinking that the putative culprit XMRV could be inhibited by anti-retroviral pills directed against HIV. It would have been a good proof, if it had! Often tumor cells from patients cannot be grown in the laboratory. Therefore, they are often transplanted and grown in mice. Indeed, cells then grew and were distributed to laboratories around the world for further studies. Yet they were contaminated with viruses picked up from the mice — and, soon, so was the whole world. Nobody knew about the mouse virus contamination for 10 years, from 1996 till 2006. A committee was formed to dig through laboratory protocols and find out what was going on. What a nightmare — I cringe at the very thought of the task of looking through all my co-workers' protocols for ten years back! John Coffin, one of the most experienced retrovirologists was the Sherlock Holmes who finally solved the problem: the "new" virus was an old animal virus, spread around the world with the "control" cells, and not a prostate-cancer virus — no Nobel Prize. Two problems, contaminated mouse cells and contaminated diagnostics tests by the polymerase chain reaction, PCR had caused the confusion. An embarrassing story — but a re-run of it is easily possible! Viruses can spread, and are indeed difficult to get rid of. Once, at the Robert Koch Institute in Berlin, the air-conditioning spread viruses through the whole building — though I must admit that that was 30 years ago. Also, the Max Planck Institute for Virology in Tübingen was once shut down completely, and we were all forced to take "vacation", because the whole building had to be disinfected. Hospitals suffer nowadays from dangerous contaminations, and recently cruise ships as well. The crew has to learn how to cope with the highly infectious noroviruses, where a few particles in an aerosol in a toilet are enough to convey an infection.

Especially mice have very many viruses. Even in-bred "healthy" laboratory mice can harbor 60 different types of viruses, and wild mice up to ten times more. In addition to virus contamination, PCR reactions also led to false-positive signals. Contaminations can be caused in surprising ways, which cannot be foreseen by regulations: Mice can escape and hide and can catch new viruses. A night-watchman at the Max Planck Institute in Berlin tamed a lab mouse with pieces of cheese during his boring night shifts. This we only noticed when the black mice had no black progeny any more but *grey* mouse babies. To everybody's delight, the fertilization act had been conducted through the cage wire. To find out what had happed in this case was easy, just by color — many other contaminations are much harder to identify.

Here is such a "hard" case: The whole of Germany participated in the detective story of "the Heilbronn killer". A murder had to be solved. Policemen took swabs from the suspect's car door handles, bedroom drawer handles and so forth. Whatever they tested for DNA by the highly sensitive PCR test system — it was positive. This went so far that a serial murderer was suspected. The culprit finally turned out to be a contamination of the

Q-tips — sticks with a cotton tip — that were used for taking the swabs for subsequent diagnostics by PCR analysis. They were contaminated with the DNA of an employee of the company that manufactured them. A control reaction should have been included, one which every scientist would automatically have performed — how could this have been forgotten about?

Another contamination turned out not to be contamination — it was the discovery of the virus HIV. During a meeting in Cold Spring Harbor in 1983, in the US, Luc Montagnier presented his data on a new virus from Paris. The retrovirologists in the back row of the audience did not believe a word of it and reacted as though they were highly amused, joking about it and mocking the speaker. After all, false alarms about the discovery of new human viruses had been all too frequent — one case I can remember had been a gibbon ape leukemia virus (GaLV), which turned out to be a laboratory contamination and not a human virus at all. However, this time Luc Montagnier was right!

5 Viruses that do not make us ill

An ocean full of viruses

Recent lectures that I have given have had titles like "Viruses, more friends than foes", "We would not exist without viruses", "Viruses — better than their reputation", "Viruses as inventors", "Viruses as drivers of evolution" and so on. Viruses do not (only) make us ill — on the contrary, some of them make us fit! I would go so far as to say: we would become sick *without* viruses. I am not being cynical — of course I know about HIV, Influenza and Ebola. However, the opposite perspective is important, and I want the reader to know about it — because it is new, because it is still almost completely unknown, and because it may one day influence our everyday life. New technologies are the basis for this new insight.

In 2013 I went to visit the "Geomar" in Kiel, the former Institute of Marine Biology of the University. In the foyer one can see a model of the research ship "MS Meteor", which offered excursions for students in the 1960s — one of which I went on, but as I always got seasick I decided to study physics instead. Even so, I am still a bit of a hobby oceanographer, so I came to listen to Curtis Suttle who is Professor of Marine Microbiology and Virology in Vancouver. Microbiome analyses — the global sequencing of all microorganisms — has in the meantime become well-known. This would normally embrace not only all bacteria, but also all viruses (the virome), but these are difficult to characterize and are still poorly known. One of the front-runners in the field is the virus expert Curtis Suttle. The lecture hall was overcrowded with students. I already knew Curtis Suttle to be an inspiring speaker and on that evening, to the students' delight, he fulfilled all expectations. Do viruses rule the world? — he asked. He was not referring to viral diseases, written about in textbooks; no, he meant the opposite. "What would the world be like without viruses?" His answer

Viruses   Stars

Not stars in the sky but phages (viruses) in the ocean with gold stain: phages (small), bacteria (bigger), and a cellular protist (big).

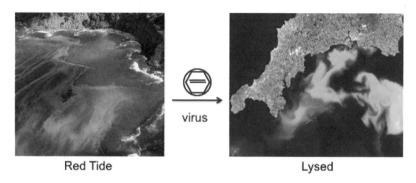

virus

Red Tide   Lysed

Algae bloom ("red tide") is terminated by giant viruses leading to milky fluid and to recycling of nutrients in the oceans.

was (and is): "We would not exist". We would not have enough oxygen to breathe, as it is produced by viruses when they clean up the algae bloom in the oceans in the fall.

His first lecture slide looked like a picture of the stars in the night sky, but it had quite a different meaning: The smallest dots were viruses, next in size were the bacteria, then came the biggest dots from protists (small single- and multi-cellular eukaryotes such as algae and some fungi). Until recently, viruses in the oceans had been undetectable. So Curtis Suttle took 200 liters of ocean water, concentrated it and passed it through a filter that kept back the bigger bacteria (and the giant viruses). The viruses in the flow-through could be seen by "epifluorescence microscopy" by staining with the fluorescent dye SYBR-Gold (an extremely sensitive cyanide dye).

The pictures revealed by this procedure were every bit as pretty as the stars at night. There are 10^{30} viruses in the oceans — more than there are stars in the sky, the number of which is estimated at 10^{25}. In total, there may be up to 10^{33} viruses on our planet. How many different virus types really exist is almost impossible to estimate, because they cannot be distinguished easily. They were first analyzed by the Global Ocean Sampling (GOS) Project in 2007, in which Craig Venter and his crew sailed in Venter's sailing boat *Sorcerer II* from the North Atlantic through the Panama Canal to the South Pacific to collect water samples, opening up as they did so a new research area. A similar ocean expedition Tara Oceans was later undertaken by EMBL members with the schooner *Tara*, which sailed for three years (2009–12) taking seawater samples from oceans around the world — a collection that will yield a mine of information on microorganisms and small animals. Evaluation of the data acquired is still in progress.

Curtis Suttle knows about salesmanship. He talks about viruses when he should use the term "phages". He hides that. "Nobody would come to my talks if I announced that it was about phages, because most people have never heard of them!" So please note: phages are viruses that infect bacteria. They are real viruses, but they are restricted to bacterial hosts. And the majority of viruses in the oceans are phages. Recently, however, giant viruses were discovered in the oceans. Nobody has the slightest idea how many of these there are, because they are big and behave like bacteria (not like viruses) under filtration. The terms "phage" and "bacteriophage" mean the same: bacteria-eaters. They do not actually eat bacteria, but it looks as though they do — in fact they lyse them. A sip of water from the Baltic Sea contains between 10^8 and 10^9 phages — and they normally do not make us ill. They are small, and belong to the group of nanoparticles. How far would they reach if we lined them up? Curtis Suttle loves such calculations: $10^{30} \times 100$ nm makes 10^{23} meters or 10^{20} km or 10^7 light years. For comparison: the Crab nebula is 4000 light years away from us. (*My Master's thesis was about cosmic rays from the Crab nebula. I measured these rays close to the Marine Institute in a bunker left over from World War II; its ceiling was seven meters thick, and was used to mimic a second atmosphere.*)

Viruses exchange genetic material with their hosts. This gene exchange, "horizontal" or "lateral" gene transfer (HGT or LGT), allows the evolution of viruses and their hosts, teaching the host lessons with new genes! The estimate by Suttle is that there are 10^{23} viral infections happening per second in the oceans, which generates an astronomical amount of HGT. This is the basis for innovation and changes in the host cells. Daily, a fantastic total of 10^{27} lysis events take place. However, among this gigantic number of transferred genes it is difficult to learn about the viruses that are newly produced. Nobody knows what to look for. The total number of phages in the oceans is about 100 times higher than the number of bacteria. 80% of the bacteria in the oceans are infected with phages.

The students in Curtis Suttle's lecture were surprised. "Viruses are cool!" Bacteria and viruses have existed for at least 3.5 billion years. Since that time they have multiplied. The microbial world in the oceans makes up the largest amount of biomass on Earth. 98% of the total biomass on Earth is contained in viruses. Every day, about 20% of the whole biomass is lysed by phages. This allows recycling of nutrients, enabling other organisms to feed. Viruses lead to increased growth rates of other organisms in the oceans, e.g. of phytoplankton consisting of plants such as green or blue algae, of which the oceans contain more than 5000 kinds. The total number of genes in bacteriophages surmounts that of all other species combined on our planet. The viruses do not even exploit all the genetic possibilities that are there.

Sequencing of viruses after concentrating them and filtering them out of seawater is very successful, but only if all of them are sequenced at once, by analyzing the water's whole "virome". Yet it is not possible to identify the individual phages, or bigger viruses. The next step is the task of dissecting the information further. It is difficult and time-consuming, even for the biggest computers. 90% of the genetic information in the sequences is unknown and unrelated to any known sequences. There are so many viruses with unknown information! This is quite surprising. Where does all the information come from? Self-made? Yes, probably, viruses try everything, and the cell cannot provide all the genetic information, because it simply does not have that much. Most viruses and bacteria cannot be grown in the laboratory, and therefore cannot be tested easily. The number of phages and other viruses in the oceans depends on the season. In the fall and the winter, many phages adopt an intracellular state, which is also named "lysogenic": they integrate into the bacterial DNA genome and reside there, or they sit outside the genome as free "episomes" — often as closed circles — just waiting or persisting in a quiescent chronic state without manifesting themselves. They grow to large lawns, biofilms. This sounds like swarm formation, where members within the swarms cannot easily be attacked by enemies from the outside — in this case, by infections. Only when bacteria are stressed by lack of nutrients or space the viruses are activated into lytic behavior, replicate and lyse their hosts. In spite of all the many viruses/phages, one can swim in oceans or lakes, and even drink the water — which may then contain 10^9 viruses per sip. Despite that, one does not become sick — not usually, anyway (but sometimes — see EHEC below).

Phages — the viruses of bacteria

Phages are the viruses of bateria. The phages were discovered by Félix d'Herelle, at the Institut Pasteur in Paris, in 1917. The story of the discovery of phages is worth remembering. D'Herelle, who was an autodidact and an unpaid worker (he never became employed at the Pasteur Institute), noticed that in a culture dish some bacterial colonies formed an unexpected lucent halo. The halo was caused by lysed bacteria. From this area he isolated, by filtration, bacteriophages, or phages — that is what he called them, because he thought the flow-through contained organisms

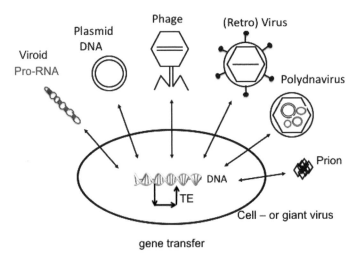

gene transfer

Infection of a cell (pro- or eukaryotic or tumor cell or even a giant virus which can also be a "cell") and gene transfer by various viruses or virus-like elements such as naked or packaged viral or foreign nucleic acids, possibly prions. Transposable elements (TEs) "locked-in viruses" inside the cell as jumping genes.

that were "eating" (Greek: *phagein*) the bacteria. The discovery of penicillin was rather similar. There again, transparent halos surrounding bacterial colonies indicated their death. And, interestingly, phages and antibiotics compete to this very day in killing bacteria. D'Herelle successfully killed bacteria that were causing diarrhea epidemics in Tunisia and other places. He verified that the phages were not dangerous, and he drank the phage-containing solution — without falling ill. He was right. In contrast, the person who also emptied a beaker full of cholera bacteria to prove how harmless they were was the chemist Max von Pettenkofer; although he was wrong, he was lucky, because they were already inactivated and dead. But d'Herelle's phages were indeed not dangerous.

"The enemy of my enemy is my friend" — whether d'Herelle used this sentence is not known, but it would be suitable. Already in 1917 he reported on the destruction of bacteria by phages. He observed that the bacteria from soldiers with diarrhea caused by dysentery disappeared in the culture dish because phages were present. Yet his academic colleagues disagreed, as they failed to reproduce his results. This was due to the fact that phages and bacteria must match one another, as both are highly selective and specialized. Not so the antibiotics. They can inhibit bacterial replication non-selectively. No prior testing of effectiveness is normally required. They can be applied immediately and are easy to handle: just a pill is needed, no biological testing is required. Consequently, the then up-and-coming antibiotics overtook the phages and destroyed d'Herelle's hopes and plans. Antibiotic treatment completely displaced the phage therapies. For many decades, phages were solely a research topic for scientists. I predict that this will soon change again, because of all the multidrug-resistant bacteria that are arising, which now also require prior testing of bacteria to find the right antibiotics. Phages will have their grand comeback!

The phages of the bacterium *E. coli,* the workhorse of basic research, were numbered as Types 1 through 7, T1 to T7. It was demonstrated that phages can bind to the bacterial surface and can inject their DNA, which is located in the phage head, through a contractile tail into selected bacteria. Electron micrographs can show the phages with their highly symmetrical heads and kinked antennae. Their appearance has become an icon of viruses, even though not all viruses look alike. Some phages exist that have no tails or antennae. Furthermore, the antennae get lost easily during preparation for the electron microscope. In fact, I had problems finding nice pictures of phages for this book! Inside the bacteria, the phage DNA has two options. It can survive as an episome outside the bacterial DNA, often in circular form, and replicate there, giving rise to hundreds of phage progeny and leading to lysis of the bacterium. Alternatively, the phage DNA can integrate into the host chromosome — a safe place for the viral DNA and also normally harmless for the bacterial host. This DNA, called prophage DNA, is passed on to subsequent generations of the bacterium. The term "DNA prophage" was much later used by Howard Temin and changed to "DNA provirus" for retroviruses, which can also integrate as DNA proviruses into the host cell's chromosome and be passed down to later generations. The retroviruses have to transcribe (beforehand) their genetic material from RNA to DNA, which is just an extra step. Apart from this, phages and retroviruses have a lot in common. What a high degree of conservation there is, all the way from phages to retroviruses! It must mean that they are related and that this mechanism is of great evolutionary advantage.

Stress for bacteria can activate the quiescent phage and terminate this so-called temperate or lysogenic state. Then the phage DNA is released from the bacterial genome and bacteria are lysed — the lytic state — and hundreds of new phage progeny are released. There are lytic viruses in mammals (including humans) which behave in a similar way, except for the retroviruses, which normally leave their host cell by budding without lysing them.

What does "stress" mean in the case of bacteria? Almost the same as for us: lack of nutrients, not enough space, too rapid growth, conditions that are too hot or too cold, or general environmental changes. Even the activation of stress factors by signaling can be compared among bacteria and other species, including humans. Signals are transmitted from the outside of the cells and result in gene regulation at the level of DNA and gene expression. Removal of repressors, or turning on of activators, can then activate cellular lysis or have other effects in bacteria or host cells.

Too much bacterial growth can often occur in oceans through the influx of excess nutrients from agriculture in the surrounding fields. Fertilizers can be washed out by rain or carried by rivers into the oceans, and there they can become enriched. High summer temperatures can further increase growth. If in the Fall the bacterial density is too high, the phages are activated, they replicate to large numbers and lyse their bacterial hosts. Then the spectacle is over. Nutrients sink to the soil, where they feed other organisms and are recycled. Thus phages are regulators of the food chain and suppliers of nutrients to animals in the oceans, and they regulate population densities of the bacteria and other hosts. Phages are specialists in population dynamics.

The ratio of phages relative to bacteria is about 100:1, and can be as high as 225:1, the deeper one gets into the ocean. The phages cause a special effect: they lyse the winner among the bacteria, as this one is the most abundant: "kill the winner" is the motto. As a consequence, biodiversity is guaranteed, because the phages prevent the winners from being the sole survivors and displacing the less successful bacteria. The phages give a chance to minorities, interestingly enough. This may make the reader think of other societies.

Lysis is supported by an enzyme called lysozyme, which would be useful as a therapeutic. Indeed, researchers — not only at the Swiss Federal Institute of Technology, the ETH at Zurich, but also elsewhere — have already mobilized their efforts to produce it and to use this lysis enzyme as a bacteria-killer. Also, very recently, the bacteria in the human gut have been investigated for the role of temperate or lytic phages — the number of lysed bacteria and free phages increases with affluence, as too rich a diet — bringing about obesity — has an effect similar to that of excess nutrient in the Baltic! This was one of the results obtained by the Tara Oceans expedition through the oceans. This is a surprising result referring to one of the smallest and largest ecosystems, the gut and the oceans.

My first lectures at the Max Planck Institute for Molecular Genetics were given in Berlin a few decades ago. At that time, in the 1970s, phage research was at its height. Heinz Schuster, one of the Directors of the Institute, was exploring the phage world, and I was teaching in my own field of research, the retroviruses. The degree of similarity between these two different systems, the two seemingly distant worlds of phages and retroviruses, surprises me to this very day, and suggests a proof of their evolutionary relationship. Do they have common ancestors? Did phage genomes originally also consist mainly of RNA, as in today's retroviruses? Most phages contain the more stable double helix of DNA. Possible reasons may be replication rates, "progress" in evolution, from RNA towards more DNA (see last chapter). Phages do not contain reverse transcriptases, and they do not use it, in spite of the fact that bacteria contain incredible amounts of reverse transcriptases, which could easily transcribe a potential phage RNA into DNA. The huge number of reverse transcriptases present in bacteria was only discovered recently, and nobody knew what they were doing there. This was especially surprising for me as a specialist in reverse transcriptase — but this information is very new. Did retrophages precede DNA phages during evolution? I know of only one single "retrophage", which harbors a reverse transcriptase gene: a phage of the bacterium *Bordetella pertussis* (which causes small children to cough terribly, frightening young mothers). It does not even carry a name. The reverse transcriptase of the phage can modify a cell entry receptor and lead to many more types of host bacteria, an interesting innovative step by the error-prone RT (causing this broader host-range).

Even today, I am still surprised that no phage symposium ever includes the retroviruses. And the reverse is also true: retrovirologists do not know about phages. I have never seen an abstract on phages at a retrovirus meeting or *vice versa* (except for mine, which have never been well received by the organizers!). Initially, phage researchers were way ahead of retrovirologists, but that has changed; no virus has ever been studied as extensively as the retrovirus HIV. And the phages have almost been forgotten about. But one day — soon I think - they will be resurrected and may save us — a case report and some speculations about their future will be presented later.

A coat for the painter and a journal for the scientist

Max Delbrück was the "pope" of phage research. Every year he came to Berlin from Los Angeles, where he was professor at Caltech, the California Institute of Technology. He was born in Berlin, where the street *Delbrück-straße* is named after his father, a historian. Max Delbrück came every year to the Max Planck Institute for Molecular Genetics in Berlin, as an external foreign member. He came not only because of the phages, but also because of his newer research hobby, phycomyces, a fungus that turns away from the wall when growing. A patient scientist Enzo Russo was investigating this together with Max. The mechanism is still unknown, and this topic would not have earned him a Nobel Prize. He had received it much earlier, in 1969, for his phage research. Max opened the door for phage research. He had the vision to use phages and their hosts, the bacteria, as a model for life, replication, mutation and immunity. The reproduction time of bacteria is, depending on the conditions of growth, around 20 minutes. Then hundreds of phages are released. The generation time for humans is several decades, far too long for fast results. But it was initially unclear whether replication of viruses and bacteria would obey the same rules of life and justify comparison with humans. We do not look like close relatives! Therefore, Delbrück's vision of a general, universal mechanism for replication of life was the idea of a genius. In one of his famous experiments he analyzed resistances of bacteria against phages and the significance of mutations in this.

Another reason for Delbrück's visit to Berlin was to meet Jeanne Mammen (1890–1976) an expressionistic painter, whom he supported. She survived World War II in a backyard flat behind the *Kurfürstendamm*. Her single room, crossed by a thick black oven pipe, was her home and her studio, with a huge chestnut tree looking in through the window. It is now a museum in her honor. The walls were covered with her paintings, which only after her death made it into major museums. She asked Max for a warm winter coat and was given his own coat. It was far too big for her, but she thankfully wore it while bombs fell onto Berlin. She portrayed Max. The portrait became a famous cubistic painting, which decorated the front page of the program booklet in honor of Max Delbrück's centenary celebrations, in 2006, at Cold Spring Harbor. Max always managed to get to East Berlin to visit Berlin–Buch during the period while the Wall was up. Every month he sent his personal issue of the *Proceedings of the National Academy of Sciences* (PNAS) to Erhard Geissler, his colleague, a phage and virus researcher at Berlin–Buch, locked behind the Iron Curtain. His pencil notes in it could still be seen. The journals first arrived in sacks, with all the back issues, and irritated the East German customs authorities — some containers even disappeared. At that time the journal was not available anywhere in East Germany, and its arrival gave a strong stimulus to phage and virus research in East Berlin. After the Wall was opened, the East German institute was re-named the Max Delbrück Center (MDC), and the affiliated café is named after his wife Manny. She came for the inauguration of the MDC early in the 1990s, with her sons, but without Max, who had passed away. In spite of her advanced age, she came loaded with a then still rather unusual rucksack, and busied herself with assembling all the many Delbrücks by making telephone appointments, working her way through a long telephone list, to go out for an ice cream.

In the Cold Spring Harbor Laboratory, which were directed by Jim Watson for many years, the famous phage courses were held annually. Max Delbrück and other scientists, such as Gunther Stent, who wrote a famous book on phages, used to teach there. They trained a whole generation of phage specialists. In that way, phage research spread worldwide. Delbrück then fostered phage research at an institute in Cologne, Germany, which was also later named after him. He was quite strict and demanding in discussions and even though he has been accused of favoring a "sloppy" approach toward research this was precisely what he regarded as the route to making new discoveries; he regarded inaccuracy as a promoter of innovation. This is proven by the innovative sloppy reverse transcriptase. He was a bit frightening or intimidating, as I remember him. He was disappointed by the discovery of the double helix. As a physicist, who had started his career in Copenhagen working with Niels Bohr on the atomic model, he had expected a new principle of life that was not merely "physical". The question "What is life?", which was made famous by the physicist Erwin Schrödinger while a lecturer in Irish exile during the Second World War, inspired a whole generation of physicists. This book has become a classic among the scientific books of the last century — a milestone of molecular biology and a wake-up call for many physicists, including Max Delbrück. The phage "lambda" was a model phage for research. Mark Ptashne, of Harvard University, has characterized it extensively and has published a book that has guided generations of students through gene regulation in lambda phage — following principles that are true for almost any biological system and not only the phages. After the discovery of the DNA double helix, phages became an interesting research topic for DNA replication. There followed research on the extremely complex topic of gene regulation by promoters and repressors — complex enough to fill a thick book. While I was giving a talk about the Myc protein, a regulator of gene expression in eukaryotic cells analogous to phage proteins, in Ptashne's institute at Harvard some 30 years ago, he got up to make a coffee — so informally that it left me with serious doubts about the quality of my presentation. However, only a few years ago, when he gave a talk in Princeton, he recognized me in the audience and stepped down from the stage to welcome me — again to my surprise. I thought he must have confused me with someone else, but no, that was not the case!

We are not alone — we are a superorganism

We are an ecosystem, teaming inside and outside our body with bacteria, viruses, archaea, and fungi, all living in a close community. We cannot remove them with soap and water — and we should not try to, because we belong together. Too many showers may do harm! This needs to be explained.

The total "microbiota" comprise all bacteria and other microorganisms that populate our body, within and without. Every year the journal *Science* lists the year's top ten scientific breakthroughs, which in 2011 and 2013 included the discovery of the microbiome. The microbiome comprises all microorganisms, bacteria, viruses, archaea, fungi, etc. — no matter whether they are known or not, whether or not one can grow them or characterize them, — only the total sequences are determined, a global approach! The motto is: "Do them all"! Sequencing techniques have improved so enormously, and have became so cheap and gained such momentum, that today we can sequence everything imaginable: the microbiome of all kinds in the guts, the birth canal, of saliva, skin, nose, lung, and whatever. We have learnt that the human body is an ecosystem. Everyone has his own microbiome, like a personal PIN code. The results were so spectacular that all newspapers reported on them. The new finding is how many microorganisms populate our body: the numbers are enormous. The data were obtained, among others, by the Human Microbiome Project (HMP), dedicated to analyzing the human microbiome. In a way, the HMP is the successor of the Human Genome Project, which was directed towards sequencing our genomes, all our genes. For the HMP, initially 242 healthy US citizens were subjected repeatedly to swab-sampling of various parts of their bodies — 5000 in total — for sequencing. 178 genomes from microbes that live in and on the human body were studied. Rare microbes turned out to be of relatively high variability, while populations of abundant ones showed less complexity: they had acquired growth advantages and could outgrow the others, so that fewer types dominated. The complexity of all this was enormous. The largest numbers of bacterial populations are present in our guts and body cavities, on the skin, in the elbow, behind the ear, and (less so) in the vagina. The microbiome of the armpits is different from that of the forearms. There are about 44 species on the arms — the richest community of the outer part of the body. The microbiome of the vagina was especially stable and of low complexity. Microbiomes from certain body parts have more in common with the same parts of another person than they do with other parts of the same body.

The metagenomic approach is thus the only one possible — whereby the word "metagenomics" is derived from "meta-analysis" (a procedure used in statistics to make different results comparable) and genomics (the complete genetic information). Imagine: 99% of all existing microorganisms are unknown and cannot be analyzed by other means.

There were several surprises found in the results of the Human Microbiome Project. First of all, most microbes do not cause diseases. Astronomical numbers of bacteria and viruses and fungi are present not only in our intestines, but also on the outer surface of our body. Many of them are essential for elementary requirements, such as supplying vitamins or digesting food that we cannot process ourselves. They synthesize vitamins, detoxify carcinogens, and metabolize drugs (we had better find out more about those!). Viruses and bacteria influence our immune system and protect us against foreign microbes that are more likely to be pathogenic for us. Thus the useful ones can control the harmful ones, which means that the known defend us against the unknown ones! Our guts are not a battlefield for fighting microorganisms; there is no war, but rather a homeostasis, a well-balanced and stable community. And our defense against outsiders!

The second surprise lay in the numbers: About 1000 to 2000 different types of bacteria can be detected in our feces, and their total mass amounts normally to about 1.5 to 2 kg. Our body consists of 10^{13} cells in total, and we house some 10^{14} microbes. Only about 10% of our cells are really our own body cells. One also has to add the unknown amount of archaea. The total number of viruses or phages in our guts can only be estimated: 500 of them have been recognized, but most of them are still unknown. Viromes, or the "Virosphere", comprising all viral sequences in a sample, are most difficult to analyze, because searching for unknown sequences is a special problem — only known ones can easily be found. Many viruses cannot be grown under laboratory conditions, so that one cannot analyze them. Perhaps there are 1000 or 5000 or even more? We need to develop search programs to find them. A recent result indicates that most phages remain inside their bacteria and are released only under non-healthy conditions — this is likely to be a diagnostic tool in the future, the number of free versus intracellular viruses in the feces — which would be an indicator of diseases if the ratio were large. This ratio can even be determined in saliva at the other end of our guts!

Then we should not forget about the fungi, which have not been investigated so thoroughly. About 1000 types of fungi are known in medicine. Only about 100 of them can be diagnosed in the fungal-diagnostic laboratory next door to my office in Zurich. In all, there are one or possibly two million different types.

Microbes contribute more genes responsible for human survival than the humans themselves do. It is estimated that bacterial protein-coding genes are 360 times more abundant than human genes. More than 8 million unique microbial genes are associated with the microbiome across the human body. They contribute to our well-being — or illness — to a degree that we cannot yet estimate. When compared with the total number of genuine human genes (20,000), this suggests that the genetic contribution of the microbiome, amounting to several million genes, is many hundred times greater than the genetic contribution from the human genome. The microbiome is our "second genome". Viruses and fungi may be our third and fourth genomes! This makes us a superorganism.

Studies on twins, comparing their microbiomes, have indicated that only identical (monozygotic) twins have similarities in their microbiomes; similarities to the microbiomes of their mothers or other relatives are much weaker. Microbiomes are so different that they could be used for forensic purposes, to identify a person by his or her individual microbiome. It may still be a bit too cumbersome to sequence traveller's microbiomes before they pass through customs checkpoints, should photo IDs one day prove insufficient! Perhaps there will be fewer legal problems with microbiome analyses than there are with a genome analysis, and perhaps it will soon be doable in almost no time — who knows?

If our saliva reflects out unique genetic complexity well enough for diagnostic or forensic purposes, then I wonder whether a kiss, or close contact among family members, might turn out to be some kind of equilibration procedure, or an adjustment of microbiomes — how unromantic! Does the microbiome influence attraction or rejection between people? Love?

Finally, if we are an ecosystem and if all the foreign genes contribute to our own genes, we may become philosophical for a moment and ask "What is a human being?" — "Who am I?" Can the microbiome even influence "free" decision? What happens if the microbiome is not healthy? Every antibiotic therapy will change the microbiome dramatically and contribute to inflammatory bowel disease (IBD) such as Crohn's disease (see below).

If we consider the timeline of the development of life on our planet, then we have to conclude that we humans are the latecomers in a preexisting world of microorganisms. They existed more than 3 billion years before we did. Naturally, only those of our ancestors who could cope with their environment survived. This automatically resulted, finally, in a peaceful coexistence. The virome specialist Forest Rohwer from San Diego describes us as "living incubators" for viruses. That does not really sound as if we were divinely created beings — it makes us sound more like service stations for microorganisms! In principle, we are not necessary at all on earth. The microbial world can exist perfectly well without us. The microbes will outlive us anyway, just as they were here before us. They are much more flexible and adaptable than we are, so they will survive catastrophes — as they have done before. The reverse is not true: we cannot live without microorganisms. We need microbes for digestion of our daily food and for protection against other pathogenic microbes. We are dependent, and not superior.

Cesarean section, milk and a "sushi" gene

The microbiomes of identical twins, and those of mother and child, resemble each other more closely than they do those of other people. However, in spite of this diversity, there is a shared "core" population of the gut microbiome which comprises about 50% of the total of 3.3 billion bacterial genes populating our body. The flora of the digestive tract of a newborn is transmitted from the mother's microbiome during delivery, from her birth canal and vagina. The initial microbiome will last for some years, and although it changes during childhood, adolescence or older age, yet to some extent it may have a lifelong influence. The first microbiome is called "pioneer species". In contrast, children born by Cesarean, C-section, are first in contact with their mother's skin microbiome (or that of a nurse or doctor!), which differs from the one of the birth canal. Germs from the environment with putative hospital-specific germs are taking on a greater importance. Premature babies are often delivered by C-section, and are therefore particularly endangered. According to recent studies, children born by C-section have a higher risk of developing allergies or asthma later in life. A healthy mother's flora protects the newborn against environmental dangers. Some hospitals or mothers prefer C-section, because surgery can be planned. The most prominent example is the practice in the Far East, where in some countries 90% of deliveries are by sectioning. Thus, allergies in countries like China may further increase strongly. It is no longer discussed whether bacterial suspensions should be applied during the birth process as a preventive step, a shower of the mother's or common vaginal bacteria as first encounter with the world — not a vision any more, but reality!

Again: the known microbiome protects against the foreign one. We cannot say that the good protects against the bad — because the bad ones may be good for others. Recollect that if we go abroad on holiday and drink the water just like the local people do who are used to it, while we may fall ill.

How long does it take for the body to adjust to the digestion of unusual food? Nature has performed such an experiment for us — it is called "lactose intolerance". About 10,000 years ago, most people in Europe and Africa were hunters and gatherers. They changed the life pattern and became farmers, with domestic animals and different food — such as milk. In order to digest cow's milk, a mutation was required. Even today, 15% of Europeans do not have that mutation — like me — and cannot digest milk. It is amusing to note that with milk intolerance I am the "wild type", and those who can tolerate milk are the mutants. Northern Europeans have the mutation and can digest milk, which helps in the production of vitamin D if there is little sunlight, as is the case in Scandinavia. 75% of the world's population cannot digest milk. The adaptation happened by mutation about 10,000 years ago. My ancestors missed out on it. There are pills against lactose intolerance — an extract of cow's stomachs, which contains the missing enzyme lactrase. It does not help me much. The knowledge of lactose intolerance is slowly spreading. Not so long ago, milk was shipped as powder to Africa against malnutrition — and it did more harm than good because Africans have lactose intolerance and plenty of sunshine.

Chinese people are reported to have lactose intolerance; however, milk has recently become an important, popular article imported from other countries to China, and in China milk production has intensified. The most surprising twist in this scenario is the new nutrition fashion: buying lactose-free milk without any need to, even for people without lactose intolerance — very surprising. Of course, it costs more! This is also happening with gluten, a component of wheat — there is a condition of gluten intolerance, coeliac disease, but many healthy people also buy gluten-free bread now — which makes me reflect on the power of good advertising.

How long does it take to adjust to a new dietary habit? One sporadic mutation occurred 10,000 years ago — a rare event. Can one speed the process up? Perhaps mutations may not be needed. The Japanese consume algae, which we Europeans cannot digest so well. Can we develop a "sushi gene"? Indeed, several of them have been found! The genes were adopted from algae and taken up by gut bacteria, so that the bacteria adapted to digest sushi. The mechanism of horizontal gene transfer is the most successful way to "learn" something new and fast. Phages are often the carriers, and they help by mediating gene transfer to a new host. This is a good example to demonstrate that viruses/phages are our gene "teachers", the drivers of evolution.

A more frightening manifestation of this kind of horizontal gene transfer (HGT) is the transfer of antibiotic resistance genes. Toxin genes, too, are being transferred by phages and caused EHEC, the biggest food scandal in Germany. Are they a real threat? — No, not as long as bacteria from animal feces are not used to decorate our salad. What about a chocolate gene, which Swiss people seem to have? Can one acquire it? That has happened already, the lactose-tolerance mutation described above.

Viruses against global warming and laying eggs

Viruses can be beneficial — this needs to be explained. Viruses cooperate and compete, and they often seize opportunities for their own advantage. Horizontal gene transfer can go in either direction; genes can be supplied to a cell, and genes can be transported out of a cell. Is the former direction the more frequent one? Then there are the "helper viruses"; one virus can help another. Defective viruses can use a protein coat from a non-defective virus present in the same cell: they share it. Beneficial viruses can even repair the genetic defects of a cell — but watch out, that is often beneficial not only for the host cell, but also for the virus itself: if a virus helps a cell to survive better, then there is more progeny for the virus as well. Also, the removal of tumor suppressors by viruses is such a case — the cell lives longer this way and will guarantee more viral progeny. Thus it is best if both have an advantage — mutualism is the right word for it. The "beneficial" viruses even include herpes viruses, adenoviruses, phages, and plant viruses. Adenovirus infections can protect against tumor growth, and herpes viruses can suppress HIV or bacteria. A boy with leukemia improved his health when he got a poxvirus infection. Such so-called "virotherapies" date back about 100 years and are based on acute, strong immunological defense reactions of the organism that are provoked by one agent but are also effective — unspecifically — against other diseases. Some unexplained cures may have been brought about in this way.

Moreover, phages can act as helpers. A host cell with radiation damage to its DNA can get support from a phage in repairing its defective DNA. Lambda phages can even help to repair dead host bacteria, making possible their (own) production of more progeny. And healthy bacteria displace pathogenic ones in our gut microbiota.

A very impressive example from plants may be of relevance for future food supply to save mankind from hunger: viruses can help against global warming! In the Yellowstone National Park there is a grass, *Dichanthelium lanuginosum*, which can grow at temperatures above 50°C, but only with the help of a fungus and a virus, in a triple symbiosis. The plant needs the fungus and the fungus needs the virus. The virus has a striking name: *Curvularia thermal tolerance virus* (CThTV). The virus induces heat resistance in the plant by acting through the fungus. So several steps are required. It is also egotistical help, because the well-being of the host in turn guarantees more viral progeny. Help by mutualism corresponds in modern business terminology to a "win-win" strategy. There are even discussions as to which is the real host of the virus, the fungus or the plant. If one removes the virus, the plant loses its thermo-resistance. How the virus can be so helpful is currently being investigated by transcriptome analysis — that is, by expression profiling of the transcripts of all its activated genes. Not surprisingly, osmosis genes seem to help the plant, reinforcing its water supply during dryness. A sugar was found to be involved in desiccation and is reduced by the virus. Also, a pigment (melanin) contributes in the fungus to stress tolerance. We will need to learn more about this, to fight against global warming in agriculture. Farmers in California may have to generate new viruses to ensure survival of their almond crops. The production of a single almond requires 4 liters of water. Often, fossil water — more than 4 billion years old — is obtained by drilling. Drilling is currently leading to subsidence of the whole state of California by 5 cm each month!

The most surprising consequence of a beneficial virus is the development of the human placenta by means of a retrovirus. It is because of a virus that humans do not have to lay eggs but can have embryos develop inside the body. Retroviruses can induce immune deficiency, a property, which caused one of the biggest catastrophes of mankind, the AIDS epidemic, caused by HIV. The ability of a virus related to HIV to suppress the immune system allows an embryo to grow within the womb of the mother without rejection by her immune system. The *envelope* protein Env of a defective human endogenous retrovirus, abbreviated as HERV-W, became part of the mother's placenta; more specifically, it formed the multi-nucleated syncytia trophoblasts and it induces a local immune deficiency which allows the embryo to grow. Mammals do not need an eggshell, as birds do, or a pouch, as kangaroos do, to allow embryos to mature in a separate compartment, outside the mother's body. The similarity of the surface protein Env of HERV-W and HIV is striking. Exactly 20 amino acids can be identified in both of them as the relevant sequence and can be detected in today's transmembrane protein gp41. Most surprisingly, a simple HIV test often used today is based on syncytia: many cells are fused and form giant cells containing many nuclei, which can be easily detected in an optical microscope. Thus, today what makes HIV a deadly virus led earlier during evolution to one of the biggest advantages of mankind, the development of the placenta. Perhaps there are more such properties for which viruses are responsible without us knowing about it. Whether syncytia in the uterus was of benefit for the ancient virus as well, I do not know; possibly there was a higher survival strategy. HERV-W is related to the endogenous sheep virus *Jaagsiekte sheep retrovirus* (enJSRV), which still occurs as an endogenous as well as an exogenous virus. The first cloned sheep Dolly died of it. This virus is not yet in equilibrium between exogenous or endogenous states — rather surprisingly, because it is believed to have existed for the last five to seven million years. Thierry Heidmann, in Paris, detected a second syncytin molecule, and similar syncytin sequences can also be found in many animal species; these sequences include the above-mentioned RELIK, the ancient virus of rabbits and other mammals such as cats and dogs. The endogenization of syncytia viruses happened between 25 to 40 million years ago. Recently even for pigs and horses viral syncytins for placentas have been described — for 18 different viruses and many species by now. They are not all related, so syncytia may have been "invented" more than once. Most recently HERV-W and its coat protein Env was found in several human diseases, such as Amyotrophic lateral sclerosis (ALS), multiple sclerosis (MS) and even type II diabetes. A vaccine against the coat Env protein is under investigation. Furthermore, one virus, the MS virus, has already been given a name: MS retrovirus (MSRV). Are these singular cases? Time will tell.

A virus full of wasp genes — is that a virus?

My favorite virus is the poly-DNA virus PDV, an insect virus, because nothing about this virus seems "correct". It does not fit into any known scheme for virus classification. First of all, it does not have a genome. It is not empty, but the DNA originates from the host and it is really lots of DNA, 30 DNA plasmids, circular DNAs. This makes PDV an exception, with much foreign DNA, and gave it its name, poly-DNA virus. Furthermore, it does not infect cells for its own replication but rather it helps the host in producing offspring; it does that even without the host. This is like job-sharing and sub-contracting — which is good for the host progeny, and ultimately also an advantage for the virus. The virus has of course also its own genetic information — but this does not lie within its particle; instead, it is delegated to the host genome, so to speak by outsourcing. The viral DNA is integrated into the genome of the host, the mother wasp, and guarantees the production of new virus particles in its ovaries. Mother wasp secretes the eggs and, with them, the virus, and she injects all into the body cavity of caterpillars. Now the viruses release the 30 DNA plasmids with genetic information for toxins, which are produced and kill the caterpillar. This results in predigested food for the young wasps. This is a perfect reversal of roles: viruses with host genes and hosts with virus genes.

This makes us wonder about the definition of a virus, which gives all its own genes away to the host and leaves the host without them. How would they ever generate viral progeny? The answer is simple: all newborn wasps carry the viral genes in their genomes, inherited from the mother wasp, and will one day repeat the cycle. This is how PDV replicates. It is a vertical reproduction from generation to generation, similar to endogenous viruses. However, these are not genuine endogenous viruses, because real exogenous particles do leave their host. I would like to include strange viruses such as PDV in the definition of viruses in general: viruses as carriers of genes for gene transfer. PDVs are not even rare, but about tens of thousands of such insect virus types exist. This seems to be of mutual advantage, except for the caterpillar, which however also shows a strange behavior. It is in mortal danger from the virus and the young wasps, but it still defends them against foreign invaders trying to attack their cocoons. Thus it helps its future murderers. This is sometimes called "motherhood" behavior. Perhaps the caterpillar can postpone its death. Gene uptake by the caterpillar has indeed been proven, in 2015, as protection against other viruses. Thus, there is some mutualism and benefit for the caterpillar too. Such an unusual virus–host interaction, a virus half endogenous and half exogenous, may be very old in terms of evolution. With even more reduction in lifestyle, the virus may never be able to leave the cell and would become entirely endogenous — or was it the other way round, did locked-in viruses become mobile?

Similar life cycles with viruses and wasps seem to exist with caterpillars and the so-called "Monarch" butterflies, and also for moths.

This example seems to be a very successful principle of Nature, which deserves imitation: Viruses as vehicles for gene transfer, not for their own but for foreign genes, constitute a genuine gene therapy experiment designed by Nature. The virus is emptied and refilled with poison genes — or curative genes for gene therapy. In exactly the same way, scientists are designing gene therapies. Gene therapy with PDV is special, because it involves a local treatment. An approach similar to that taken by PDV is also being pursued in *ex vivo* gene therapy with local (topical) application for humans. Thus this approach existed long before researchers and gene therapists thought of it, and so far we have not been nearly as successful as PDV has been.

Prions — can do without genes

Viruses with genes from the host without their own genes are strange, but even more surprising are viruses with no genes at all, without any genetic information. They are at present not defined as viruses but as prions. This term prion is a fusion of "protein" and "infection" coined by Stanley Prusiner. These exceptional entities are possibly another variant of viruses, consisting of proteins only, without any nucleic acids. Most scientists exclude them from the category of "viruses" — personally I think that they are just another variation of one theme, namely strange viruses!

Even when Stanley Prusiner received the Nobel Prize for the discovery of prions, his results were highly controversial. Infectious particles without nucleic acid? — that seemed totally impossible. He was describing his results in a conference in Bethesda in 2001, when the first tower of the World Trade Center was hit by an airplane, on "9/11" — and he stepped down from the podium. We all were petrified.

Evidence that prions do not contain infectious nucleic acids with ultrasensitive methods came from Prusiner together with Detlev Riesner, supporting their "protein-only" nature. But many researchers tried to disprove this hypothesis. One of them was Charles Weissmann in Zurich. He underwent a Pauline conversion and finally came to believe in prions. At the age of 80 he showed, as founding chief of the new Scripps Florida Institute, that prions can acquire mutations which can be inherited. This brings the concept of prions close to that of viruses. Weissmann had earlier been very critical, and apparently wanted to convince himself. Many years previously he had knocked out prions in mice and showed how healthy they looked, and concluded correctly they must have more than one prion gene. Weissmann also showed that the silver needles used in brain surgery could transmit prion disease, because prions stick strongly to the needles and withstand all disinfection procedures. Prions tolerate heat: to inactivate them they need to be heated in a steam autoclave to 120°C for two hours. Alzheimer's disease may also be transmissible in a similar iatrogenic fashion caused inadvertently by medical treatment — at least according to recent speculations.

Everyone has heard of "mad-cow disease", bovine spongiform encephalopathy (BSE) in which prions are also very likely to be involved, and many will remember the TV news showing cattle being burned in Great Britain. Spreading BSE among cattle was a man-made process, a result of giving them the wrong fodder! Diseased animals were milled and fed to healthy cattle, thereby transmitting BSE, because suppliers had simplified the production procedure by reducing the temperature for inactivation of potential contaminants. Some customers thought that they had got a bargain and imported cheap cattle fodder, but they also imported the disease. This has been discussed as a possible source of BSE infections in Switzerland.

Prions and BSE may lead to Creutzfeldt–Jakob disease (CJD), kuru in humans, and scrapie in sheep. The generic name is "transmissible spongiform encephalitis" (TSE). The disease kuru was discovered by Carleton Gajdusek in Papua New Guinea and was also attributed to unhealthy food: he could link cannibalism to the disease. Natives had consumed humans, whereby the brains were specifically reserved for men — but precisely the brains were preferentially contaminated with prions, leading to disability, wasting, dementia and death mainly of men. Gajdusek traveled to remote places and was able to demonstrate this connection between cannibalism and death. As a consequence, cannibalism was forbidden. He brought 60 boys to the US from his journeys, educated them, sponsored their University degrees, and sent them back to their home countries to take leading positons. One day a picture of the Nobel prizewinner appeared on the cover of a US magazine — handcuffed. One of the boys had accused him of sexual abuse. Robert Gallo paid a high sum in bail, which, however only helped Gajdusek transiently against imprisonment.

Scrapie was transferred to sheep in Iceland by imported German sheep, even though this disease had never appeared in the sheep of the Northern German marshland, with its green landscapes and dykes. The German sheep had also gone through the required quarantine. Yet in Iceland they were crowded together just before shearing, which spread the disease. We should all bear this in mind: crowds increase the risk of epidemics.

Prion proteins are present in every organism and are present at high concentrations in the brain. They play a part in the normal development of the brain, during neurogenesis and possibly in long-term memory. They only cause diseases if they are misfolded. Wrong folding is perpetuated: the wrongly folded proteins serve as a template and the "wrongness" is transmitted as "contagious", leading to increased numbers of misfolded proteins and ultimately to insoluble aggregates (amyloid fibers). This phenomenon we can sometimes see when looking at crystals, where a wrong structure continues to be formed.

(*The healthy prion proteins, PrP (or cellular PrPc) have structures different from those of the disease-causing scrapie prion proteins (PrPsc);*

they contain more beta sheets and fewer alpha turns, while the reverse ratio is found in healthy prions.) The disease was first observed in sheep, and "scrapie" describes mental illness, a symptom of which is continued scraping.

About 180,000 cattle contracted BSE in the space of 30 years, and 220 people died of it. The number was much lower than originally anticipated, so the European Union decided not to require inspections of cattle meat any more.

Prions may be a model for other brain diseases, such as Huntington's disease or Alzheimer's disease (AD). There are genes involved — but also infectious agents, possibly prions: according to recent case reports, transmission of AD might have been caused by growth hormone extracts from human brains given to children who did not grow properly. The extract may have been contaminated with "AD proteins" — almost like kuru! They are also designated as "abeta prions". Sterilization procedures, and even blood transfusions, will now be inspected for such contaminants.

In summary, there are two extremes: proteins without nucleic acids (the prions), and nucleic acids without proteins (the viroids) — and PDV with foreign genes only. All of them are variations of one theme for the term "viruses": the various combinations that lie between the two extremes, "proteins only" and "nucleic acids only", are the textbook viruses.

6 Viruses — "giant" as cells

Giant viruses of algae and a swimming ban in the Baltic

The Baltic Sea was finally warm enough for a swim — which, however ended with a shock: the water was turbid as far as one could see, and swans were stirring up dirty, brown mud. A notice board warned us that the algal bloom could kill dogs and be hazardous for people's health. The cause was unicellular algae, small eukaryotes belonging to the plant plankton (*phytoplankton*). They can be green, brown or red, and they can grow explosively during hot summer periods, well nourished by spills from the over-fertilized farming fields that line the shore. The algae form solid layers reaching almost to the horizon. Red tides are mentioned in the Bible as a sign of the Apocalypse — could that have been algal bloom instead of blood? Not impossible!

A contaminated ocean — how does it get cleaned up? By viruses! Lack of nutrients and of space activates the viruses inside the microorganisms, and they then destroy the algal bloom, leaving behind white milky suspensions in the sea, which can even be observed from spaceships. The debris then serves as recycled food for fish or shrimps.

The viruses discussed here are now "real" viruses, not the phages, the viruses of bacteria described above. They are in fact *really* big ones, giant viruses. Algae and viruses have co-evolved for 3 billion years, and it is thought that these viruses belong, in evolution, to the oldest ones we know of. Some of the algae look very pretty, such as the "chalk algae" (*haptophytes*), with their little collar-like lamellar decorations. They are common, and they populate oceans from the Arctic to the equator. They prefer the light water surface. Their name is *Emiliania huxleyi*, or *E. hux* for short, and their viruses are designated accordingly EhV or one specific virus EhV-86. It is a giant virus with double-stranded DNA of 400,000 base

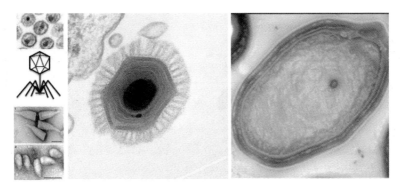

Various viruses: HIV (*top left*), phage (schematic), two archaea viruses (*left*) and two giant viruses (*middle and right*) (not to scale).

pairs corresponding to about 500 genes. (For comparison, a retrovirus has 10,000 base pairs and about 10 genes and the phages described in the previous chapter have sizes ranging from 5,000 to 150,000 base pairs). The algae were named after Cesare Emiliani and Thomas Huxley. This algal virus, with a DNA genome, is also called *Coccolithovirus*, with the Greek word *lithos* for stone, referring to the stony calcium or "chalk" algae that are its hosts. They were selected as "algae of the year" in 2009 because of their influence on the climate and the environment. They are only one of 300 types, and in total 40,000 different algae are known. Calcium carbonate is produced by the virus EhV after destruction of the algae and accumulates at the bottom of the sea. If this goes on for millions of years and the water level sinks, one can see the shiny white coast of Dover or the rocks of the Rugia Island in the Baltic. Nobody seeing this beautiful coastline will immediately think of viruses. Neither did Caspar David Friedrich, when he painted his famous picture of Rugia Island.

A few years ago I had to evaluate a substantial grant application: the goal was to "fertilize" the ocean with iron, so that algal growth would increase and the algae would remove the greenhouse gas CO_2 from the atmosphere, sinking with it to the bottom of the sea. Fertilizing a whole ocean? That seemed risky to me, when I remembered the Baltic Sea, which got out of balance only because of a hot summer. That was a warning. We do not know enough yet. All the reviewers were skeptical, but the idea keeps coming back.

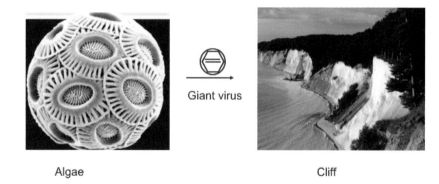

Algae   Cliff

Viruses as "architects": Calcareous algae (*Emiliania huxleyi*) lysed by giant virus built the white chalk cliff.

Lysis of algae by viruses also generates the typical smell of the oceans, the chemical dimethyl sulfide, which again nobody attributes to the action of viruses — not even a virologist would. The smell is not only caused by phages, which regulate CO_2 emission, but also by algae viruses, which influence cloud formation, rain and chalk cliffs. And nobody would think of viruses — let alone of giant viruses — while watching the weather forecast in the evening news.

Giant viruses such as the Chlorella viruses prefer green algae, whereas EhV prefers chalk algae. Chlorella algae got their name from chlorophyll. Indeed, these algae helped Melvin Calvin, whose lectures I attended at UC Berkeley, to solve the chemistry of photosynthesis. We found to our surprise the first chlorella virus in human guts — and asked the person whether she liked sushi food — no. (Read more about it in Chap. 10, feces transfer.)

An expert discipline dealing with the virology of the ocean did not exist until recently. The phages in the oceans, with their astronomical abundance, came as a surprise, but giant viruses are still not widely recognized. The reason is simple: one cannot find them easily, and they do not cause diseases. Filtration is the standard procedure to separate viruses from their host cells. However, giant viruses are of similar size to the bacteria, so one cannot separate them with filters; indeed, to this very day nobody knows how many of them there are. Some researchers think there are many! Furthermore, single-celled organisms such as algae were not even believed to be hosts of viruses. This seems incredible if one remembers that the number of algae viruses arising newly every single day may amount to 10^{19}. In comparison, phages replicate at a rate of 10^{24} new infections per second, and 10^{30} of them populate our planet. The total number of viruses is estimated between 10^{31} to 10^{33} — and the number of stars 10^{25}, and grains of sand may amount to 10^{28}, depending on how one does the calculation!

These giant viruses contain double-stranded DNA, with about 500,000 base pairs and corresponding to about 500 genes, contained in icosahedral protein structures. This is of similar size to the parasitic bacteria, which are small bacteria. "Real" bacteria, such as the well-known laboratory strain *E. coli*, are bigger, and comprise about 5 million base pairs. If one compares these giant viruses with small bacteria, they are of similar size. The DNA of giant viruses can even contain rare sensitive single-stranded DNA regions, and some stretches of it are non-coding, that is, without genetic information for proteins. That makes them appear to be very ancient; indeed, their origin, as estimated by their discoverers, dates back 2.7 billion years.

If we compare the number of giant viral genes with those of HIV or influenza, which have about 10 genes each, they are indeed gigantic. These giant viruses are even small among the newest giant viruses. The biggest new ones are in fact five times bigger, and contain up to 2500 genes. Most of the genes of the giant viruses cannot be found in the gene database, and they are almost unrelated to one another. Of two different algae viruses with 1000 genes each, just 14 of them are related. All at a sudden, there is a whole new world of genes, pathways, and metabolism — of life!

Amoebae viruses can tickle

The novel giant viruses are often viruses of amoebae, a host not many virologists ever thought about. These viruses too were completely unknown until recently. While searching for the dangerous *Legionella* bacteria, which cover the insides of waterpipes as biofilms, and can spread during a shower and caused Legionnaires' disease, the giant viruses were overlooked by researchers. In 2003 the first giant viruses were described as viruses of amoebae. In 2008 further giant viruses were discovered in the water of cooling towers in Paris. They were designated as "mimiviruses" which means they mimic bacteria, which are similar in size! This should not be confused with "mini" — small — because they are really quite the opposite. They are so gigantic that they were overlooked as viruses. Giant viruses are "almost-bacteria". One can see them in the light microscope, which was impossible for viruses known up to then. Virologists use electron microscopes, not light microscopes. But they were not accepted as bacteria either, because they could not be detected in the standard assays for bacteria, the plaque tests. They were furthermore so heavy that they sank to the bottom of culture dishes in the laboratory and thereby escaped detection by bacteriologists.

Shortly after discovery of the mimiviruses two other giant viruses were discovered, *Megavirus chilensis* near the beach of Chile and *Marseillevirus* close to Marseille. All three giant viruses are known as amoebae viruses, comprising a million base pairs each, corresponding to about 1000 proteins. Today, new giant viruses are constantly being detected. The genetic material of amoebae viruses is composed differently from all other known genomes on Earth. They are so big, with more genes than seem necessary: the gene of the smallest living species, the bacteria *mycoplasm (Mycoplasma genitalium)* comprises about 500,000 base pairs, corresponding to about 482 genes or proteins. Craig Venter generated early in 2016 the first fully synthetic independently growing cell with 473 genes, topping a former minimalization effort with an "almost living" synthetic entity (comprising 382 genes) which had to be placed into an empty cell which supplied some help. Many bacteria such as *Chlamydia* or *Rickettsia*, or any of 150 other bacterial types, are smaller than amoebae viruses. This is particularly true of bacteria that became intracellular parasites, receiving support from the infected host cell. The smallest bacterium is *Hodgkinia cicadicola*, with 145,000 base pairs, coding for 169 proteins. It hides inside insects and is shielded there from the outside. Such parasites are degenerate and specialized, and they delegate functions to the host cell. Mitochondria are former bacteria, and so are chloroplasts in plant cells. Both of these symbionts are highly specialized and dependent on the host cell. They are out-competed in size by far by the giant viruses.

Amoebae viruses take up new genes by horizontal or lateral gene transfer from the inner part of the amoebae, where also other bacteria and other viruses exist, which are normally taken up and digested as food by the amoebae. There is DNA floating around, which giant viruses incorporate — by horizontal gene transfer. DNA from amoebae they disdain, perhaps because the DNA of the amoebae is located inside the nucleus and is therefore difficult to reach from the cytoplasm. Of the total genome of amoebae viruses, 56% is derived from eukaryotes, 29% from bacteria, 1% from archaea, 5% from other viruses and about 10% are unknown. Such a genetic complexity must indicate that the giant viruses interact with their environment and undergo significant gene exchange. One of the giant viruses, the Marseillevirus, contains a chimeric RNA-DNA genome, which is very unusual. Was some RNA somehow left over on its evolutionary path to DNA? These viruses do not fit into the viral landscape.

And there is another big surprise. The giant viruses harbor genes for protein synthesis. Protein synthesis is considered as the most important privilege of living cells and completely out of reach for viruses. Thus, the discovery of giant viruses containing components of the apparatus for protein synthesis completely set aside our previous understanding of the world of viruses. These viruses do not possess the whole set of components for protein synthesis, so they are defective in that respect, but even so...! Did they stop evolving at some half-way point on the evolutionary road to becoming living entities like bacteria?

Some mimiviruses have strange large hair crowns, which are used to contact host cells. These are really fibrils, composed of collagen, and make the giant viruses look even bigger. These long extensions make me wonder what they are for, uptake or protection? Mimiviruses have icosahedral nuclear structures, the nucleus measuring 500 nm across and the layer of hair is about 140 nm thick. The collagen hair resembles our connective tissue (and not our hair). Viruses almost always have surface molecules, used for recognizing, binding to and entering the specific host cells for which they are specialized. With these hairs giant viruses seem to tickle their host cells to gain entry. Complex receptors do not seem to have existed in this early world. Entry by touching or tickling is a simple variant compared with the highly specialized docking sites that HIV uses for entering lymphocytes, imitating the ligands of host-cell receptors. Gigaviruses seem to get away with mere mechanical irritation of their host cells.

Almost as though it had been ordered specially for the football world cup in Brazil in 2014, a new giant virus was discovered in the rain forests of the River Amazon. It is called "Samba" and also has long hair sticking out. Soon, all these many new viruses will no longer seem surprising. The host cells, the amoebae, seem ancient. They swallow other viruses, especially herpes viruses, in one gulp and digest them. In order to protect themselves against the fate of being digested, the giant viruses out-trick the amoebae, because they can be taken up without being eaten up. They move into vacuoles inside the amoebae and merge with the vacuoles' membranes; these are a safe place where the giant viruses can even replicate. Replication takes place in a separate compartment, a separate factory, independent of the cellular nucleus, designated as a "virus factory". It becomes bigger, the more viruses are produced. The viral progeny is then secreted out of the factory.

In the laboratory, an experiment was performed in which giant viruses were passaged from one amoeba generation to the next, in all 150 times. After these passages the giant viruses had changed significantly — they had lost some genes. Among these was the loss of their "hair", which was apparently not essential in this setting, by reductive evolution. Loss of genes in the culture dish was striking with the giant viruses. It will be discussed in the last chapter, because loss of genes can go back almost to zero — if a rich environment supplies the minimal needs. Evolution in reaction vessels has become a favorite experiment in the laboratory, although it requires lots of patience.

Sputnik — viruses of viruses

The giant viruses have another surprise up their sleeve. Yes, they really exist: *viruses of viruses*. Giant viruses can be infected by other viruses and allow their replication as real hosts. This property of giant viruses is highly exceptional. Normally, viruses infect cells — but these giant viruses can be like cells for other viruses, which in turn, after all, places giant viruses close to cells. Giant viruses are "almost-cells". The viruses of viruses are termed "virophages", by analogy to the bacteriophages, the viruses of bacteria. They have names such as Sputnik or Ma-virophage. The latter one is derived from Maverick viruses, which are destructive computer viruses. Sputnik and Ma-virophages depend for their replication on the giant viruses. They have about 20 genes, and thus about 20,000 base pairs, and only three of these genes are derived from the giant virus. They are rather small viruses, but potent, because they can destroy their viral hosts. Sputnik occupies the viral factory, where the giant viruses multiply, and replicates its own progeny on the expense of the giant virus; it even hijacks the host's proteins for its own reproduction. In this process the 20 times larger mimiviruses die. So, here, one virus is the host of another virus, but not in the sense of being a helper virus but more like a deadly computer virus. This is quite unusual.

The Ma-virophage infects another type of giant virus, called Cafeteria roenbergensis virus (CroV). What a name: it is called after a café in the Danish town Rønbjerg, not the name of the discoverer — I had a hard time to find that out until I called Matthias Fischer, a specialist on virophages in Heidelberg. Cro is a microflagellate, a unicellular, eukaryotic, bacteria-digesting organism, not like amoebae or algae. It is widespread in the oceans, which it penetrates to a great depth. Its giant virus CroV causes the breakdown of the flagellates and thereby regulates their population density. The coffee-shop-virus, CroV is under investigation by Matthias Fischer, who is one of Curtis Suttle's former students. It is composed of 730,000 base pairs and 544 gene products from bacteria, eukaryotes and DNA phages. Samples of this virus containing the virophage Ma were collected by Craig Venter (not next to a coffee shop). After finishing the sequencing of the human genome, Venter combined his research with his sailing hobby. Passing through the Sargasso Sea (in the Atlantic Ocean) he took water samples and handed them out to other researchers for sequencing and further analysis. This is now keeping numerous bioinformaticians busy. A whole genomic world is opening up, in which most of the genes and metabolic pathways are unknown. Life at the bottom of the oceans is very different from ours! What a surprise.

The Ma-virophage replicates just like the other virophage Sputnik in the virus factory inside the amoebae, without becoming digested. Only four of Ma's twenty genes are identical to those of Sputnik. The Ma virophage even carries retroviral genes, such as the viral integrase. (*Retroviruses influence everything! Ma also behaves like a retrovirus, because it integrates its DNA into the flagellate DNA, which guarantees its inheritance.*) On the other hand, Ma uses the coffee-shop virus CroV as a host for replication and then kills it — as phages do. Thus Ma is a retrovirus for one host and a phage for the other host, a half-and-half. This is unique, as far as I know. Thus it is an interesting intermediate of evolution. It is exploiting and helping with social and with egotistical behavior, both at the same time. Nobody could have imagined such a complex relationship — a *ménage à trois*!

The organization "Global Ocean Survey" is searching for other virophages. Perhaps virophages are not even rare. In the meantime, half a dozen has been identified. One is called Antarctic Organic Lake Virophage, another one the Phaeocystis Globosa Virophage. The new

Samba mimivirus also harbors a virophage. Now we can start to compare the compositions of the genes of the virophages — they are puzzles of dozens of gene sources.

Do not try to remember all this, gentle reader. It is enough if you are surprised!

XXL-sized viruses — the pandoraviruses

Even before its completion the "giant virus" chapter has had to be extended.

From now on, one has to add the date of discoveries — so fast is progress in the detection of new viruses. Two new viruses have been reported since 20 July 2013. Until now the largest giant virus was the *Megavirus chilensis* from amoebae with 1.25 million base pairs. Let us call it XL. Now there are two new even bigger ones, almost twice as large, super-giant "XXL" viruses. Their DNA amounts to 1.91 and 2.47 million base pairs, corresponding to 1900 and 2500 genes or gene products. Jean-Michel Claverie, in Marseilles, isolated the new viruses: first the pandoravirus *P. salinus*, the biggest of all known giant viruses in sediments on the coast of Chile, and the other one, *P. dulcis*, in a pond in Australia. Claverie was invited to give a talk at a university in Australia, where he walked over to a pond on the campus, taking along a glass bottle which he filled with water and mud from the pond. At home he isolated *P. dulcis!* The simultaneous discovery of two such giant viruses at two so distant regions of the globe cannot be a lucky coincidence. These viruses must be common.

The newly discovered viruses have the appearance of a Greek amphora or — a bit less poetically — they are egg-shaped, not icosahedral. Yet it was not their shape that gave rise to the names, but rather the optimism of their discoverer that there will be more such viruses described by the Greek legend of Pandora's box — full of viruses, an optimistic assumption. Perhaps there will be even XXXL jumbo viruses. You never know! Perhaps they will be hardly distinguishable from bacteria, and may still be overlooked!

The two viruses are genetically totally unrelated to each other or other giant viruses. They replicate in amoebae like other giant viruses, but without signs of the energetically favorable icosahedra, no membrane, no hairs — rather naked entities in fact. Of the 2370 genes of *P. salinus* only 101 resemble the genes of eukaryotes, 43 those of bacteria, and 42 those of other viruses. Almost incredibly, 93% of their genes are unknown. Yet the shape of the two is related, they look similar, from Chile all the way to Australia. Their egg shape is composed of several layers. Also their DNA is unusual, made up as it is of linear double strands with repetitive ends. Just like other giant viruses they have quasi-cellular properties: genes for DNA replication as well as for protein synthesis, which is normally supplied by the host cell and missing in viruses. The discoverers Didier Raoult and Jean-Michel Claverie have proposed the existence of a new domain of life, a fourth kingdom in addition to the three existing ones, bacteria, archaea (both prokaryotes), and eukaryotes, and the new gigaviruses (or, for short, "giruses"). That sounds like good salesmanship for higher ranking of their discovery. They go even further by placing the giruses below the three domains, at the bottom of the tree of life as common ancestor. They estimate the age of these viruses to be 2.7 billion years.

Viruses located at the roots of the tree of life? More people think like that. But I believe more in viroids, because giant viruses are far too big for a beginning! The giant viruses could be incomplete bacteria, which are on their way to becoming real bacteria and stopped in a side branch and remained unfinished; or, alternatively, they could be degenerated bacteria that have lost some of their genes and their independence. In any case, the giant viruses are transitional forms between viruses and cells. The borderline is not sharp, but continuous. How did evolution proceed — did small viruses become bigger and progress to becoming giant viruses and, further down the path, cells? This corresponds to an increase in complexity and size. Or is the opposite more likely — did loss of genes and regression lead to viruses? I believe the first is more probable, and this will be discussed later. I wrote a scientific article on viruses as drivers of evolution, as suppliers of genes and inventors, participating during the early development of life, helping to build up all genomes and even antiviral defense — which I tried to publish in peer-reviewed international journals. This led to an unusual number of anonymous referees. One editor sent out my article for appraisal by 7 referees, which I had never experienced before. The answers ranged from enthusiastic ("a new view", "unexpected concept", "innovative", "visionary") — all the way to questioning my knowledge and background in basic virology. One can read the article "What contemporary viruses tell us about evolution" and a short version: "Are viruses our oldest ancestors?" Others had written about this before, such as Luis P. Villarreal (who wrote a whole book on it) and more recently, Eugene Koonin, a biomathematician at the National Institute of Health (NIH), and Carl Zimmer, a journalist who wrote a small booklet on "Mammals made by viruses". Koonin used the abbreviation LUCAV, alluding to LUCA, the last universal common ancestor, and added V for viruses, thinking of an ancient virus world or "virosphere". LUCA must be very complicated, since the first "minimal" artificial cell produced by Craig Venter needs 473 genes for life. But life must have started in much simpler fashion. The innovative viruses tried many alternatives and did not start from a single virus but a mixture — or quasispecies — of viruses, a cloud, a population composed of many different sequences. They could have been the beginning. Then horizontal gene transfer helped to build up higher complexities. A single simple "Ur-Virus", a pre- or pro-virus, would never have had a chance.

My passion for the giant viruses must have been infectious, because I was invited to the studio of a radio station to explain the newest twist about giant viruses: there was the predicted "XXXL"-virus, *Pithovirus*; its discovery was announced on 3 March 2014. This one too was discovered by Jean-Michel Claverie. *Pithos* is a Greek word and in ancient times meant a reservoir, as indicated by the authors. In contrast to the pandoraviruses the pithoviruses do not look like amphoras but have, like the mimiviruses, an icosahedral structure containing circular DNA and a coat, exposing the surprising "hairs". (*Pithoviruses are three times bigger than the two pandoraviruses, yet they harbor only one-fifth as many genes, "only" 596 in total. The virus is some kind of a combination of mimiviruses and pandoraviruses. It also replicates in amoebae, but this time inside the nucleus.*) This is just another variant. Again, pithoviruses have features of living bacterial cells, can transcribe DNA to RNA and contain some genes for protein synthesis. Are they really viruses, or bacteria? Not only the reviewers but also the readers wanted to know. This is an important question at the borderline between viruses and bacterial cells. Only bacteria can divide and replicate (duplicate themselves), while viruses are unable to split into two halves and grow. Typical of viruses is their life cycle: within 20 hours they replicate so that the amoebae burst and release thousands of viruses at a stroke. That corresponds to a classical viral replication as it is known today. (*I personally believe that at the very beginning viruses were viroids and did not need a cell; energy is required, but it could initially have been thermal or chemical energy — and then they could have doubled their RNA near the black smokers deep in the ocean, as described more than once in this book!*)

Pithoviruses have a specialty. These giant viruses were recovered from the permafrost in Siberia and are therefore called *Pithovirus sibericum*. They were found in the hollow ice-sampling drill-tips. With help of the radiocarbon method it was possible to calculate their age: 30,000 years. At that time Neanderthal Man was running around. The most incredible thing is that after this long period of frozen storage the viruses are still biologically active, which means that they can infect amoebae in the laboratory and can replicate. Every virologist stores viruses in a deep-freezer, even for thirty years — that is the normal way of keeping them, and they retain their infectivity, especially DNA viruses. However, 30,000 years of permafrost is a duration hard to imagine. Perhaps the journal *Science* also had some doubts, and did not believe in such long preservation in permafrost, because it rejected the paper on pithoviruses. Claverie and his colleagues performed the drilling horizontally, not down into the depths. Can this guarantee 30,000 years of permafrost? With these viruses nothing had to be repaired or mutated in the laboratory. The same was the case for the 100-year-old influenza viruses of a soldier who had been buried in the permafrost since World War I from whom the virus was recently isolated and reactivated. However, the influenza genome consists of RNA, which is more labile than DNA.

The isolation of the pithovirus raised some concern: what else could have been buried there for so long? The authors are careful and define their virus as a "safety indicator" for other, potentially more dangerous, viruses — a sort of index fossil. So far there has not been any indication that any disease is caused by giant viruses, and they are considered safe. The authors, however, pointed out — to every reader's shock — the poxviruses. They had been considered extinct, but they are somewhat akin to the newest giant viruses. Poxviruses had previously not been included in the "giant virus" family, yet their large linear DNA genomes, their icosahedral structure, and their mode of replication inside the cellular nucleus make them potential members of this category. Could they also have been hibernating in the permafrost in Siberia for 30,000 years, and could they potentially be reactivated? Are there other, similar viruses? Perhaps some we do not know any more or not yet? That would be frightening. (*Some other rare viruses such as Asco, Irido, Phycodna or Asfoviruses are not too distant from the poxviruses or gigaviruses.*)

The biggest poxvirus is the *Canarypox virus*, and it is used extensively in gene therapy as a vaccine strain. It has about 300,000 base pairs and about 300 proteins.

The discovery of three so different giant viruses within a few months of each other, at three locations as far apart as Australia, California and Siberia, cannot be a pure accident. These viruses must be much more widely distributed and quite common. Thus, a new era in virus research has begun.

The hunt is on, in full cry — in the Fall of 2015 the effects of global warming revealed another 30,000-year-old new virus isolate, *Mollivirus sibericum*, with about 500 genes — this story of discoveries is going to be continued!

Two Guinness world records: the biggest viruses in the biggest cells

Amoebae suddenly attracted attention as the hosts of giant viruses. Many of us had not previously been acquainted with them. They are huge, unicellular eukaryotes of many different kinds. Their preferred habitats are sweet water, rivers, humid soils and moss. Lacking a real structure, they move with pseudopodia and replicate by cell division. Their genomes are gigantic, one type, *Amoeba dubia* comprises the biggest genome known on our planet with 700 billion base pairs. This is not a typing error! It is about 200 times as big as our genome. Second in the hierarchy of giant genomes is the plant *Paris japonica* with 150 billion bases — and we have only 3.2 billion bases.

Other Amoebae comprise some 290 billion base pairs — "only" 100 times larger than ours. The biggest cells also host the biggest viruses, the

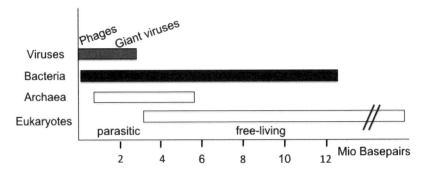

Genome sizes: giant viruses are bigger than many parasitic bacteria.

giant viruses. How does this double superlative come about? Not a single gene of the giant viruses originates from the genome of the amoebae, but only from the rich gene pool within the amoebae by lateral gene transfer, genes from other microorganisms, which are digested inside the amoebae. There is plenty of "food" for viruses to become giants on. The genomes of the amoebae contain numerous gene duplications, pseudogenes, which do not seem to be used. In this respect amoebae resemble some plants such as maize, wheat, bananas or tulips, which also have gigantic genomes with many gene duplications, and which can easily throw out genes without significant consequences — almost half of their genes at once! This paradox about such huge genomes are being discussed as consequences of cultivation, but in the listed examples there are already exceptions.

Amoebae normally engulf viruses and digest them — except for the giant viruses, which hide in a separate compartment (the virus factories or the nucleus). Amoebae are related to circulating macrophages in humans, the phagocytes of our immune system — flexible cells that migrate through our body as guardians to seek out pathogens and destroy them in compartments called lysosomes. They are filled with enzymes for degradation. They collect garbage in our body, such as cell debris, bacteria, viruses and archaea — they can even swallow gold particles. The latter ability is useful for detecting macrophages in electron microscopes. Their variety of "food" uptake makes them a mixing vessel of all kinds of genes. This may explain why the biggest cells harbor the biggest viruses. Amoebae are present not only in the circulating water of cooling towers and air conditioners, but also in the drinking water from our water pipes. Most of them are harmless, unless they are infected with bacteria such as *Legionella*; if so, then they can be very dangerous when they are distributed in small droplets during a shower, in room humidifiers, or in whirlpools! One type of amoebae, *Entamoeba histolytica*, can cause amoebic diarrhea (or amoebiasis), which can be dangerous and can even cause epidemics, especially if waste water and drinking water are not kept well apart, as is the case in some underdeveloped countries in Africa and Asia. My father once contracted amoebiasis during a long journey. This disease needs to be reported to the health authorities, and his medical office was about to close. Even though there are many kinds of amoebae, the *Acanthamoebae* — the hosts of giant viruses — do not cause any known disease.

Can viruses see?

Stupid question? Do the giant viruses of algae and amoebae really allow us to ask it meaningfully? Do they have eyes? What would they need, to be able to see? Nobody is likely to take this question seriously! However, viruses are indeed equipped with a gene that is known to be required for vision. They must recognize something, because they react to light and trigger their host to swim towards light, even though they have no lenses and no brain. How viruses steer their host cells is not known. Giant viruses can achieve this — the plankton DNA-viruses (*Phycodnaviruses*), activating their hosts, the unicellular eukaryotic algae. The giant viruses of amoebae also "drive" their hosts. Giant viruses harbour a precursor to a light receptor, proteo-rhodopsin. This is a so-called seven-transmembrane receptor, a widely encountered signal transducer, which winds itself through membranes in seven helices. Humans use them for transmitting light signals. However, in the case of the giant viruses a point mutation is present in the third helix; this mutation allows light to be accepted but prevents the signals from being transmitted further. That does not come as a great surprise, because where are the signals supposed to go anyway? The light quantum activates the virus, and the virus impels the host to swim towards the light, to the surface of the sea, where rich food is to be expected. This mechanism is called phototaxis. It is not only of an advantage for the host; it is also in line with the viral principle of generating as much progeny as possible: the healthier the host, the better it is for the virus. There is a preference for green light — even that has been discovered. Is that the color of the plankton, promising a rich food supply? The gene is a mixture of related genes; 30% of it is similar to the well-known rhodopsin genes of bacteria, protists, yeast and other eukaryotes including ourselves. There are surprising "eyes" found in Nature, such as some tiny black dots at the rim of jellyfish, which are also thought to be precursors of eyes. In total, five kinds of viruses contain rhodopsin precursors — so such viruses are not even rare. Where did the "proto-eye" genes come from? Have they been transmitted from viruses to hosts, or vice versa? Did they co-evolve with the viruses? Nobody knows.

The many different eyes known in Nature were only invented once, according to Walter Gehring, the renowned eye researcher in Basle, Switzerland who studied the evolution of eyes. He published 50 colorful pictures of eyes in 2012 in a magazine, as part of a debate on how such diverse structures could have one single origin. To demonstrate the autonomy of the eye genes, he even put an eye onto a leg of the fruit fly *Drosophila*, using some developmental molecular tricks. The chimeric flies looked frightening and scared people — "horrible genetic-engineering monsters" and may have cost him the Nobel Prize. He had never heard of virus eyes. That came as a surprise to him, and it would open up a new research topic, as he said when he heard me talk about it during a seminar at Klosters in the Swiss Alps, organized by Manfred Eigen. I gave him the publication by Eugene Koonin — and he missed dinner reading it.

Archaea like it hot and salty

Some regions on our planet are extremely hot, extremely dark, extremely cold, extremely salty, extremely acidic, or extremely basic. Life under such extreme conditions is possible for the "extremophiles". They can tolerate water temperatures of up to 150°C at high pressure in the deep sea, strong acids at pH 2 or strong alkalis at pH 12, or high salt concentrations in salt lakes. Even in permafrost or in drill-tips sampling deep regions in the soil, where nobody would expect any living organism, one can find living extremophiles. One would expect it to be impossible for life to exist under such conditions. The question about the border between life and death does not apply to extremophiles, as discussed for the viruses, because these organisms really are alive: they metabolize and replicate in a manner similar to bacteria. They can metabolize under completely unexpected conditions, and use unusual energy sources, for example, taking up free electrons from electrodes and using them for redox reactions — it sounds like children licking ice cream for sugar uptake. Then there are salt crystals, which are hygroscopic, attracting humidity from the air, from fog in regions of the world where there has been no rain for 200 million years, the Atacama Desert in South America — yet still cells can live there inside salt crystals, in small "fluid inclusions", little droplets. Some of these archaea were revived after an estimated 34,000 years, as demonstrated at the American Museum of Natural History in New York. This has stimulated visionaries, film producers and the NASA to speculate — or even hope — that there may be life on Mars or other planets. Indeed, fluid inclusions have been found in Martian rocks.

Archaea are no longer referred to by their historical name "archaeal bacteria", but simply as archaea, since too many differences have been found between them and bacteria. They were, however identified as a separate domain of life, independent of bacteria and eukaryotes. In spite of this, they show some similarities with bacteria in respect of size, mode of replication, and the ability to host viruses, using antiviral defense mechanisms similar to those of bacteria and harboring reverse transcriptases. The name archaea suggests that they may be the oldest form of life, the beginning of living cells. Today they are placed somewhere between the bacteria and the eukaryotes in the evolutionary tree, and they are related to both of them. When Carl Woese discovered the archaea in 1977, the differences between the domains of life were not so clearly defined. Woese chose as reference the RNA of ribosomes and discovered sequence differences, which led to the organization of life into three domains. The sequences are specific for all types of bacteria and are used today for the identification of millions of different bacteria, a procedure that is surprisingly reliable. (*The RNA is called "16/18S ribosomal RNA (or rRNA)" in pro- and eukaryotes. The corresponding DNA is used for sequencing and classification of bacterial types.*) Early in 2015, a Danish group applied a much more refined sequencing analysis to the archaea and placed them much closer to the eukaryotes within the tree of life, not further down to the bottom but further up, away from the origin of life. They found numerous eukaryotic-like genes, more than anticipated. These archaea are called *Loki archaea* after a deep water hydrothermal vent with this name — Loki is also the name of a Scandinavian god. So the debate about what is older is ongoing.

Archaea are remarkably good survivors. Extremophiles have adapted extremely well to their highly specialized life conditions — but they are also surprisingly inflexible. They may have taken a long time to adjust slowly to new living conditions, since they often had to develop novel metabolic pathways, many of which are highly specialized and still unknown. As a consequence of their inflexible lifestyle they cannot easily be transported, or grown under laboratory conditions. Thus, paradoxically, the heat-loving archaea die during transport under warm conditions, because they can only tolerate a small window of temperature change. Other oxygen- refusing archaea prefer the combination of low temperatures and oxygen during storage; they can survive for more than ten years in the cold room. This raises the question whether such abnormal laboratory conditions may lead to wrong results, ones which do not reflect the real lives of these micro-organisms. Analysis of their microbiome could help, by sequencing of all sequences of a population of archaea directly, without transport or manip-ulation steps in the laboratory. Archaea are as abundant as bacteria, but much less well-known and studied, for these very reasons.

The German researcher Wolfram Zillig, who passed away recently, collected archaea in distant areas of the earth: in Iceland with its geysers and in Yellowstone National Park, which is almost like a window into the inner parts of our world. He even climbed into the craters of volcanoes, and was an adventurer and driver of this research. His samples are still being used today. Karl O. Stetter is studying them, specializing especially in their parasites. The archaea harbor parasites with a ball-like appearance that stick to the surface and help archaea to survive.The host archaea are called *Ignicoccus hospitalis*, and the tiny parasites, which travel with the archaea, are called "riding dwarfs" or *Nanoarchaeum equitans*. I would call them "jockeys"!

Could this strange "jockey parasite" be a virus, possibly a giant virus? It looks that way to me, but that may be just wild speculation. Then Stetter would have discovered the giant viruses! Perhaps he should think about it. Bacteria do not seem to have such jockeys, but hydra and a worm will be mentioned — and what about us?

Archaea have special membranes, containing ether molecules, and may thereby have an advantage for survival under extreme conditions. Archaea do not normally cause diseases; they are not familiar with the metabolism of our cells, and perhaps that is the reason why they do not disturb anything. Or perhaps they have coevolved with us for so long, as they are very ancient.

A former student of Zillig's was David Prangishvili, now at the Institut Pasteur in Paris. He studies the thermophilic archaea together with their viruses, which prefer temperatures above 80°C. Some of them even survive a normal laboratory sterilization procedure, which we assume kills all these microorganisms — but we do not need to worry, since they do not cause diseases. He has so far detected 24 virus families, which are so diverse in shape and properties that they cannot easily be classified. They do not look like viruses at all: one consists of a two long arms with hooks at either end and a thicker central region, the *Acidianus two-tailed virus* (ATV). Others look, in electron micrographs, like lemons, sticks, spirals or bottles. Even in the remotest regions of the world, the structures of these strange viruses look alike, if they encounter similar environmental conditions. This means that the special surroundings must result in such structures. The maxim "form follows function" is apparently also true in Nature, and not only in the arts, or in architecture and design.

All archaea, and most of their viruses, contain double-stranded DNA, which is more stable than RNA. Possibly, extremophiles do not accommodate RNA viruses, which may be too labile in hot environments. The reverse transcriptase responsible for transcribing RNA to DNA can indeed be found in archaea, but what is it there for? Produced by retrotransposons, I suppose, jumping genes, for DNA changes by "copy-and-paste" mechanisms. Some viral sequences in the archaeal viruses are distantly related to today's herpesviruses — whatever that relationship may mean. Recently, an unknown viral structure was detected, in which a protein is wrapped around DNA like a helix and the DNA is less tightly packaged — perhaps an advantage at high temperatures?

There are also viruses of archaea with icosahedral heads and tails, which resemble the phages of bacteria. They occur in lakes containing high salt concentrations, but not in hot springs. Why and how salt contributes to such structures is not really known. So different types of archaea harbor different types of phages.

Archaea do not always live under extreme conditions. They are also found in the feces of humans, or of ruminants such as cows.

Archaea have a strong immune system, an antiviral defense system, which is directed against the lytic life cycle of the viruses. They also have the CRISPR/Cas9 defense system against DNA-enemies such as the phages. All this is very similar to bacteria and their phages.

Could archaea and their viruses also be successful in the event of catastrophes bringing about new extreme conditions? Life can adjust to extremes. That is what the archaea are teaching us. Are they then also the best survivors? This may depend on how quickly changes occur. Perhaps not, they may be too slow to adapt! Survival under new extreme conditions requires evolution of new metabolic pathways, and that takes time. Can humans learn from the archaea how to cope with new environmental changes? Adaptation would require much time. Survival cannot happen fast — it did not in the past, as far as we know. So humans would probably die out before they became extremophilic!

Many viruses of the archaea have unknown genes. Where did they come from? Probably not from the cells, where they cannot be identified. The spectrum of genes in the viruses is much larger than that of all cells combined. Viruses are fast in developing novel genes, also DNA viruses, thanks to their error-prone replication. Viruses are the drivers of evolution. They are the inventors, they test the many possibilities and supply novel information to the cells, not vice versa. So will viruses help us to become thermophilic and survive longer? They probably did so in the past.

7 Viruses as fossils

Viruses inherited

Howard Temin not only detected the enzyme reverse transcriptase, described in Chap. 3; he also discovered endogenous viruses. He was the most imaginative pioneer in the field of retroviruses. Against fierce resistance from the public, and even from his own co-workers, he postulated that there must be an inheritable form of RNA-containing retrovirus with a DNA intermediate in DNA genomes. This is how he, together with others, discovered not only reverse transcriptase but also the endogenous viruses. Endogenous viruses need a reverse transcriptase to transcribe viral RNA into DNA. I was one of the non-believers and failed to detect the reverse transcriptase.

Viruses are normally transmitted "horizontally" as exogenous viruses, which can spread inside the body and also from person to person. However, such viruses can also sometimes infect germ cells; they are then passed down "vertically" from generation to generation. This is normally a very rare event — but it can happen, because a viral infection involves trillions of viruses in the body. Retroviruses normally integrate into somatic cells, but some also make it into germ cells, as DNA proviruses, and they are then passed on as such to the progeny. Can we inherit viruses from our parents? We should not — and that is why gene therapy does not permit the use of replicating viruses, which might enter germ cells and reach future generations. In research, that is an absolute no-go area.

Most of the inherited "endogenous" viruses are from retroviruses. But, in fact, other viruses can also end up in our genomes, even RNA-containing viruses which never go through a DNA stage take advantage of one of the many reverse transcriptase molecules floating around in our cells, and make DNA copies, thus allowing integration. This is currently

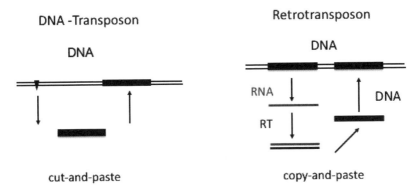

Jumping genes or transposable elements: DNA-transposons or retrotransposons lead to movement or duplication of genes in the genomes of cells.

being studied in a new field of research, paleovirology, which came into being through the new powerful sequencing methodologies. Researchers looked eagerly for viruses in our genome and found them.

Often, viruses do not produce real virus particles any more: they are defective. The inherited endogenous viral sequences stay integrated at fixed locations and, by this, they can be distinguished from newly occurring acute infections which lead to integration more or less at random. This is a simple distinction and is easily made.

The transition from exogenous to endogenous viruses will be discussed using the Australian koalas as an example — because in that instance one can observe the process of endogenization almost in real time. And the reader may be surprised to learn that even Ebola viruses can enter germlines — which *unfortunately* has not happened in humans, but has done so in bats, pigs and kangaroos. *Unfortunately?* This needs to be explained in more detail. This is the message: endogenous viruses protect against exogenous viruses: No entry!

Exogenous viruses, primarily the retroviruses, survive within the genome of a (somatic) cell for as long as the cell lives. In the case of HIV, the infected lymphocytes have a lifespan up to 60 days. Then the cells will die, together with the integrated DNA proviruses. By then virus-producing cells have shed 10^{10} virions per day in a patient. Yet such survival can be perpetuated almost forever if the DNA provirus integrates into a germ cell, as germ cells are passed on from generation to generation. Thus, retroviruses in the genome of a mother or a father are inherited by the children as their own genes, as "self", not as foreign material. This has been going on for about 100 million years, or maybe longer. In contrast to somatic cells, germ cells do not normally produce progeny, no new virus particles. 100 million years is not a sharp boundary, but longer periods of integration are difficult to prove because the viral sequences undergo mutations, deletions and changes, which at some point accumulate to such an extent that it is no longer possible to recognize the viral origin of the DNA.

Why did endogenous viruses accumulate in our genomes? They must originate from real viral infections. Why are they preserved in our genomes as DNA proviruses for millions of years? By chance? By accident? Is integration a safe harbor for viruses, hiding as cellular genes and evading immune recognition by the host? Is an endogenous virus protected against the immune system of the host? — If it is, then integration would be of great benefit for the virus — but also for the cell, as we shall see. All this is true.

The cell can almost never get rid of an endogenous, integrated DNA provirus. Some residual sequences always stay behind. These are the ends of the DNA proviruses, promoters for viral genes, the *long terminal repeats* (LTRs). They are identical at the two ends of the DNA provirus, because of the mechanism of replication by which they were generated. They can recombine after integration and thereby eliminate the sequences in between, cleaving lariat-like structures and leaving only one LTR behind. This leads to numerous solo-LTRs and only a few full-length retroviruses. (*In contrast to DNA proviruses in mammalian cells, the DNA prophages in bacterial cells can be kicked out. Are phages superior in this respect? This process is induced in bacteria by stress, whereby the host bacterium is normally lysed, which means it dies. In humans there is normally no cell lysis.*)

The endogenous virus dies when the cell dies. Also mammalian cells counteract the endogenous viruses by shortening them, if they become useless with time. "Useless" means that they are of no advantage, no longer protecting the cell against outside viruses (see below). Normally, only what is left has any function.

This is how so many crippled viruses accumulated in our genomes, a process which has gone on continuously for millions of years. Often, among the first genes lost are the ones coding for the envelope proteins "Env". Env proteins are necessary for packaging and release of intact virus particles. Without an Env-coat, viruses are "naked", cannot leave a cell and cannot enter a new cell. But, here again, there are two possibilities: Env can not only be lost, but it can also be acquired during evolution or borrowed from a helper virus.

Moreover, endogenous viruses can be inactivated through mutagenesis — this mechanism can at least functionally get rid of the endogenous virus. "Stop" codons occur when the viruses are no longer selected, no longer helping to defend a cell against other viruses, so they start deteriorating. Endogenous viruses can indeed lead to surprising advantages for a cell. They can protect their host cell against superinfection by other viruses. If the cell is "occupied", busy with replicating the first virus, then it blocks a second virus from getting in. This phenomenon is also called "viral interference", which reminds us of "interferon", an antiviral defense molecule in mammals, which will be discussed (Chap. 9). Also phages can occupy a bacterial host cell, preventing other phages from infecting it. For bacteria, the phenomenon is known as "superinfection exclusion" — no entry. Even the evolutionary ancient bioelements, the viroids, which only consist of naked non-coding ncRNAs, destroy competitors by cleaving them. This mechanism, called "silencing", is a very general phenomenon in immune defense. It is based on RNA, but viral proteins too can keep competing viruses out by occupying receptors at the cell surface, which are often the anchors for viral entry. This is just one of many viral defense mechanisms, which can vary widely. In general, viruses defend themselves against other viruses by supplying an antiviral defense to the cell, generally known as a cell's or organism's immune system. All known immune systems were built by viruses. And they are built against viruses! Both, viruses resemble antiviral defense!

This is a strong statement — but it can be proven. There is mutual benefit for the virus and the cell. Thus, viruses — even crippled viruses — may still be useful for their host, and therefore they may not have disappeared, but instead have lasted for billions of years inside our genomes.

It came as a shock to the scientific community worldwide when it was discovered that our human genome is composed of almost 50% of retroviruses or virus-like elements. They are often not intact viruses, but degenerated retroviruses. This has been one of the most spectacular scientific results of our millennium so far. It is true for every eukaryotic organism: mammals, plants, insects, yeast with their spores, etc. Another surprise is, that even integrated defective retroviral elements can move around — but only inside the cell, without ever being able to leave it. Yet they are passed on to later generations as "cellular" genes and are difficult to eliminate.

However, a very surprising protection operates during fertilization. A new embryo starts from scratch. It inherits parental genes with all the endogenous retroviruses. Yet they are normally inactivated by "silencing", by functional inactivation. This is a fantastic arrangement by Nature to shut down parental leftovers and guarantee a new start for the next generation. This silencing mechanism was only discovered recently. The silencing of endogenous viruses is based on chemical modifications of the DNA or the chromatin, the packaging protein of the DNA. Designer genes, intended for use in gene therapy, are also silenced in diseased cells as potentially dangerous — much to the disappointment of researchers, physicians and patients.

Do endogenous viruses exist only in eukaryotes? What about bacteria? Yes, indeed, they also exist in bacteria: up to 20% of bacterial genomes consist of DNA prophages. And only recently it was discovered that many of the prophages are fragmented and interspersed by other spacer sequences. This is then the immune system of bacteria against phages, designated CRISPR/Cas9, the most exciting breakthrough in modern molecular-biological research. In summary then, integrated DNA prophages or DNA proviruses protect their hosts against phages and viruses, respectively. In the one case the hosts are bacteria/prokaryotes, and in the other case eukaryotes — including us.

Phoenix from the DNA

The human genome codes for about 40,000 human endogenous retroviruses (HERVs), integrated as DNA proviruses. Some of them are sufficiently intact to be activated artificially in cells in culture, by the use of appropriate environmental conditions such as ultraviolet light, starvation or cellular stress, or by signaling molecules such as interferon. One group of these human endogenous retroviruses, HERV-K, is still active and can produce proteins — in rare cases even intact virus particles — which can be seen in the electron microscope. However, they normally cannot spread to other cells. Some of these viruses were found in the human placenta, in the brain of mouse embryos and some may play a role in cancer, such as malignant melanoma or diseases as arthritis. Whether these observations will be of general importance for diseases is too early to say.

Often the HERVs are not fully intact and cannot replicate. How do we know that these endogenous viruses are derived from real viruses, that our genome is indeed a cemetery for fossil viruses? This was such an important question that two groups tried to prove it, which strongly supported the observation. Thierry Heidmann, in Paris, analyzed viral sequences of the human genome and aligned nine such fossil endogenous retroviral sequences. All of them were full of stop codons and each of them was unable to express proteins, full virus particles or replicate. However, the stop codons were not evenly distributed, so that he could find regions with open reading frames and put them together. He reconstructed a full-length retroviral genome, synthesized the sequence as DNA and transferred it into cells. Indeed, a replicating retrovirus was produced and could be identified in the electron microscope as a typical retrovirus, one that could newly infect other cells, including human cells. This was a striking proof that the "dead" endogenous retroviruses, populating our genomes in large numbers, indeed originated from real intact viruses — most of which had in the meantime become extinct. That was a totally unexpected resurrection of an ancient virus — and it was frightening. It would have been worth some criticism, even massive public protest, because a virus that existed 35 million years ago and was active until about 1 million years ago could possibly be still harmful today and might even cause some unknown disease. To my surprise, there was no protest whatsoever from the press or the public. Did they overlook the potential risks, or did nobody understand the experiment? Of course the authors took all the necessary precautions. The virus was named Phoenix, resurrected — not from the ashes, as in ancient Greek mythology, but from our genome — to become a real virus again.

Independently of this, Aris Katzourakis in England compared the sequences of the different endogenous mammalian retroviruses and assembled a consensus sequence from them, which allowed him to reconstruct a replication-competent and infectious virus, which he called "HERVcon" (con for consensus). He could infect cells and even animals with it. Katzourakis was also interested in reconstructing other ancient viruses from existing genomes. One of them was a rabbit retrovirus from an endogenous rabbit virus. He designated it, punning a little, as "RELIK", standing for rabbit endogenous lentivirus type K. It was endogenized about 12 million years ago into the genome of rabbits and is a complicated lentivirus of a type similar to HIV. These viruses are more complex, with more auxiliary genes, than the often-detected retroviruses. Until RELIK was discovered it was generally thought that such complex lentiviruses cannot become endogenous viruses. Well, they can. This is also a proof that an HIV-type virus existed a long time ago. It may even be older than can be concluded from the experiment. Why he chose to analyze the rabbit genome is not clear. Did rabbit breeders sponsor the experiment?

Could we search for other viruses in the genomes? Could one find unknown viruses, too? A virus-search algorithm is needed to find unknown viruses. Not an easy task. How can one determine the age of these ancient viruses — what kind of clock is ticking? For age determination one can compare the ends of viruses, the viral promoters, the abovementioned long terminal repeats (LTRs), which flank the HERV-DNA provirus. They are identical during DNA provirus synthesis. With time, they can accumulate mutations — as can any other gene. Assuming a certain mutation rate, this can lead to different mutation patterns from which one can infer how long ago the two LTRs might have diverged. Some HERVs are 100 million years old. For one of them, HERV-K, which we analyzed, we determined an age of 35 million years.

How koalas survive deadly viruses

Koalas belong to the favorite animals in Australia. Many of them die in car accidents. Therefore, special hospitals exist. Hunting them has also increased their chance of extinction. To save them, many were transferred to some islands surrounding Australia and kept in protective custody. However, there they contracted the gibbon ape leukemia virus, a monkey retrovirus, and many of them died — but some nonetheless survived. More detailed studies indicated that the unexpected resistance of the survivors was due to integrated monkey-virus sequences in the genome of the koalas. Viral sequences were present in all koala cells, at the same integration site in each cell, indicating that the viral sequences had been endogenized at that location and stably inherited. In contrast, newly infecting viruses do not have a preference for any single integration site in the genomes and integrate at many sites at random. This is a simple but very informative test.

The process of endogenization must have taken place during the hundred years in which koalas had been evacuated from the mainland, where

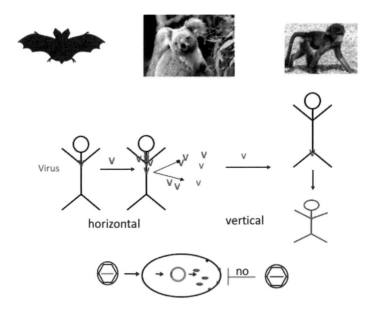

Top and middle: Bat, koala, monkey, and human can acquire endogenous retroviruses by vertical transmission (inherited by viruses in germ cells) protecting against other virus infections (no entry), horizontal viral transmission spreads via somatic cells.

Bottom: Viruses inside cells secrete factors, which keep other viruses out, "no entry", thus viruses cause antiviral defense — and make the host immune.

no such sequences have been detected in the koalas' genomes. The virus was apparently able not only to replicate in the somatic cells of the koalas, but also to enter their germline, which resulted in viral integration at fixed sites in their progeny. Thus the retrovirus sequences were inherited as endogenous viruses — in the surprisingly short period of 100 years. This was endogenization in real time, from exogenous to endogenous viruses — and surprisingly fast. Researchers started to study the process as a "reality show" for endogenization. The endogenous viruses have become the organism's "self" and no longer cause any harm. The important result is that the endogenous viruses protect the koalas against the exogenous viruses — a phenomenon that is known in other cases: the monkey virus SIV, which led to HIV in humans, does not cause death or even disease in monkeys. Also, Ebola viruses do not kill bats any more once they have become endogenous viruses. The point is "not any more", because the viruses probably killed their hosts originally, but after sufficient time, a coexistence finally evolved, making the survivors resistant. How do the viruses get into the germ cells, what are the receptors, how long does endogenization take? These are some of the important questions.

Endogenization in 100 years would correspond to about 20 koala generations. One would have expected thousands of years to be needed. Can one expect a similar time frame for HIV infecting humans? Would humans then be protected against HIV? Ten generations of koalas would correspond to 300 or more years in humans. That is a terribly long time. Yet our genome is full of retrovirus sequences that accumulated during evolution. They are much, much older. RELIK in rabbits was endogenized 12 million years ago. It is a complex virus, similar to HIV. For HIV it is not even clear whether the virus is able to infect germ cells. The receptor for viral entry may not exist on our germ cells, and if it does not then HIV may not be able to infect them. Only with such infection could HIV be inherited. Can HIV change, to enter germ cells? Not impossible. Recently two publications indicated that HIV is indeed capable of entering the human germline. So we can wait and see what will happen — yet protection of humans against HIV, even by a fast endogenization process, will be much too slow and will cost many lives.

There is a precedent for an endogenous HIV: the foamy viruses, which are quite a surprising example. They resemble HIV in having many auxiliary genes and exhibiting complicated regulation of gene expression. However, they are not pathogenic for humans. "Foamy" comes from the appearance of the lung of infected animals or cell culture. Both monkeys and humans can be infected, but they do not fall ill. The virus can only replicate in monkeys, not humans, so that humans are like a "dead-end-street" for this virus. It is harmless in humans. Foamy viruses are the oldest known complex (HIV-like) retroviruses, 100 million years old. How long does it take for a virus to become harmless? We do not know, it does happen. Obviously. SIV does not kill monkeys — any more — so since when has that been the case?

Perhaps there are more viruses of this kind. They are difficult to detect, because they have a DNA genome and thus, at first sight, do not resemble retroviruses. Also the sheep virus JSV (*Jaagsiekte Sheep Virus,* equivalent to the human placenta-forming virus, HERV-W) exists as an exogenous as well as an endogenous virus.

Fish viruses can also exhibit such dual roles, retroviruses replicating as exogenous and also integrated as endogenized viruses. Very little is known about fish retroviruses, yet the numbers of fish-farming facilities are currently increasing to cope with the demand for food. Perhaps a young virologist may want to study them and go to the German Nobel laureate Christiane Nüsslein-Volhard, who hatches more than 20,000 zebrafish in her aquariums, which are piled up to the roof at the Max Planck Institute in Tübingen. These fish became the stars of research, because they are transparent and one can "see" the effects of their mutations. She was honored for this by a standing ovation in the US. The queen of flies has become the queen of fish. They surely have viruses — we just do not know much about them yet.

Paleovirology

How many different types of endogenous viruses are there in our genomes? So far the possibility has been discussed that retroviruses integrate in two ways: horizontally and vertically. Virus particles are spreading from cell to cell or animal to animal by horizontal transmission. Then there are the vertically transmitted retroviruses, which infect germ cells and integrate into the germline. With these cells they are passed on to the next generation. When the human genome was sequenced, not only retroviral sequences were detected in our genome but also the (DNA) sequences of DNA viruses and DNA from RNA viruses, the Borna- and Ebola viruses. These two viruses never normally adopt intermediate DNA. They must have "secretly converted", by some unusual mechanism, to being DNA viruses. A reverse transcriptase (RT) could have been borrowed inside the cell and have helped to reverse-transcribe the RNA to DNA, so that the RNA viruses could be integrated as DNA copies into the genome. Our cells are full of RTs, as we only learnt recently. They may originate from other retroviruses or retrotransposons and then become hijacked. This is how the Borna- and Ebola viruses integrated into animal genomes. The sequence of their genomes has not changed much over time, leading to the assumption that they must serve some purpose. They indeed protect against new infections by the same virus, by interference. Bornaviruses are present as DNA in many species, in humans, monkeys, marmosets, elephants, lemurs, mice, rats and birds — but not in horses. This is how humans may have become resistant, with their endogenous Bornavirus sequences against infections by exogenous Bornaviruses. No endogenous Bornaviruses have been detected in horses, and they indeed get sick by exogenous Bornaviruses, coming down with depression — a disease in horses. In humans the endogenous Bornaviruses express some highly conserved viral nuclear proteins, which protect us against *de novo* viral infections. Depression in humans have other causes.

Ebola viruses can be traced back in animal genomes to about 40 million years ago. Some may even be 100 million years old. They are surprisingly intact and even today express some viral proteins that are related to a present-day Zaire strain. Ebola and the related Marburg Disease Virus, both filoviruses (a name referring to their thread-like appearance in electron microscopic pictures) entered the genomes independently. Filoviruses present in genomes can be found in bats in North America, in Asian primates, in rodents in South America and Australia, and besides bats they have also been found in swine. In contrast to Bornaviruses the Ebola viruses have never become integrated into human genomes, so we are not protected. Why not? Maybe infection occurred too recently, because endogenization takes time, as we discussed above in connection with the koalas. It could also be possible, that human germ cells cannot be infected by Ebola viruses. This should be tested! A good research project.

About 10 other virus types found their illegitimate way into mammalian genomes. Among them were even rare viruses such as circoviruses, with their single-stranded DNA genomes. They are frequent in pigs, chickens and pigeons. One has to conclude that almost all virus types, with different replication strategies, can make it as endogenous viruses into mammalian genomes. Integration is preferred by viruses that replicate in the nuclei of cells. Thus, integration of viruses into the human genome is ongoing and is happening under unusual conditions and protects the host.

Furthermore, animal species with endogenous viruses may produce viruses that are infectious for other species and could possibly cause diseases there. Has anybody looked to see whether a swine Ebolavirus can infect people? This may be worth testing.

Crippled viruses

About 50% of our genetic material consists of endogenous retroviruses or retrovirus-related genes. During evolution we experienced numerous retrovirus infections, and they all have left traces in our genome: Phoenix in human and RELIK in rabbit genomes have been cited as proof of this. With time, the integrated viruses mutated and may in some cases not be

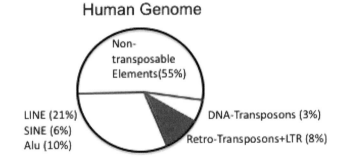

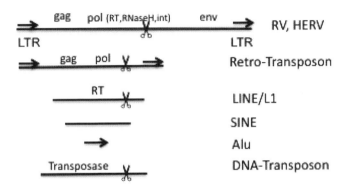

Human genome with transposable elements, truncated retroviruses (RV) make around 50% of the human genome and get shorter with evolutionary age.

recognizable as viruses any more. Such degenerate viruses cannot move or migrate or penetrate into other organisms. They stay within the cell, or — even more housebound — stay within the nucleus of the cell. Old viruses become domesticated and prefer to stay at "home"! Scientists call them "fossils" and our genome a "graveyard" of former viral infections. The graves are at least 35 million, and some probably up to 200 million, years old. Others are as "young" as one million years old. In spite of that, these viral relics can do a lot. They can move from one site of the DNA to another, giving rise to the term "jumping genes". They can get in and out of the genome — but not out of the cell. These "jumping genes" are transposable elements (TE) or transposons.

Even more athletic are some degenerated viruses that can perform a "salto" in between, whereby DNA is transcribed into RNA and then back to

DNA by the reverse transcriptase. Then this DNA copy integrates into the genome. These transposable elements are designated as retrotransposons.

To visualize it simply, think of the text-handling program on your computer: "cut-and-paste" for transposons and "copy-and-paste" for retrotransposons. In the former case a piece of DNA is cut and moved to another location in the genome and integrated, which hurts the DNA twice. In the second case the transcribed DNA is integrated, with only a single genotoxic event. The genetic text gets longer. It grows by exactly the size of the piece of DNA transcribed. The gene is duplicated — a mechanism that is extremely important for our genome to acquire new functions, by mutating one copy and leaving the other one intact as a back-up. The human genome "forgot" about the simple "cut-and-paste" procedure long ago, yet plants use it extensively — "still", one may have to add. Plants are in this respect somewhat more old-fashioned, perhaps because they metabolize more slowly and are less able to adapt, so that two genetic defects are more innovative than one. The "cut-and-paste" procedure was abandoned by our genome about 35 million years ago, when it switched to "copy-and-paste". One should be precise and differentiate between DNA transposons and retrotransposons; the two are often subsumed under the term "transposable elements" (TEs). The retrotransposons cause only one lesion by integration, as a less dramatic change or variation of approved and successful properties.

Viewed as a whole, the principle is simple — transposons behave like viruses, but with the restriction that they move only within a cell and not between cells; that is, they stay endogenous and are not exogenous. The imprisonment of jumping viruses could also be interpreted differently: this may have happened very early during evolution, in which case endogenous viruses may have preceded exogenous viruses. Perhaps the first viruses could only jump around in the genomes, and only later learnt to migrate from cell to cell or organism to organism. An argument in favor of this letter notion is that the coat protein, the envelope "Env" — protein of retroviruses, which is a prerequisite for packaging of viruses and moving over longer distances, was acquired later than the other genes were. Only when the virus is protected by a coat can it leave one cell and enter another. However, the envelope protein can also get lost. Probably both mechanisms operated: acquisition and degeneration — trapped viruses started to migrate and moving viruses become prisoners. Neither can the possibility of a common ancestor be excluded.

How all these ex-viruses in our genomes were discovered has been shown in one of the most spectacular papers I have ever seen. It was published just at the beginning of our millennium and it opened up a completely new research direction, posing enough questions for decades. When as a student, about 100 years ago, Max Planck was told that he should not study physics, because physics had come to an end and was dead. That was exactly the time at which quantum mechanics started to open up a new type of physics, which remained at the forefront of research for half a century. The search for an answer to the question of what the sequences in our genome may mean could again keep us busy for half a century. If that is sufficient!

The exciting paper I refer to was published in *Nature* in 2001. It was the summary of the International Human Genome Sequencing Consortium, with Eric S. Lander as first author and contributions from some 20 research laboratories from all around the world. Its title was "Initial sequencing and analysis of the human genome", a huge paper, comprising 60 pages and more than 40 figures, the biggest I have ever seen in that journal. Nobody could have dreamt what our genome is composed of: viruses! More or less complete viruses, ex-viruses, or virus-like elements! 50 percent or even more! 80%? 100%? That has been the main reason for me to write this book — I wanted to tell others about it and try to explain it.

The authors divide progress in biology in the 20th century into four quarters of 25 years each. The first quarter was the re-discovery of the chromosomes and inheritance in the context of Gregor Mendel's laws. The second quarter was the discovery of the DNA double helix as the basis of inheritance and information storage. The third quarter brought recombinant DNA technologies and the cloning of genes, and the fourth quarter the deciphering of genes and genomes. As of 2001, about 599 viruses and viroids had been sequenced, 205 naturally occurring plasmids (double-stranded circular naked DNA), 185 organelles, 31 bacteria, 7 archaea, a fungus, two animals and one plant — and two human genomes of J. Watson and C. Venter. Then the abovementioned paper appeared, and since 2001 things have really started moving with incredible speed. About 95% of the human genome was sequenced in just 15 months, containing ten times more sequences than all the previous ones together. Our human genome comprises 3.2 billion nucleotides. What do they say? Most of it is unknown: we recognize the four letters but not what they mean, at least not most of it. The first shock was that we only harbor 20,000 genes, coding genes — i.e., genes coding for proteins (the authors calculated this to be 30,000 but the figure has since been corrected and reduced, now again from 22.000 to 20.000). On the basis of the total number of nucleotides, one had expected ten times as many. It was the belief that humans are special — that is what humans think! Now it turned out that humans have not more genes than many other organisms. What makes humans humans? To air a secret: our genes are longer (that was the reason for the miscalculation), and they are fragmented into pieces, described above as exons and introns, where exons, about seven on average per gene, can be recombined by splicing. The ability to combine the pieces is what makes us great. Combinatorics. The next big surprise was that our genes are not unique for us humans, but are a potpourri of genes, chimeras from all kinds of sources: bacteria, archaea, fungi, plants, and nonsense, garbage, "junk" — the latter many of us do not want to believe. Indeed, we do not know whether there is really junk in our genome. Most genetic information in our genome is not uniquely "human" but has arisen by horizontal gene transfer (HGT), from all the many other living organisms around us. About 10–20% of our genome is identical with that of bacteria, and almost 50% with that of retroviruses or retrovirus-like elements; 5% results from fungi; the contribution of archaea, plants or other viruses cannot be quantified yet. Other genes have not been correlated with known functions, they are simply unknown — many of the non-coding genes are currently at the focus of intensive research. We are all relatives, "brothers and sisters" — all living beings. I always remember this if I happen to step on a rain-worm and kill it. This was also remembered by the composer Thomas Larcher, who dedicated a piece of music to me in his collection of pieces "Poems": "Don't step on the Regenwurm"!

That spectacular result about the composition of our genome and the relationship of all living entities on our planet even prompted the then Pope Benedict XVI to ask scientists for an explanation. A while ago, the *Nature* paper would probably have ended up on the Vatican's index of forbidden books. We are less unique than we thought. Yet humans are not completely pushed off their throne; after all, we are the most complex living beings.

Almost 50 per cent of our genomes consists of fossil viruses: the older, the more degenerate and the shorter. A simple classification follows from that, a hierarchy based on their lengths: from long to short, there are names such as: HERV, LINE, DNA-Transposons, SINE and Alu. DNA transposons do not quite fit because they are not necessarily shorter. Another rule is noticeable: The shorter the transposable elements, the more abundant they are, perhaps because they do not require much space.

These viral elements are not only pieces of our genomes; they are also present in similar ways in all genomes of all species: chicken, mouse, platy-pus, opossum, fungi, plants, flies — and in lower species all the way down to Elba worms (see Chap. 10)) and jellyfish; there are no exceptions. Yet there are some differences. Humans have more LINEs, chicken have no SINEs, while platypus has more of them; rice and fungi have no retrotrans-posons. So far, nobody has found out why this is so. I asked one of the co-authors of the genomic analysis of platypus, Michael Kube (then at the Max Planck Institute in Berlin). He could not offer any interpretation — at present nobody can. How did we get such a genome — and all the other species also? We got it, at some time in our history, by gene transfer, HGT, and by diverse virus and other microbial infections. A virus infection is a great gain and innovative push for a genome. Many new genes are inserted in a single shot. That is rewarding for a genome. Viruses are the most versa-tile inventors. They are the motors of evolution. The reader should note the direction in which this is taking use: Viruses and microorganisms made us! That is a new view of our world.

Cancer or geniuses from viruses?

Here are some details, which the reader can omit, because a summary was just given above. But perhaps the following sections will also be informative.

The largest retroelements in our genomes are the Human Endogenous Retroviruses, HERVs. They make up about 8% of our genome, and there are 450,000 of them. Yet only about 40,000 are fully intact. They have two LTRs as promoters at either end, and *gag* and *pol* genes for replication. Pol proteins comprise a viral protease, the reverse transcriptase, an RNase H, and integrase, all the enzymes necessary for replication of a complete ret-rovirus. The envelope protein Env is not always present. Most endogenous retroviruses are inactive and full of mutations. Some of them can leave the cell as particles, but may not be able to infect other cells. The endogenous viruses which we can still detect, prefer integration between "real" genes, into the introns, which is less harmful for the genome than integration within one of the 20,000 real genes, the protein-coding genes. Apparently, only the cells with such harmless integrations have survived until today; others can no longer be found. The integration event of one particular HERV-K, which we analyzed, took place according to our calculation 35 million years ago. One may ask what happened at that point in time. Meteorites, solar winds, supernovae, changes in the axis of the earth — whatever, nobody really knows.

A co-worker of mine, Felix Broecker, was able to show that an integrated endogenous retrovirus HERV-K can lead to cancer. The integrated retroviruses can insert in the "right" or in the "wrong" direction, which means parallel or antiparallel to the orientation of neighboring genes. The read-through products from the virus into the neighboring genes can result in either stronger or weaker expression. If a neighboring gene is essential (such as a tumor suppressor gene) and the antiparallel expression of the virus inactivates it, then the cell may become a tumor cell. We were able to demonstrate this under laboratory conditions in cultured cells. Is this a cause of cancer in humans? Where, in which tumors and how often might this occur? Thorny questions of that kind were asked by the referees of our manuscript. We do not know. Even specialists in informatics from the Institute for Advanced Study in Princeton did not have data sets from patient-derived tumors that contained enough data to answer the question. I do not think that it will be possible to give the answer, even in principle. Even a large consortium on cancer was unable to conclude from their data whether HERVs are generally involved in tumor formation or only sporadically. HERV-K may contribute to sporadic amyotrophic lateral sclerosis (ALS) in some patients — but the question remains as to how general this is.

There are many degenerate HERVs, which can shrink all the way down to one single promoter LTR. This is left if viruses are eliminated by recombination of the two identical LTRs: all the rest forms a "lariat" or loop, which is then cut off. One LTR is the smallest possible leftover, the footprint of a retrovirus. Yet influencing neighboring genes by turning them on or off — that is what these residual promoter-type "solo-LTRs" can still do, which is a lot!

The next rather close relatives of retroviruses are the retrotransposons or retroelements. Their main characteristic is the transposition of DNA by an RNA intermediate through the reverse transcriptase by "copy-and-paste". Retroelements are rather similar to the retroviruses, in that they code for a reverse transcriptase, the molecular scissors RNase H, the integrase, and an RNA-binding protein for transport and protection of the RNA. They also have the long terminal repeats (LTRs), which are the viral promoters required for integration and regulation of gene expression. With Env proteins they can leave the cell, without they cannot. These are often called endogenous retrotransposons.

The retrotransposons come in a second "flavor"; the second one is even shorter and has also lost the promoters LTR, called LINEs (long interspersed nuclear elements), with only two sets of proteins, "Orf1" and "Orf2", used for transport and the reverse transcriptase together with an endonuclease for transposition. LINE elements, or short L1, are transposable TEs that can jump (transpose) by the "copy-and-paste" mechanism and thus increase the size of the genome by gene duplication. They make up 21% of our genome, and each is 10,000 nucleotides long. The human genome harbors about 850,000 copies of these LINE elements. LINE-elements have duplicated and amplified themselves over 80 million years and influenced the human genome. Some of these are still active today, about 100 in the brains of embryos, or, according to other estimates, once in 10,000 cells. It has been shown that some of them can "jump" in the brains of mouse embryos and can cause significant changes. In a human brain disease, Rett syndrome, jumping is more frequent than in healthy brains, as shown by Fred Gage, a brain researcher in the USA. About 60 cases of this have been reported, which could be correlated with new insertions that may result in diseases such as hemophilia, brain diseases, autism or cancer. In the most recent discussions even the age-related Alzheimer's disease was related to transposing elements by Fred Gage. In adult humans on average about 14 jumps occur during the lifetime of a neuron and nobody knows any consequences.

Activation of retroelements is supposed to occur by cellular stress. Such stress could be UV light, chemical substances, poisons, or DNA-damaging agents. There are plausible reasons for activation of TEs. If a cell is under stress, it may have to change its properties to cope with the stress — and retrotransposons can quickly lead to new features and guarantee better survival. I repeat my "mantra": viruses are drivers of evolution and the inventors of new properties. We would not survive without them!

To conclude, if the number of LINE-associated diseases is not the tip of an iceberg, then they are extremely rare in diseases. This is my suspicion, but I may be proved wrong by more results in the future.

The "copy-and-paste" will lead to only a single point of damage to the genome. To avoid this, somatic cells keep the jumping procedure under stricter control than embryos. There, jumping may have significant consequences. Can then — apart from diseases — also a Mozart, a Newton, or an Einstein arise — a cancer or a genius? The possibility of autism and conditions such as Asperger syndrome (which "sages" have with exceptional skills) developing by this mechanism are being debated — or whether jumping genes led to the geniuses of Newton and Einstein.

Retrotransposons may have contributed to the ability of humans to speak, to develop social behavior and to have a conscience. Even plants are considered perhaps to have learnt in this way how to attract insects for replication and gene exchange. "*Lotus*" is a beautiful name for a retrotransposon in flowering plants. The number of transposable elements, TEs in cultivated plants is gigantic. In maize it amounts to 90%, and in wheat to over 80%. For comparison, about 42% of the human genome consists of TEs, some estimates go as high as 66%. TEs are not trivial to find in genomic analyses, because they are not always located there at the same position and do not fit by alignment onto the reference genome. These comparisons are normally based on conserved properties. This is why jumping genes and integrated viruses are easily overlooked or even sorted out as unimportant in sequencing analyses. Therefore, their relevance is underestimated, despite their contribution to efficient innovations in the genomes of many species.

Retrotransposons lead to gene duplication. This is frequent in humans and is correlated with properties that make us special. The evolutionary specialist Svante Pääbo in Leipzig, analyzed gene duplications in human genomes and discovered endogenous retroviruses and LINE elements also in the Neanderthal Man's genome. Neanderthal Man lived somewhere between 200,000 and 40,000 years ago.

Rather like a Greek statue, a gene duplication is characterized by a "supporting leg" and "free leg"; thus one "leg" can try novel things without risk, because the other one guarantees a secure back-up. Therefore, gene duplication may be especially important for expanding genomes and new adjustments. Duplicated genes can also degenerate and be left behind as pseudogenes without any real function. Even whole genomes — and not only individual genes — can be duplicated. The genome of yeast doubled in size about 100 million years ago. The record is held by wheat, the genome of which has increased sixfold and therefore is now huge, bigger than ours. And tulips reached ten times the size of our genome, with three related chromosomes. Such genome repetitions may be a man-made consequence of cultivation.

In human genomes, gene amplification is highest in tumor cells, where many oncogenes increase at least twentyfold in number. The myc gene in particular is amplified in brain tumors ten-thousand-fold, which is probably the reason for the high aggressiveness of this particular tumor.

Who built the DNA — viruses?

I would have liked to gloss over the small viral leftovers in our genome, yet there are so many of them, and they may be doing something. Do they belong to the much-discussed "junk" DNA — garbage, forgotten leftovers? There is no clear answer yet. Some readers may wish to jump ahead to the next chapter.

The next smaller group of transposable elements comprises the even shorter SINEs, Small Interspersed Nuclear Elements, which are much smaller and can do almost nothing anymore. They wander around in our genome and influence other genes. That is their characteristic behavior. They hijack the reverse transcriptase, RT of the larger LINEs, because they do not have their own. Thus they owe their existence to foreign reverse transcriptases. SINEs arose from short RNAs, which were reverse-transcribed into DNA and integrated as SINE DNA. SINEs also lack the viral LTR promoters and thus they do not have any long-range disturbing effects. They have no endonuclease, but use simply DNA breaks to enter DNA genomes. SINEs contribute up to about 13% of the genome, yet each of them is only around 100 to 300 nucleotides long. They are very abundant, with a total of 1.5 million copies. Historically, the SINEs were thought to be "junk", but they can lead to novel functions.

The smallest TEs are the Alu sequences, 300 base pairs long with repetitive DNA sequences. Alu sequences I would have preferred to omit here, because they make the subject-matter rather complicated for the reader. Yet there are about one million of them in our genome. These 10 per cent of our genome cannot be ignored. They only occur in the genomes of primates, and they do not code for proteins; rather, they contain an internal promoter, not an LTR but a shorter one, which affects neighboring genes. They got their name from a restriction enzyme Alu, derived from the bacterium *Arthrobacter luteus* (the abbreviation has nothing to do with aluminum! — though that does help one to remember it). The DNA fragments carry the sequence recognized by the Alu enzyme, which cleaves it and thereby allows identification of the elements. Alu elements can only jump if they borrow the reverse transcriptase, just like the SINE elements do. The RNAs of these elements contribute to the regulatory RNAs in the human genome. Regulation here is local, not long-range. And humans have the greatest number of SINEs/Alus. However, no living animal is as rich in SINEs as the Australian platypus — and nobody knows why!

Why around 50% of our genome comprises retrovirus-like elements, nobody understands. Could it have consisted completely of retrovirus-like elements, up to 100% and not only 50%? Perhaps we only notice virus-like sequences of the 50%, and the rest is no longer recognizable as viruses and has disappeared in genetic "noise".

Furthermore, 50% is not a limit, but a statistical middle-of-the-range value. Some human genes consist of 80–85% retroelements, as shown by Felix Broecker, who analyzed such a gene (*the protein kinase inhibitor beta, which inhibits the kinase PKB/AKT*). The retroelements are not arranged linearly: instead, they overlap, can integrate into each other. The step from 90% to 100% is still a huge one, but it is not a fundamental hurdle. Thus I would like to ask: was our genome once 100 per cent viral? Is there an upper limit to how many jumping genes there could be? If we assume that there were 1.5 million transposable elements, each 10,000 nucleotides long — corresponding to the size of a retrovirus today — then the total length would exceed the size of our whole genome by a factor of five. Perhaps the "viruses" were not all there at the same time and were not full-length.

Was our genome once composed solely of retroviral elements? This is what I suspect — but cannot prove it, and I am sure some colleagues will doubt it. Yet one colleague to whom I mentioned this idea immediately discussed it in his next talk: Luis Villarreal, a pioneer and one of the first to point out the role of viruses at the beginning of life and during early evolution. Sufficient information would be available from the viruses to build our genomes. There is more information in viral sequences than is used by all living creatures together. I find it easy to imagine that retroelements are precursors of retroviruses, or truncated retroviruses, and built our genomes. I find it difficult to turn around the argument of colleagues that viruses are satellites that escaped from the cells. Where did the cells come from? What came before the retroviruses will be discussed in the section on RNA — viroids or ribozymes (see Chaps. 8 and 12).

Endogenous retroviruses cannot be followed back further than about 100 million years. "Hominids" existed about 200,000 years ago. Much older is Lucy. She is perhaps our oldest mother and lived in Ethiopia about 3 million years ago. She was 1.10 m short and her arms reached down below her knees, but she walked upright on two legs. That counts for being our relative — yet life began 3.9 billion years ago. What happened in between? The gap is huge. There is no reason to believe that in the meantime no retroviruses or any of their relatives were around to fill the genomes of bacteria, archaea, plants, animals and humans. Even in hydra, platypus, yeast, the small wild flower *Arabidopsis thaliana,* everywhere we can find similar genetic compositions. *Everywhere* simply means in all eukaryotes. Computers fail with their search programs and cannot correct a text with too many changes. Phoenix, a retrovirus reconstructed from our genome, dates back 35 million years.

Freeman Dyson, to whom I tried to explain, that I think viruses built us, that viruses do not make us sick but fit, that viruses may be our oldest ancestors, listened, sitting in his office at the Institute for Advanced Study in Princeton with books stacked from the floor to the ceiling, and laughed about my discourse. He was immediately interested: the more unexpected and contradictory arguments go, the more he likes to discuss them. "Write this down by any means," he said in German. He speaks German well, but hides it. He encourages me — even though I think he often holds a different opinion from mine. After all, he is the author of a book with the title: *The Scientist as Rebel.*

The next group of transposable elements, are the DNA transposons. They are not considered to be related to the retroviruses, or so we are told in textbooks. Really not? Perhaps they are related after all! They lack a real reverse transcriptase, RT, that was the argument, but their enzyme, the transposase, is closely related to the RT. DNA transposons follow the "cut-and-paste" mechanism; they cleave and paste with a nuclease and an integrase that are related to the retroviral RNase H and integrase. So there are the family ties.

DNA transposons jump in plants and other organisms, but not in humans, where they do not jump any more. DNA transposons are rare fossils in the human genome, amounting to only about 3%, corresponding to 300,000 copies. But they are inactive. DNA transposons are frequent in many plants at almost 50%, in maize at 85% of the respective genome. The ratios of DNA versus retrotransposons greatly vary, many plants have also many retrotransposons. DNA transposons do not multiply, but only change the integration site without doubling and increasing their copy numbers. But some transposons can also replicate. This is not well known, but it is important because of the many gene duplications that result.

Starting and landing of transposons before and after the jump hurts! It can cause lesions, genetic defects in the genome. But these may likely be useful as toxic for the cell. The cell may lose or learn something new. Thus, the interaction with the cell must be optimized: too much jumping destabilizes the genome, and the cell will die. On the other hand, not enough jumping is not informative and innovative enough for the cell, and the cell may also die. In the eukaryote yeast researchers have succeeded in determining the frequency of jumping. A yeast transposon Ty-1 jumps in one cell out of 20,000. The jump of a DNA transposon with "cut-and-paste" can be innovative because of insertional mutagenesis, the abovementioned genotoxic events, which can lead to cancer as well as producing geniuses. They are twice as "dangerous" or "useful" — i.e., innovative — as the retrotransposons. Thus DNA-transposons are probably the most important force in driving genomes to adapt to new conditions. This may be most important for organisms that otherwise have little variation. It may be particularly true for some plants, or even trees. They cannot move, cannot run away, they just stay where they are for a lifetime — up to 9500 years, which is the age of the oldest tree we know of. They are visited by insects, beetles, birds, fungi, etc. which may convey new information. Maybe not enough new information, because plants have the greatest number of active DNA-transposons. Recently a poisonous plant, *Paris japonica*, was described with 150 billion basepairs, that is 50 times the human genome and the biggest giant genome known, topping the tulips — I bet it has the highest number of DNA transposons.

Yeast (*S. cerevisiae* or *S. pombe*) in contrast has no DNA transposons (but many retrotranspons). Why two types of amoebae vary extremely, one of them containing almost only retrotransposons (*Entamoeba (E.) histolytica)* and another one almost only DNA transposons (*E. invadens)* is enigmatic to me. Thus, it is difficult to find rules about the ratios of the various transposable elements.

Transposons are not only very active in plants but also in flies, where they have specialized in jumping to the ends of the chromosomes as protection against erosion during cell division. They are thus relatives, or precursors, of the telomerases (indeed, they have similar structures), which elongate our chromosomal ends during embryogenesis. During cell division they protect our genomes from deletion of essential information.

Now comes a surprise: transposons generated our immune system. How? Transposons defend a cell against related transposons. That is called antiviral defense — and is our immune system, which defends a cell. The mechanism is simple: any virus (including transposons) prevents a similar virus from entering a cell. This guarantees the virus successful replication and high progeny within the limited resources of a cell. Thus, defense of a cell by the monopoly of a virus is for the benefit of viral replication. Viruses have built up an antiviral defense system, which ultimately became a general immune system of the host cell. The viruses "taught" the cells how to keep viruses out. The immune system is directed against its "inventors" or initiators by a simple negative feedback mechanism.

Not only eukaryotic cells, but also bacterial cells defend themselves against too many transposons. Transposons have to be inactivated over time, and cells need to accumulate inactivating mutations for the maintenance of successful genetic acquisitions. It should be pointed out that there are defense mechanisms by cells against transposable elements, because too much jumping can do harm. Jumping is counteracted by "silencing" mechanisms (see Chap. 9).

"Mrs. Mendel's" maize

The geneticist Barbara McClintock studied maize kernels and was surprised by all their different colors. The laws of inheritance of genes, propounded by Gregor Mendel, failed when she analyzed the colors of Indian corn. She could not explain them. This led her to the discovery of jumping

Epigenetics

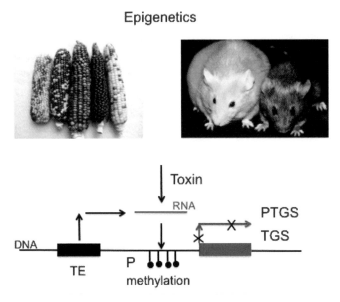

Epigenetic changes are transient, caused by environment such as toxin, or viruses or transposable elements (TEs) which can cause color changes in maize or mice by modification (methylation) of the promoter (P) of a color gene. Mechanisms also involve transcriptional gene silencing (TGS) or post-transcriptional gene silencing (PTGS) (no RNA).

genes, the transposons in the 1940s. She overthrew a universally accepted concept about the stability of our genetic material, even before the double helix of DNA was discovered in 1953. She discovered instabilities in the DNA of maize and attributed it to jumping genes, which influenced the colors. Already in 1944–45 she observed the phenomenon of movable genes within cell chromosomes, causing mutations and color changes in Indian corn. When she reported it to the scientific community in 1951 she was confronted not only with complete lack of understanding, but even with hostility and rejection. Her results were felt to be heretical. She was 50 years ahead of her time.

Jumping genes, or DNA transposons, were then observed throughout the next 20 years all over the place — in yeast, in bacteria, in all living species — including humans, though in a slightly more complex way. Retrotransposons dominate in our genomes, while the DNA transposons became inactive in humans a long time ago.

One can "see" jumping genes — surprisingly — but only if they affect color genes.

DNA transposons can jump into the vicinity of color genes and change their regulation, which McClintock defined as "control genes". Other people thought that pathogens were causing the color changes. This indeed is also true, viruses caused the tulipomania (see Chap. 8). Jumps can occur anywhere, but by looking one can only detect changes in the regulation of color genes. It is rare that one can see such complex mechanisms with the naked eye. It required a Barbara McClintock, though, to solve the mechanism — at least in part, because the mechanism is still not fully understood. During meetings in 2015 her photograph was shown repeatedly.

Barbara McClintock discovered the phenomenon "epigenetics", beyond genetics, because explanation of the colors of maize by Gregor Mendel`s laws on genetics failed. For her discovery of the jumping genes one may well name McClintock "Mrs. Mendel", so importantly we must rate her discovery. Today we know that epigenetics is based on chemical modifications of the DNA or the proteins packaging the DNA, which then deregulate gene expression. The effect is due to environmental changes and is normally not inherited. Two major mechanisms are known by now, brought about by chemical modifications, such as the attachment of methyl groups to the DNA or changes in histone proteins (chromatin) by acetylation. Methylation of a promoter — which is her predicted "control gene" — shuts off gene expression.

McClintock was not understood, but luckily she was ultimately not left out. She missed a first Nobel Prize. She had contributed significantly to our understanding of control regions, not only in maize but also in bacteria. The so-called "operon model" in bacteria was honored with a Nobel Prize in 1965, which went to two French colleagues from Paris, François Jacob and Jacques Monod. She was at least a mental inspirer.

McClintock had to go through the bitter experience of having her scientific papers rejected. So she gave up publishing. Normally that would have been the end of a career — in science anyway. But the leaders of the Cold Spring Harbor Laboratory, Fred Hershey and later Jim Watson, noticed and possibly foresaw the importance of her research. She was given a permanent laboratory. There she stayed until the end of her life. Her office was full of papers, piled up to the ceiling. If visitors came she would not be unfriendly, but she was not at all communicative. Visitors would have to raise new questions or topics continuously — because otherwise the conversation would soon flag and die out. She would say almost nothing, and was friendly, but relieved, when the visitor left. In 1983 she received the Nobel Prize "for her discovery of mobile genetic elements", the discovery of the jumping genes. People started to understand the implications of her results also for viruses and cancer cells. She was the third woman ever to receive the prize and one of the few laureates who did not share it with others — she received it alone.

She was then 81 years old. No-one had ever sat with her in her car, she said in an interview, such a loner — or an autist? — was she. Shortly after her 90th birthday she passed away. She lived in one of the white wooden houses that probably originate from the times of the whalers, who came to Cold Spring Harbor to get fresh water. Above her flat in Hooper House were a few guest rooms for participants at meetings, as spartan as monastic cells. They were always my favorite address during symposia. In the little rooms time seemed to stand still; one toilet door had a defective lock and the occupant was forced to keep the door closed from the inside with one foot — for decades! Everything was simple. Some rooms allowed a view down to the Canadian geese and the harbor with its changing tides. In the mornings McClintock crossed the wet lawn, thick socks on her feet. This is how one might imagine the ascent to Olympus or the goal of pilgrims with the *spiritus loci* of a genius. It is interesting that McClintock would not have been able to discover the phenomenon of jumping genes in the human genome — since there is no jumping any more, at least not in the way that DNA transposons jump around in maize. Plants have the most mobile genomes, and maize was ideal, with its colored kernels. Maize is still full of surprises, because it can lose a third or half of its genes at a stroke. The genome of maize is as big as the human one, and 85% of it consists of DNA transposons, whereas the human genome has only 3% transposons, all of which are inactive. The "cut-and-paste" mechanism of transposons was analyzed by McClintock not only in maize but also in bacteria with the help of phages, where genes jump from the phage genome to the bacterial genome, rapidly back and forth, like ping-pong balls. This flexible little phage is called *Mu*, because it *mu*-tates the genome of its hosts. Exploiting Mu as a faster model for studying and understanding the mobile elements was another stroke of genius by McClintock.

Not the famous maize, but the almost unnoticed and boring-looking field weed *Arabidopsis thaliana* was selected first for sequencing of a whole plant genome, because its genome is only 130 million base pairs long, which is small compared with that of maize or wheat. The numbers of jumping genes and of repetitions are similarly small, which makes sequencing easier. Perhaps the low frequency of jumping explains the small genome or did nobody care to cultivate it?

The tools required by Mu for jumping resemble those of retroviruses, especially the integrase. Mu phages always integrate, just like retroviruses — distinct from other phages which do not integrate in the lytic phase, which Mu however does. It was Merck Co. that exploited the integrase of the phage Mu in a high-throughput test for screening possible inhibitors of the HIV integrase. Mu as model for integration of retroviral DNA into a bacterial genome replacing HIV! I could not believe it! Speed was a major advantage, and it did not come with any safety-laboratory requirement. Indeed, it resulted in one of the best clinically relevant HIV-integrase inhibitors, the drug Raltegravir. The highly surprising similarity between the phage Mu jumping around in bacteria and the jumping genes in maize — all the way to retroviruses such as HIV — indicates a staggering degree of evolutionary conservation: from a phage in bacteria to maize to HIV in humans! I am surprised about it myself. I also admired the researchers at Merck Co., who had this thrilling idea, one that indeed resulted in a big therapeutic success. There is no more convincing similarity than this one, between a retrovirus and a phage, HIV and Mu.

Another surprise was that Mu could also serve as a model for gene therapy, namely for the mechanism of how a foreign gene can be integrated into the germline cells: by a staggered cut (a step with dinucleotide overhanging ends), which has to be filled by the foreign gene. The processes of integration of phage Mu and retroviruses resemble each other down to the smallest details and were studied in the less dangerous and faster Mu system.

Almost at the same time as McClintock's discovery of epigenetics, another phenomenon was detected in maize: in 1956 R. Alexander Brink, a Canadian geneticist, also noticed a totally unexplained inheritance, surprisingly patterns also with the colors of the kernels of maize. It lasted longer than one generation without following the rules of mutations, so he called it "para-mutation", genetics without genes! That has remained almost unnoticed, even till today. It leads to the inheritance of acquired properties, including learning — a topic that once inspired political discussions. Again, this phenomenon does not involve direct genetic changes of genes — but in contrast to epigenetic changes these are long-lasting: they are maintained over many generations. Now scientists start to remember, analyzing the surprising inheritance of changes without mutations. It got a name: transgenerational inheritance. They got alert from the observation that fat fathers have fat grandsons, also true for fat mothers and fat granddaughters — keep reading why! I lift the secret: The entities inherited are special regulators, called piRNA and transmitted in the sperm! This is a future perspective and will be discussed in the next chapter and at the end (Chap. 12).

In front of the lecture hall in Cold Spring Harbor Laboratory, there is a little glass box in memory of McClintock: it gave me the impression of her minimalistic lifestyle — almost poverty: a pair of tweezers, a graying ear of maize, a picture postcard, her glasses… close to nothing. Her notebooks are in the library. She would probably have appreciated the modest exposure. It is her thinking which survives, not these goodies. It is astounding what can be seen by looking at maize kernels. A glance at my tin plate with colored maize ears — a present from a friend, purchased at the weekly farmer's market in Graz, Austria — reminds me of McClintock's jumping genes. All these belong together: transposons, retrotransposons, retroviruses all the way to antiviral defense and immune systems. These billion-year-long stories are told to us by a few colored kernels. Everybody should save some of these beautiful witnesses of the history of life home — the colored maize from Graz of course, and not the yellow stuff in a tin can from the food industry where there is nothing to be seen!

Any reader who has got as far as this sentence should write to me and let me know if he has found a mistake, if he thinks anything needs to be changed or corrected: he will then get a free copy of the book with my dedication.

Poisonous toys and epigenetics in Agouti mice

Young mothers, please watch out: Colored mice can identify dangerous baby's dummies. This is equivalent to the colored maize in mice. The mice are not as colorful as maize, but they are also colored, and the color is an indicator: normal mice are black and healthy, whereas sick mice are yellow and fat. Black mice suppress diseases and obesity, diabetes and even some tumors. This is why they are used as living test models for various environmental influences. The mice are called *Agouti* mice after the name of the color gene *Agouti*. Looking at the two mice, which appear totally different — one yellow and fat, the other one black and thin — one cannot believe that they are really identical twins with identical genes. Epigenetics is the explanation.

They differ by epigenetic effects, which means by acquired transient modifications of their DNA and not by mutations in the DNA. The epigenetic changes occur through environmental factors. In this case carcinogens can lead to chemical modifications of the color genes by methylation of the DNA promoter and/or acetylation of histones. Thereby, genes are newly regulated, which can be observed, with normal color gene mice becoming yellow, whereas modified color genes leads to black fur. Intermediate changes result in brown colors.

Now I would like to draw the attention of young mothers: if one feeds the Agouti mice with carcinogens such as bisphenol A, which is present in many baby's toys, in dummies, or in plastic baby bottles from the Far East, then the next generation of mice will be yellow-colored because of activation of the Agouti color gene. This indicates "danger"! Agouti mice are test-mice or biosensors of environmental alterations, such as carcinogens and toxins, and also fine dust, chemicals, psychological factors or simply age, all of which can lead to a yellow color. Feeding of the yellow mother mouse with a second compound in addition such as vitamin B_{12} or folic acid, the mice progeny reverts to "black and healthy". One nutrition or environmental factor can neutralize another one. Here, the time point and the developmental stage are also relevant, but the principle is surprising. However, I would not recommend vitamins as baby food to counter the effects of toxic toys; rather, buy less toxic toys!

How can poisons cause color changes? The answer comes from a totally new research area and is a multistep process: toxins lead to small regulatory RNA, a methylator (*RNA-dependent DNA methyltransferase*) which then methylates DNA and regulates color genes.

McClintock showed that jumping genes (or RNA) can induce epigenetic changes and multicolored maize kernels. Mice with changing colors were not yet known to her. She may have been surprised about similarities between maize and mice. Yet the principle is similar: epigenetic changes of control genes regulate the appearance of new colors. When you go for a walk, you may notice dogs and cats with white tips on their ears, or in the lower leg, or with white toes. Horses can also show white regions and Kois with color patterns. They are frequent– this book is full of them!

But you will be surprised to read later at the end of the book — that some of it is due to "paragenetics".

Interestingly, one RNA molecule can neutralize the effect of another, so two nutrients or even two poisons can cancel each other out. What does all of this have to do with viruses? Jumping genes, transposons, or retro-transposons are virus-related and affect control genes, the promoters and thus colors. Transposons are like locked-in viruses, which cannot leave a cell, imprisoned into the cell because they lack a coat in which to move around outside.

The beauty is that white tips of ears or feet are indicators of compli-cated epigenetic mechanisms, environmental changes of genes. I do not know whether the offspring of your dog or horse will also have white-tipped ears — if so, then transient epigenetic changes may have been turned into permanent mutations by some ambitious breeder!

Sleeping beauty, ancient fish, platypus, and kois

Some examples of what transposons do in various species will now fol-low. Again, there was the resurrection of a dead body: a transposon from salmon, with the suitable name "*mariner*", was revitalized from an inactive fossil transposon and was called "sleeping beauty" (SB), after the figure in the fairy tale of the Brothers Grimm. It was reactivated by a procedure similar to that described above for Phoenix, the reactivated retrovirus, which became infectious again after 35 million years of "dormancy". This sleeping beauty slumbered on for over 10 million years. Then researchers came to wake it up. Again, stop codons were replaced in the laboratory and the transposon was repaired so that it was able to jump again. The salmon SB-transposon was selected as the molecule of the year in 2009 in the US by an International Society for Molecular and Cell Biology, which has given it quite a reputation. Now everybody is waiting for success with this transposon in gene therapy without (or less?) side effects, because it does not integrate as retroviruses do. Indeed, there is some progress in 2016, the transposon was used to create a green cow, even more surprising a green cow with glowing red eyes, and now people shall get transgenes or therapeutic genes, not color genes, to cure eye diseases. Wait and see.

An exotic animal full of jumping genes is the *African coelacanth*, a fish as long as a dinner table. For a long time, it was considered extinct until a museum director in South Africa recognized one amid the catch in a net. Since then about a dozen of them have been encountered. Its picture decorated the front cover of *Nature* in 2013. Coelacanths have been around for the past 40 million years and are considered to be the oldest living fish — living fossils. The fish grows very slowly, with an increase in length almost immeasurable over 20 years, and it lives for more than 100 years. This can be determined from its special organs for equilibrium, almost like "tree rings". Coelacanth's genome is 2.9 billion base pairs long, which is unexpectedly large, and it is full of repetitive sequences, which are amplified (up to 60%) without much diversification. Axel Meyer in Konstanz, a specialist in *cichlids* (tilapia) fish from Lake Victoria in Africa, is co-author of the genome analysis of this ancient fish. Even he does not know why the genome of the fish is so constant. Did it have no natural enemies? The fish is especially interesting because it can swim and run, which should be detectable in its genes. I would like to add this fish, with its so many fossil-type viruses, to my curiosity collection, my *Wunderkammer* of strange creatures.

Another exotic animal is the platypus, an unusual-looking animal from Australia which is becoming extinct. Its genome was fully sequenced by colleagues at the Max Planck Institute for Molecular Genetics at Berlin, who found that 50% of the platypus genome is made up of transposable elements, some of which are very repetitive, and there are extremely large amounts of SINEs, which must have multiplied significantly. No other organism has so many SINEs. The reasons for all that are not known. The platypus is half-way between birds and mammals — flat-footed, with the beak like a duck's, thick fur, and a huge tail in which it can store fat; it lays eggs and produces milk. The male can spray snail venom. It can be dated back about 100 million years. It is as strange as science fiction. Its genome is a mixture from diverse species.

Then there are some fish, the *kois*, which suddenly died in fish hatcheries. Kois are Japanese ornamental fish that are shipped around the world to serve as decoration in outdoor ponds. These fish are carps with symbolic character, a Japanese version of the holy cows from India. The farming of these special fish, with their colorful patterns, has a long tradition. Perhaps an attentive reader is already asking whether the colors are caused by jumping genes as the maize kernel colors. This is what I suspect, because they trouble the fish-farmers with colors constantly changing from one generation to the next and not being properly inherited. This rings a bell — it sounds suspiciously reminiscent of effects seen with maize and Agouti mice — and with striped tulips (see Chap. 8): by epigenetic changes?

Why did the kois die all of a sudden? They died of koi herpesvirus. Was the population density too high? Did a newcomer fish bring in a virus and contaminate the others? No, humans changed the environment; they straightened out the borders of the lake, and they removed the fish ladders, so that the fish could not swim upstream and lay eggs. Stress for the kois! This affected their immune system and resulted in activation of herpesviruses. In Germany, salmon-ladders have become obligatory now, so that the fish can jump upstream. Our carps have similar habits and problems, too. And stress is well-known to activate herpesviruses on our lips, and cause cold sores — the same phenomenon. Environmental stress acts on real viruses and epigenetics.

A fence with empty spaces

How surprising: only two per cent of our genome codes for proteins. What about the rest, the other 98 per cent? It guarantees that nothing goes wrong with the protein-bearing 2%. Regulatory or non-coding ncRNAs originate from the 98%. For years, the Australian scientist John Mattick has emphasized the importance of this kind of ncRNAs. Mattick showed earlier, on a graph in a *Scientific American* article, how the amount of regulatory RNA increases in higher organisms — from bacteria, unicellular organisms, fungi, plants, vertebrates — from zero to about 98% of the total RNA. We rank at the top. Why is there so much more ncRNA in animals with higher complexity? The answer is simple: to allow higher complexity.

Our genomic DNA codes in total for 20,000 genes. Yet we are surpassed, perhaps surprisingly, by bananas, which have about 35,000 genes, while maize have as many as 50,000 genes, as do wheat and tulips. For wheat, three chromosomes have been identified which are related to each

Split genes

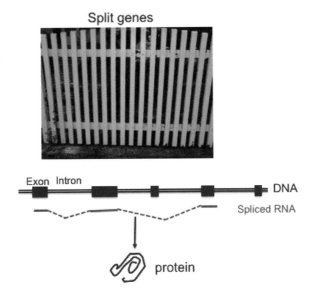

Genes are like a fence corresponding to exons and introns in the DNA, combinations of exon RNAs (by splicing out introns) lead to various proteins by combinations.

other, looking as though some kind of multiplication procedure is at work. The flour from wheat with only one of the three chromosomes was shown to be unpalatable in bread and was difficult to bake. Giant genomes may possibly arise as a consequence of long breeding and cultivation by humans. Even though there are exceptions.

The field flower *Arabidopsis thaliana* has 28,000 genes — probably nobody has ever been interested in cultivating this "useless" weed. So its genome has stayed small. Then comes the worm *C. elegans* with 23,000 genes. In respect of the number of genes, humans do not rank at the top. With our 20,000 genes we have even fewer than the desolate-looking field plant *Arabidopsis thaliana,* even fewer than amoebae do.

The number of genes in monkeys and mice is similar to ours. This sounds arbitrary. The number of genes certainly does not correlate at all with intelligence! A public dispute raised the puzzling question: Why do humans have so few genes? Our genes are not more numerous, but they are longer. All other species have shorter genes. Humans have the longest DNA, with an average of 160,000 base pairs per gene. Thus maize, bananas and tulips have more genes than we do, but these genes are shorter and "less informative": rice and maize have shorter genes of about one-tenth or one-fifteenth as long as ours, 10,000 to 50,000 base pairs per gene. Bacterial genes comprise about 1400 base pairs and many viruses have the smallest number, about 1000 base pairs per gene. The small numbers for viral genomes are worth remembering in the context of this book, because there are exceptions, the giant viruses (Chap. 6).

Here is a strange surprise: there is a Scandinavian fir tree that has a sixfold bigger genome than ours and is the oldest tree known on our planet, 9500 years of age. Yet the number of coding genes is as small as that of the field flower. Many of its genes are again repetitive, with nothing new. Does a boring life last longer? I shall add this fir to my curiosity collection.

Pretty effective processes for cleaning up unnecessary genetic material have been described in flies, where scientists even speak of gene "house-cleaning". Cleaning up (deleting) genes costs energy. Which is cheaper, or more cost-effective: storing unnecessary DNA in a genome, or removing it? Is the stored DNA equivalent to what is often designated as "junk DNA", DNA full of unknown or potentially meaningless sequences? Removal of redundant DNA is not well understood.

If I think of the basement of my home, where I urgently should clean up, I nevertheless prefer not to throw anything away, because the cost of keeping everything is not that high. I do not know what the garbage may be useful for one day. Curtains were used to make dresses during the War, when there was a shortage of everything. Perhaps excess of DNA becomes useful in periods of scarcity, as a back-up. Cleaning up would be much more cumbersome.

What purpose could the extremely large genes in the human genome serve, if not for protein-coding information? The answer is: for regulatory information, management, administration, combination of gene pieces to form new genes!

The large genes are full of introns interspersing the exons, they are the gaps in between the exons. Above, we used a fence as a model for introns and exons, with gaps and posts. The introns express regulatory non-coding or ncRNA, which is not for proteins but for regulation of the proteins. Human genes consist on average of seven exons/introns, whereby the introns can be very long, a hundred times longer than the exons, corresponding to a fence with big gaps (or an open gate). Now comes the important point: exons can be combined by splicing and generate complexity, especially since our genes are the largest ones. That makes humans more complex than all other species.

When DNA was discovered, the idea was that one gene codes for one protein, and that was as far as it went. That is incorrect for us humans — though not for viruses, because viruses do not have introns. In spite of that, they can also perform splicing but only by newly combining areas of exons — again one of the minimalistic, economical properties of the viruses. Yet viruses too need regulation of gene expression. Splicing was discovered in viruses by electron microscopy, where pieces of a viral genome were "visibly" looped out. Splicing makes possible the exploitation of different reading frames, a principle that operates at all levels from viruses to us. For viruses the amount of non-coding ncRNA is minimal, almost zero; for bacteria it is about 10% (in some reports one can find zero), for yeast 30%, for *C. elegans* 70%, for flies 85% and for mice and humans 98%. One can also express it the other way round: for viruses the protein-coding information comprises almost 100% and for us only 2%. So one might say that 2% of our genome is regulated by the other 98%, corresponding to the executive and the legislative in a political system. (*Executive correctly reminds us of the word exons!*). Does that mean we have a terribly brain-driven legislative? No, it is just like in politics. In Switzerland seven people represent the executive and 200 the legislative/regulative. Thus it seems that our gene regulation corresponds to the proportions in one of the best democracies in the world, at least in the Swiss view!

The legislative — the introns define the programs, the co-ordination in time and space, development, differentiation, specialization, growth and cellular death (apoptosis) programs, along with adaptation, immune response, innovation and peak performance. To conclude: combinatorial linkages of genetic regions by splicing make humans special. That is our strength!

ENCODE for understanding "junk-DNA"

What is going on in the 98 per cent of our human genome? Two per cent codes for proteins and the rest we did not know. Therefore, a study was initiated to follow up the Human Genome Project, called ENCODE (Encyclopedia of DNA Elements). The ENCODE project aims to find out what takes place in the "desert regions", as the authors term the unknown non-coding (nc) "empty" spaces in between the coding regions. Only two percent of the DNA can be attributed to proteins. Most of this 98% is expressed as non-coding ncRNA, which can be ascribed to regulatory functions to organize the rest, expression of the proteins. The "desert" regions contain short- or long-range regulatory elements such as enhancers, repressors, silencers with and without inactive and active sites, insulators, promoters in gene proximity, instructions on processing (that means trimming of RNA), chemical modifications such as methylation of DNA and acetylation of chromatin (probably also phosphorylation and other modifications), and the RNA transcripts of genes. The role of "insulators" is a recent discovery: they do what their name implies, blocking regions located between promoter and an enhancer. How they do this is still unclear. It is the goal to learn how and when a gene is active. There will also be analyses of signal intensities (how active a gene is), of how much is expressed, of evolutionary conservation. That again excludes the viruses, because they are not conserved — so there will certainly be more research required.

Recently handling of huge amount of data is supported by astrophysics, also mentioned by a Dutch genomic analysis. Imagine a pile of paper 15 m high, 35 km in length, and covered with only the four letters A, G, T, and C in an order nobody understands as yet, comprising 15 terabytes (or 15,000 gigabytes) this needs to be understood as text, and broken down into sentences, chapters, headings, or paragraphs. The total length of the human genome is 3.2×10^9 nucleotides.

More than 2500 publications are now beginning to emerge, many years after ENCODE's initiation in 2003. Surprisingly, some of them are puzzling. They describe 70,000 promoters, the start sequences and regulatory regions in front of genes — even though we have only 20,000 genes. Why are there so many more promoters than genes? Also, 40,000 enhancers were determined, regions more distant from a gene, with very long-ranging effects across "boundaries", regions without proteins. In the "desert" regions, where plenty of regulatory ncRNA is produced, mutations for diseases seem to occur. Some defective regulatory elements were discovered which could cause diseases. Deregulation of gene expression often means cancer. Indeed, the diseases appear to be based in the non-coding regions.

So far only 10 percent of the DNA has been analyzed. By the end of 2014, 440 scientists from 32 groups working according to standardized protocols with 147 cell types had performed 1648 experiments.

So is there now a conclusive answer as to how much of our DNA is junk? No, not really. The "desert" region of the DNA turned out to be 80 percent active functionally and 75 percent of it is transcribed onto RNA, not to make proteins but regulators.

Thus most of the DNA is doing something; many viruses are crippled to what may appear as junk, but they are often still doing something, their LTRs were mentioned as residual sequences, promoting other genes. Others may be on their way to disappearing, so there may be indeed some junk around. Evolution is ongoing, it never ends! If one wants to perform genetic diagnostics, then one has to take into account the fact that two "normal" people differ in their genomes in 0.5 percent amounting to 16 million base pairs differences — this number has recently increased by fivefold — out of the total of 3.2 billion. It is against this background that one has to define sequences associated with diseases. Which of the single-nucleotide polymorphisms (SNPs) indicate diseases and which mean nothing at all? That is the challenge to find out. In order to overcome this signal-to-noise ratio new technologies have to be developed.

Perhaps there will again be some short-cuts to find answers to all or some of these questions. Craig Venter found cheaper and faster ways to analyze the human genome; he saved mankind ten years of work and reduced the cost by several orders of magnitude. Will he or someone else be able to repeat this with the ENCODE project? Let us hope so. All the data arising from it are publicly available, and courses have recently been announced to train people in using the data. Everybody can help — if he can!

8 Viruses — our oldest ancestors?

In the beginning was RNA

Life started not with Adam and Eve, not in Paradise — I think — but down in the oceans, around volcanoes, where it was hot — more like hell than heaven. There is also another possibility being debated: ice! Life may have started in fluid channels inside ice. If one wants to quote the Bible, one could refer to the first sentence of the Gospel of St. John: "In the beginning was the Word." Johann Wolfgang Goethe, too, reflected in his *magnum opus* "Faust" on this sentence from the New Testament: word, or sense, or force — which came first? This is a surprising congruence with our present notion about the origin of life: letters and words were first. The first biomolecules, from about 3.9 billion years ago, were composed of short RNA, and RNA consists of nucleotides as building blocks, which can be described in short as four letters A, U, G and C. Only four letters are required to encode the complexity of life on earth — that is stunningly few. The bases are parts of the nucleotides, and their names provide the abbreviations. Where did these first four letters come from in the beginning? Perhaps there were only two in the beginning. Did they arise by lightning and thunderstorms in a primordial soup at the bottom of the oceans around volcanoes or hydrothermal vents, the "Black Smokers"? There temperatures can range from 0 up to 400°C. (*Yes: because of the high pressure, temperatures can exceed 100°C.*) Researchers are trying to recapitulate the conditions that could have led to the first RNA molecules. This was so difficult that some scientists used it as an argument against the role of RNA at the beginning of life. Did RNA originate in permafrost? Indeed, RNA is stable at low temperatures. Researchers at the Max Planck Institute in Gottingen store RNA in deep freezers to find out whether it can grow, replicate,

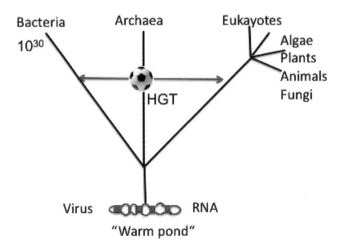

Tree of life, the football symbolizes horizontal gene transfer (HGT) by viruses, we are all related!

and mutate. Is a start in ice more likely? High temperatures would degrade the RNA.

If we need an explanation for something unknown, we tend to think of extraterrestrial origins or meteorites. Joseph Kirschvink, at the California Institute of Technology, suggested meteorites as having donated components of life, at the celebration of Freeman Dyson's 90th birthday. Was he being serious? Probably yes, however, it depends on the kind and size of the "components" — iron yes, but RNA? If this discussion about the origin of life is not quite your cup of tea, then jump to the salad section — you will only miss some details. Otherwise, for those still interested: Recent experimental results have been spectacular! John D. Sutherland and his colleagues in the U.K. published the very surprising outcome of an experiment in 2015. Within a single reaction vessel, containing solely HCN, P, H_2S and H_2O (hydrogen cyanide, phosphor, hydrogen sulfide and water) and with ultraviolet light as an energy source, they produced all the three main building blocks of life: nucleotides for nucleic acids, amino acids for proteins and lipids to form drops, all together in one single test-tube. Thus, RNA could well have been produced on earth — a notion that had frequently been questioned by biochemists. Amino acids are also considered by some specialists possibly to have been the primary building blocks of life, yet they do not replicate. Ultraviolet light could be replaced by other energy sources. Light was not necessarily the first energy supplier, since it cannot penetrate more than about 200 meters below sea level. Life needs energy. There are no "perpetual motion machines". Yet energy does not necessarily have to come from the sun or from a cell. Chemical energy might suffice. Cells came so much later. This leads us to the question: What came first, the virus or the cell? Generally, it is assumed, viruses need cells and therefore cells were first on the scene. Yet viruses need energy — not necessarily a cell's energy. Deep in the oceans there exists chemical energy, which is present in chemical compounds; and there is thermal energy. Some chemical reactions could be performed by early forms of life, which can be so specialized that they can "lick" at electrodes to pick up electrons for chemical reactions and reduce carbon dioxide CO_2, as was shown in 2015 in an exhibition at the American Museum of Natural History about "Life at the Limits". Water may indeed have its origin in the Big Bang, and thus be 14 billion years old, having arrived on Earth as "dirty snowballs" from extraterrestrial space. Other elements, too, came from outer space, where dust gathered under the pressure of gravity and fused into heavier elements. Life is indeed made of "stardust".

Today viruses have "minimalized" their lifestyle and depend on cells. But at the beginning this may have been different. (How this might have happened is discussed in Chap. 12.) A start has to be simple — Charles Darwin thought so too, but not all of my colleagues agree with this.

If we assume that RNA came first, which is widely accepted to be the case, then the first RNA molecules must have been short. The longer an RNA molecule becomes, the more unstable it gets. Manfred Eigen calculated that the frequency of errors during the doubling of RNA is balanced by the length of the RNA: error rate × length equals approximately 1 (*e.g. 100 nucleotides and an error rate of 1% gives an average error rate per RNA molecule of 1; or the copying of the 10,000 nucleotides of HIV would allow about 100 mistakes.*) Too many mistakes in copying lead to an "error catastrophe" and terminate development, while too few mistakes do not create enough new information. Indeed, Eigen tried to kill HIV by artificially increasing the number of mutations. The sequence space for RNA is immense. If we assume a length of 50 nucleotides based on the four nucleotides, then 4^{50} or 10^{30} different sequences are possible. Poliovirus, with its 7500 nucleotides, can in theory choose between 4^{7500} sequences.

That is more than there are atoms on our planet or stars in the sky — of which there are "only" 10^{25}. Thus more sequences are possible than could ever be used in Nature.

There is another special feature of RNA: RNA is never a single defined species, but always a mixture of diverse sequences, a swarm. Eigen introduced the word "quasispecies" for it. This is a result of the high mutation frequency of RNA molecules undergoing replication, which does not lead to one single type of molecule but a population of many closely related, yet different, ones. The quasispecies is what makes the innovative potential of RNA viruses so strong, and superior to other species. This mixture also makes the RNA viruses — such as HIV and influenza — so problematic, if we treat one variant another takes over. DNA replication is less error-prone because the second strand in the double helix limits changes in the first strand, and cells have developed proof-reading mechanisms, a sort of "correction paste" — all of this is missing for single-stranded RNA. Which sub-population should best be treated by antiviral therapy? Fortunately, some viruses are fitter than others and may predominate in the population. They are treated first. However, once they have been suppressed successfully, other less fit variants catch up. These often grow slightly more slowly. If they are suppressed by therapeutics, the earlier frontrunner comes back up. This can happen very rapidly — within 8 weeks in the case of HIV, a discovery that surprised many scientists. The "winner" can then be attacked again with the first drug. How can cells be put at the origin of life, if they need a minimum of 473 genes for replication, protein synthesis, and growth in the test-tube, as shown by Craig Venter? This artificial mini-bacterium, which only grows rather slowly, corresponds to 531,560 nucleotides. A natural living cell, such as the bacterium *E. coli*, has 4300 genes, which corresponds to 4.6 million nucleotides — too big for a start. How small can organisms be? Viruses are at the borderline of life, even though the giant viruses, with up to 2500 genes, can be five times larger than some bacterial cells. Poxviruses, which are almost giant viruses, have 500 genes, herpes viruses 100 genes, the retro-, influenza, or polioviruses about 10 genes each. Smaller still are plant viruses, such as tobacco mosaic viruses TMV with only four genes. Many of them have dual functions in one molecule — the typical minimalistic viral trick: one molecule with two short different ends is sufficient for conferring several properties upon one molecule. Are there entities with no genes at all? Yes, indeed, there are.

They are my favorite candidates for the origin of life. They are small and simple. It is incredible what they can do — almost everything. Please continue reading (see viroids, p. 199f.)!

First chicken or egg? — Neither nor!

There is an RNA molecule with very special properties, catalytic RNA, also designated as "ribozyme" — a word combined from "ribo-" and "enzyme", since the RNA behaves like an enzyme, even though it is not a protein (which enzymes normally consist of). This ribozyme can cleave other RNA repeatedly without being used up, typical of catalytic activities of enzymes in general.

Only recently, it was shown in California that one can put such RNA molecules into a test-tube, and there they can do surprising things. They can cleave, join, replicate and even mutate to evolve, and defend themselves against related RNA. RNA replication in a test-tube was demonstrated by Jerry Joyce in La Jolla, who had searched for many years to find the appropriate conditions for such an experiment. He repeatedly selected from an RNA pool the faster-replicating RNA for another round of replication, which resulted in ever faster replication rates. This corresponds to evolution in the test-tube, where RNA can multiply and improve itself — i.e. evolve. This may not yet be real life, but it is getting close. A really dead grain of sand cannot do that. Thus, a ribozyme is at least very much more alive than a sand corn.

Catalytic RNA was discovered by Tom Cech and Sidney Altman. At the time, catalytic activity was thought to be the privilege of proteins, and nobody gave a thought to RNA as a possible catalyst. Today, ribozymes may be regarded as the most important building blocks at the beginning of, and even throughout, evolution — up to the present. The catalytic RNA can even perform self-cleavage and then repair the cut, which is similar to the mechanism of splicing, discussed earlier. Ribozymes, small RNA molecules, can do it all: cleave, join, replicate, evolve — while mammalian cells need more than a hundred proteins for only one of the same procedures, for splicing. Of course there is no real comparison, as we are much more complicated, also in respect of splicing. Yet ribozymes are so simple and versatile, so don't they appear very ancient? Some more support for this will follow.

Now there is a new answer to the question of which came first, DNA or proteins — a molecular-biological version of the well-known question about the chicken or the egg. The answer is simple: neither of them. It was RNA! This is even mentioned in the Nobel Prize committee's press release for Cech and Altman in 1989, when they were honored for their spectacular discovery of catalytic RNA, the ribozymes. Protein enzymes, however, did not fall into total disgrace — because most of them are orders of magnitude more efficient than catalytic RNA. Protein enzymes are the basis of our entire metabolism, in all our cells. But they came later. And protein synthesis needs ribozymes! Wait and see. Some ribozymes developed during evolution to protein enzymes. The trend from RNA to proteins is a general progress in evolution and can frequently be observed.

Viroids — the first viruses?

Heinz Ludwig Sänger was Professor at the Institute of Virology at the University of Giessen. Sänger's father owned a plant nursery and Sänger had noticed that some leaves of plants looked dried-out, in spite of the fact that they had no lack of water. He doggedly pursued the cause of this and finally found the viroids. I would like to include the viroids in the group of viruses, and will explain that later. Sänger became a plant viologist and almost lost his job, because for years he failed to produce viroids from diseased plants. That was in the 1960s. The German Science Foundation ran out of patience and was close to terminating his grants. The head of his Institute had to step in and provide an eloquent defense of this type of research, and it is lucky that he did. Finally, Sänger raised the temperature of his greenhouse. That sounds trivial, and too simple — but it was absolutely essential, and it finally allowed the production and isolation of viroids. His laboratory smelled terribly of phenol, which was required to isolate the viral RNA from the plant leaves — as if viroid research consisted solely of phenol extractions — which it indeed did for quite some time. Ultimately, enough viroid RNA had been collected, so that Sänger and his colleagues could determine the sequence of the RNA. That was long before high-throughput sequencing became possible, and sequencing 365 nucleotides was a technical record. It was done with great stealth, so that the helpful but rather dominant Director of the Institute could not broadcast the news prematurely. The RNA was not only sequenced, but was also identified as a ring. This was concluded from the unusual migration properties of the RNA in the gel-electrophoresis machine, which separates molecules by size using an electric current. Also Sänger's competitor (or colleague) Theodor O. Diener in the US, noticed a circular RNA in the electron-microscope pictures. However, rings of RNA were so unusual that the idea was dismissed as an artifact. They even cut out the ring structures and did not publish them. They must have been annoyed with themselves later, having missed a spectacular novelty.

All that was new, and ahead of its time. It could have meant a Nobel Prize for Sänger and Diener or even other groups. They discovered the viroids independently and simultaneously. Sänger became Max Planck Director in Munich — but later turned to mysticism, away from science.

Detlev Riesner, who later founded the Qiagen company, continued the analyses of the viroids by purifying and trying to isolate the smallest possible amounts of RNA. Riesner was later on searching for RNA in prion preparations together with the prion specialist Stanley Prusiner in the US. Till today, no RNA has been found in prions. Up to the possible level of sensitivity. As nucleic acid expert Riesner initiated Germany's most successful biotech company, Qiagen. Wherever in the world I entered a laboratory — Korea, China, Egypt — everywhere, in more than 20 countries, the blue Qiagen boxes could not be overlooked. Also, my co-workers wanted to buy Qiagen products for everything, sometimes even for the most trivial and even meaningless experiments! This brought me to the limits of my patience and I refused to sign the purchase orders. Even little bottles with distilled and sterile water could be found in the boxes, as for an "instant cake": add water and stir — that was all. But the one-way disposable boxes were the reason for the incredible success of the company, because no contamination with even the most minute amounts of nucleic acid could ever occur — which is otherwise a frequent artifact. Riesner is still an enthusiastic entrepreneur, far beyond retirement age. He is just starting a new company with ribozymes made from DNA, to be put into service against neurodegenerative diseases such as Alzheimer's disease, with the power of a Golden Business Angel. He heads the Board of Directors of about 15 biotechnology companies, and is closer to biotechnology than to his hobby of vintage cars. I was a co-founder in one of his companies. Did he miss the Nobel Prize? — "More than once" was his answer.

Viroids — illiterate all-rounders

Cadang Cadang — this is what natives from the Philippines call the disease of their coconut palm trees. It is caused by viroids. What are viroids? Aren't they viruses? They behave like viruses, but do not look like viruses, they are naked and miss a coat — so the compromise name given to them was "viroid". I think they should be included in the family of viruses, but not all virologists agree. If we see a person in a coat and call this a human, is then the naked person a "hominoid" or still a human? So, I would like to include the viroids into the family of viruses. They simply are naked RNA. I wrote an article about it "Viruses are our oldest ancestors" which I published, albeit followed by a modest question mark! If viroids are the oldest biomolecules, then viruses are indeed our oldest ancestors.

We were sitting at the breakfast table in Busan, South Korea, during a World Summit of Virology, at which there were at most 50 participants. This was certainly not the summit that had been announced, and not quite up to the expectations of the participants. The Koreans may not have foreseen how few participants would show up. It resulted, however in a surprising consequence: the very few slightly frustrated participants talked to each other more than 25,000 people do at an HIV World Summit — something to think about.

So we listened to the 80-year-old John W. Randles, who had been asked more than 50 years ago by politicians from the Philippines for help against whatever was killing the palm trees. He was shown the devastated forests from an airplane, and he found the solution: natives climb up to the top of the trees, where they cut off the fruit always using the same knife — thereby carrying a putative infectious agent from one tree to the next. Indeed, in the sap of the trees viroids were present, later called Coconut Cadang Cadang viroids. They can do enormous economic damage, since they infect palm trees, pineapples and chili — and also potatoes in Europe. Contaminated knives for the palm trees resemble the contaminated syringes and needles that transmit HIV among drug abusers or in ill-equipped hospitals that cannot afford disposable needles. HIV infections were spread in hospitals at the beginning of the disease outbreak in Kinshasa, Africa. Similarly, hepatitis C virus, HCV was unfortunately spread by injections against schistosomiasis (same as bilharzia) with contaminated needles throughout Egypt. This abolished the tropical disease caused by the paired worms in the urinary tract, but the contaminated needles infected millions of people with HCV, leading to hepatocellular carcinomas. Similar to Cadang Cadang viroids in palm trees.

Viroids are catalytically active ribozymes, plant pathogens, which infect plants, replicate, cause diseases and behave just like viruses — but that is not what they are called. Nobody has had the courage to designate them as viruses, because they did not obey the rules for viruses: naked RNA without a coat does not fit into the definition of viruses that was current in the 1960s, and neither does it do so today. So scientists tried to find a compromise, and called them "viroids", virus-like entities. Viroids are circular, often catalytically active RNA, consisting solely of RNA. Therefore, even today they are still not regarded as "real viruses". Viroids do not use a coat, not only that, they cannot even read the genetic code, which codes for amino acids, even though the four nucleotides are available, yet not organized as triplets, a prerequisite for recognition of encoded amino acids. Thus viroids must have existed before the code developed. They are, so to speak, genetic illiterates. And — most surprisingly — they have stayed that way right up to today. And they are frequent on our planet. Viroids are present mainly in plants and seem to me as witnesses and indicators of the beginning of life. They belong to the most ancient representatives and relics of the early RNA world — even though they are not much appreciated as such today. They must be older than the genetic code, because their RNA is non-coding; they preceded protein synthesis. But, here comes a twist: they made protein synthesis possible, as there are no proteins without ribozymes! They themselves lack all genetic information for amino acids or proteins. Could their RNA sequences then be random? Never! Certainly the RNA of viroids contains information — if not as coding sequences, then for sure in its structure, which also somehow depends on the sequence. Structural and electrical charge information leads to the formation of hairpin-loop structures, required for replication, for cleaving and joining and interaction with the environment.

Some of the viroids are up to 70% self-complementary, which means that they are almost double-stranded with single-stranded loops in between. They allow interactions with other molecules in their surroundings. Such structures are more stable than single-stranded RNA — which is well known by any biochemist in the laboratory, where many sleepless nights are caused by the high rate at which RNA is degraded. All instruments and solutions (including water), have to be treated with special reagents (*such as DEPC, diethylpyrocarbonate*), to make them not only germ-free but also nuclease-free — free from RNA-destroying enzymes. Better still would be a completely separate laboratory without any air exchange. Even Charles Darwin warned of "nucleases" (though he did not use that term), predicting that under the conditions we have today on our earth, the origin of life cannot be repeated. It would be too hostile here. Nucleases are "hostile". Furthermore, we will never know what kind of conditions initially existed on our planet. But Darwin also said that one cannot exclude the possibility that all living entities on earth may have a single precursor as their origin. Perhaps the viroids? They are simple enough. They are the only ultra-simple, self-replicating biomolecules even in the present world. At the beginning of life, protection against environmental threats was certainly necessary; but how was it afforded — by hairpin-loop structures, compartments, niches, or by lipids? We do not know.

It is noteworthy that viroids are ribozymes, collapsed closed circles, partially double-stranded, and catalytically active. Ribozymes may be 50 nucleotides long and viroids today are about seven times as big. Both of them are closed circles. Later during evolution, the viroids sometimes lost their catalytic activity. The much-studied *Potato Spindle Tuber Viroid* (Pospiviroid or PSTVd, the small d indicates the viroid) is enzymatically inactive. But the loss of a function indicates that the activity was no longer needed. The virus became "lazy". The cells took over some tasks. In order to replicate they then needed help from the host plant cell. We know about two forces in biology, increase as well as decrease of complexity — whereby decrease is often possible if the lost ability is supplied from somewhere else. So the loss of viroid activity could be due to a change in lifestyle, depending on the host cell.

A reader may wonder how potatoes die of viroids. The pathogenicity of the viroids is not well understood. Does it operate through RNA–RNA or RNA–protein interaction inside cells, or by viroid activity? Viroids may activate the silencing siRNA machinery and induce gene silencing via the RNA in the silencer complex, RISC. There is a constant sequence, a "core RNA" region in viroids, is it a pathogenicity region — what is that? Certainly nucleic acids can interact also with proteins inside the cells, which could cause pathogenicity.

By the way: I include viroids in the virus family. If so, then I would have to conclude: "Viruses are our oldest ancestors" — because viroids are the most ancient viruses, about which I published, albeit followed by a modest question mark!

RNA belongs to the origin of life, to the transition from chemistry to biology. RNA could fulfil two functions reminiscent of what Freeman Dyson refers to in the title of his book *Origins of Life* — which means at least two origins, as he uses the plural. These origins are "software" and "hardware". Dyson was inspired by John von Neumann, one of the founders of computing, who was one of his colleagues at the Institute for Advanced Study at Princeton. RNA would fulfil both his requirements since it is the only molecule on earth I know of with such a dual function. RNA is a carrier of information — software — as well as a machine, an enzyme — hardware — for its own replication. What a unique prerequisite for an origin, for a beginning.

When I asked Freeman Dyson in Princeton whether he still believed in two origins of life, he answered enigmatically: "I did not claim that two origins exist, it is just a possibility."

Originally, he had been thinking of genetic information and metabolism as the two origins.

Circles of RNA

Ancient RNA is having a comeback. It closely resembles the newly discovered circular RNA, abbreviated to "circRNA". Relatives of circRNA have been described above as the (evolutionarily) very old viroids and ribozymes. Their family grew enormously all of a sudden.

circRNA was overlooked almost for decades, because researchers were looking for RNA molecules by searching for their ends. But circular RNA has no ends. This circRNA is non-coding and has a regulatory role. We are now getting close to a dozen known regulatory ncRNA. The world of RNA is more at the center of our world than we anticipated. What precisely is this circRNA good for? It is a backup for other regulatory RNA. It is a "chief regulator" of other regulatory RNA, a regulator of the regulators. Why is such a boss necessary? Presumably regulation is not allowed to be error-prone and needs a two-step security mechanism. circRNA are not even rare — our cells are full of them. There are about 25,000 RNA circles per cell in our body. They have also been detected in archaea, in fungi and unsurprisingly in mice. They are most abundant in the brain, at neuronal junctions, at synapses, where they may influence development as they do in flies, or accumulate during aging. This immediately raises questions about Alzheimer's disease — could it be due to the failure of a chief regulator?

At 1500 nucleotides these circles are slightly bigger than the viroids (which are also closed circles) — about three times their size. Our present-day cells are filled with "relatives of viroids". What a surprise! Viroids are found mainly in plants, while circRNAs are found in many species. Are they also in plants? How are the circles generated? "Cyclase ribozymes" made of RNA can produce circles. Many questions still remain unanswered.

Circles have significant advantages, they are very stable with their partially double-stranded, hairpin-loop-like regions and because they cannot be degraded from the ends. Each circle can titrate out hundreds of other regulatory RNA, such as silencer siRNAs, reminding me of my grandmother's sticky stripes of flypaper hanging from the ceiling of the kitchen. The authors describe the circRNA as a "sponge", not sticky stripes, which absorbs the regulatory microRNAs.

In mouse brains, a microRNA of this kind, miR-7, was removed by a circular circRNA with potential consequences for cancer — which is presently being investigated. The novel circRNA are completely non-coding "illiterates", as I call them. Yet they are perfectly capable of fulfilling functions with their structure — which is ultimately also based on sequences, but not on the genetic code, so we are not able to "read" them. CircRNA is produced in a special way: not by splicing out introns but by exons, in a mechanism called "back-splicing" or "reversed splicing". For years, this was thought to be a mistake in splicing. Isn't it indeed very surprising that the oldest type of RNA in our present-day cells is so dominant and so relevant? No, perhaps not, because a circular RNA is so well designed, so unbeatable, of such high resistance and such perfect technical quality inside cells that it did not disappear. The circular circRNA survived the dangerous beginning in the primordial soup, and after that the danger of degradation was far less acute! Cells are safe harbors.

From today's circRNA one may perhaps even learn how viroids function, about their interaction with other RNA or other types of factors, or about their "sponge-like" behaviour. Recently, ribozymes/viroids or circRNA have unexpectedly obtained new family members, the piRNA a very surprising non-coding regulatory RNA involved in transgenerational inheritance (Chap. 12).

What a pity that the viroid researcher H.L. Sänger in Giessen did not live to learn about the general consequences of the viroids, which become a principle of biology 50 years after their spectacular discovery. A present-day circRNA researcher from Berlin did not seem to remember the viroid RNA, or at least did not mention it in his publication or plenary lecture. When I asked him why not, the answer was — oh well, that was *sooo* long ago — very amusing, even in a public presentation. The papers by Sänger and Diener were not quoted at all. Is that a bad habit, or forgetfulness, or even misconduct? The knowledge of history of science is short-lived, sadly so. Some people find it boring!

Ribosomes are ribozymes — viruses make proteins!

How are proteins made? Protein synthesis is a hallmark in evolution and separates two worlds: the living world from the dead world. But the twist in this story is that the viruses often considered as members of the "dead" world, or more precisely viroids, are the most, or perhaps even the only real, essential elements required to produce life! They are the key to protein synthesis: viroids are required for life.

T. Cech and S. Altman (Nobel Prize 1989) placed ribozymes at the center of the biological world. And then there was another Nobel Prize for ribozymes: for Tom Steitz (2009). He observed that ribosomes, the machines of protein synthesis, do not function as everybody thought. Especially in Berlin, at the Max Planck Institute for Molecular Genetics, it was assumed that the major players in the game of protein synthesis were themselves proteins. Ribosomes consist of about one hundred proteins, which were the focus of intensive, large-scale research. One co-worker was employed in Berlin for each protein, and there were hundreds of anti-body-producing sheep. They kept the grass short at Berlin's Tempelhof Airport and were good for antibody production. I smuggled in some of my sheep, that produced antibodies against oncoproteins, and bribed the shepherd to look after them. Protein sequencing, too, was a major effort to characterize the hundred ribosomal proteins and the structure of ribosomes. It was not all wrong — but the key players were not the proteins, but RNA, or more precisely: catalytic RNA, well hidden in the inner part of the huge multi-protein complex inside the ribosome. Tom Cech coined the slogan: "Ribosomes are ribozymes". The RNA is the workhorse, not the many proteins, which serve merely as scaffolding to keep the RNA's conformation correct. All the proteins were crystallized by a trick: they were isolated from archaea — the first time in my life I heard about them, now 30 years ago. The thermoresistance of archaea makes them more stable and easier to crystallize for structure determination. This finally resulted in a Nobel Prize in 2009 for Ada Yonath and others. Yonath, from Israel, was a frequent guest in Berlin. She discovered the tunnel through the center of ribosomes, which can be occupied by the newly formed, growing amino-acid chains — a picture that once decorated the front cover of the journal *Nature*. As children we had a red wooden knitting-spool shaped like a mushroom with four nails, and used a crochet-hook to knit a cord, which then grew out of the bottom of the spool — like a ribosome releasing a growing polypeptide chain. The knitting hand corresponds to a ribozyme! The rest to the ribosome. Ada Yonath pointed out in her Stockholm Nobel lecture that the hole was made of RNA, not of proteins. It evolved from a more primitive proto-ribosome during evolution. Here too, the beginning was RNA, a holy grail at the center of the machinery of protein synthesis. The RNA is catalytically active — and is reminiscent of the viroids. A virus-like element — a viroid — at the center of ribosomes — that is quite unexpected, and nobody seems to notice it. A hundred ribosomal proteins only serve as scaffolding, or servants, to help the ribozyme do the job properly. The ribosomes are different in every bacterial type and one of their RNA components (*called the 16S rRNA, or rDNA equivalent*) is the basis for their identification and classification. The ribosomes are the targets of the life-saving antibiotics, some of which occupy the hole and thereby block the protein synthesis. So the bacteria then die and we survive. Resistance against antibiotics is a major concern today, and mutations to block the hole again by antibiotics are one approach to overcome it. If it works — fine! Ada Yonath is trying it, and she was right before!

The pacemaker of this research in Berlin was Heinz-Günther Wittmann, who has now passed away. His life achievement was the basis for this success. He delivered all the proteins to Hamburg, where Ada Yonath — at DESY (German Electron Synchroton) as well as in Israel — performed the crystallographic analyses to identify the structure of the ribosome. I remember when archaea were introduced for that purpose at the Max Planck Institute in Berlin; they can tolerate 100 degrees for growth, and it was hoped that their ribosomal proteins would be more stable and crystallize better — which turned out to be true. Even a flight into orbit around the Earth was tested as a possibility for better crystallization of ribosomes under low-gravity conditions, but that failed. A sample of my reverse transcriptase was admitted as a passenger on that flight, but it too failed to crystallize. (*Indeed, it was later crystallized from the thermophilic archaeon Pyrococcus furiosus.*) I should have thought of isolating the enzyme reverse transcriptase from archaea for crystallization myself, but I didn't. I simply missed it! Crystal structures are the basis for drug design, which is why they are so important.

Surprisingly, a relationship between ribozymes and proteins can be shown even today with some proteins, which carry a little RNA tail, as if it had been forgotten there! Is it a leftover from evolutionarily distant times? The first protein synthesis started the other way round, with RNA and some amino acids attached to it (I think). This later resulted in proteins "only". But perhaps somewhere on the way there was this strange protein with a little RNA fragment attached to it. One of these protein-RNA chimaeras is vitamin B_{12}, and another is acetyl-CoA. They are both very important molecules for cellular metabolism, and even today they tell us how they may have been generated. Proteins with conserved RNA tails — that reminds us a bit of our appendix. Such transition states are fascinating and informative, because they tell us about our past and about evolution, from RNA to proteins. I would like to add these two chimaeras to my collection of curiosities and store them in a *Wunderkammer*.

How did DNA get started? Protein synthesis does not necessarily require DNA, but RNA. Protein synthesis is *the* hallmark of life — but early life did not need DNA. DNA did not precede RNA or proteins in our world. How is it possible to remove an oxygen moiety from RNA to generate DNA (*deoxy*ribonucleic acid)? Nobody knows! An enzyme can do that today, the ribonucleotide reductase. When such a protein enzyme arose is not known. With the right building blocks, my favorite molecule, the reverse transcriptase could have produced DNA from RNA, but when was that? (*Today an RT would need coding DNA for mRNA and protein synthesis.*)

There are deoxyribozymes, small catalytic DNAs, which can cleave and join just like ribozymes without proteins, but they do not occur in Nature today. The often-quoted Jack Szostak looks for them. He demonstrates that RNA can grow without proteins such as polymerases, how did DNA arise in a protein-free early RNA world? He may find out. The researcher Patrick Forterre from Paris has conceived of three RNA-containing cells — for bacteria, archaea and eukaryotes — which he terms "ribovirocells", as all three are virus-producing cells. How did then DNA arise? Thus, the question about the origin of DNA remains open — surprising, for such an important molecule. One thing is for sure: the reverse transcriptase was a major driving force for the transition from the RNA to the DNA world. In 2015, the reverse transcriptase was discovered as the most abundant protein in genomes, cells, in various organisms including ocean samples. The reverse transcriptase is such a frequent enzyme as part of the universal retrotransposons, jumping genes. There must be a lot of jumping going on. Some researchers think that it was the most important, key enzyme for life during evolution. This is now supported by data! Note: others think so, and indeed I do, but I am biased as I have been a reverse transcriptase specialist for the past 45 years! Amazing, from the first discovery of the enzyme to the most abundant protein in the world!

Clover leaves

A prominent molecular RNA structure is the clover leaf, with three loops around a stem — that is the model that one can draw on a sheet of paper, not what is detected by crystallographic structure determination. This RNA is very stable. Could it be that we therefore find this structure everywhere from the molecular to the macroscopic world, the cloverleaf in green meadows? That is a serious question! It is furthermore amusing to note that clover-leaf RNAs are used extensively by plant viruses. Clover-leaf structures are located at the ends of their often single-stranded RNA genomes in order to protect them against degradation. Telomeres at the ends of our chromosomes have RNA pseudoknots as a similar protection, not totally unrelated to the RNA at the ends of viruses. Also phages, and RNA viruses such as hepatitis C virus and polioviruses, have terminal looped structures of that kind. This is a general measure taken by viruses to protect their ends. There are even viruses that consist of a single clover-leaf RNA and a single amino acid — a virus! Less is almost unimaginable. Another virus, the narnavirus (which has RNA in its name) from baker's yeast, *Saccharomyces cerevisiae,* consists solely of half a clover leaf and nothing else. This is very minimalistic and can hardly be reduced further.

RNA viruses developed such clover-leaf structures, probably from three hairpins. These RNAs are 75 to 90 bases long and rigid because of some double-stranded regions. Now I must finally unveil the real name of this structure, "tRNA", standing for "transfer RNA", because it transfers individual amino acids to the ribosome for incorporation into a growing polypeptide chain. This is how they were discovered and defined. One tRNA and one amino acid evolved finally to 60 kinds of tRNAs and 20 amino acids in our cells. They are the important components for protein synthesis by ribosomes today. Their beginning may have originated from viruses, because they used the tRNA structures with their RNA genomes before there was protein synthesis. The tRNAs are noncoding ncRNAs anyway, which makes them appear ancient. The viruses from the yeast *S. cerevisiae* seem like early stages of protein synthesis. Retrovirus replication, too, is reminiscent of an early protein synthesis. Thus, one may think of viruses as designers or precursors of protein synthesis machineries.

One can also turn the argument around and suggest that viruses have "stolen" the tRNAs and amino acids from the cell. Indeed, that is what the reverse transcriptase does today to initiate retroviral replication. It picks a tRNA from the pool in the cells. Each virus type has a different tRNA preference. The reverse transcriptase needs a handle to hold on to get started and uses a tRNA for this purpose. All textbooks of virology consider viruses as thieves of cellular components. I think it is the other way round! My view is that viruses were the inventors of cellular building blocks, the body-builders of cells — and today they appear to be reaping the profits of their former achievements. It may be interesting to note that it is the ability to synthesize proteins that many virologists regard as the criterion that separates life from death. If no protein synthesis is possible, the system is not alive — which is true of most viruses today. But, please do not forget the giant viruses — they contain tRNAs and ribosomes, hallmarks of protein synthesis, but only incomplete sets or fragments, not all of them. Did they lose something, or did they remain unfinished? Still, that provoked their discoverers very recently to proclaim that the giant viruses are the origin of life. It is now left to the reader to reach his own conclusion or to wait for more results of more experiments.

RNA has become an important tool in biotechnology. RNA can exhibit, just by folding, properties similar to those of the much more complex proteins made of 20 amino acids — in contrast to the four nucleotides required for RNA synthesis. Synthetic RNA is therefore much cheaper than proteins, and faster to synthesize. It can assume complicated structures and can fit onto receptors, replacing proteins, even mimicking specific protein ligands. These RNAs are called aptamers. Such RNAs can be synthesized in the laboratory from only four nucleotides, while proteins require 20 amino acids, and they can even undergo evolution and selection in the test-tube to improve their structure and composition; this method is called SELEX, an acronym describing the selection and optimization procedure. It was Tom Cech who developed this approach. Manfred Eigen, the Nobel Prize Laureate in Chemistry from the Max Planck Institute in Göttingen, founded two companies, Evotec and Direvo. The names already indicate what the goals were, evolutionary chemistry, "directed evolution". He developed artificial evolution applied to chemical compounds as new optimization procedure in the test-tube. This resulted in protein enzymes with much higher catalytic activities, never found in Nature, more efficient processes, better exploitation of biomass, and novel biologically active compounds. The special feature of Eigen's work was that he applied the evolutionary principle also to the non-biological world of molecules, in order to improve chemical products, dead matter. Originally a physicist Eigen was far ahead of his contemporary colleagues — by establishing a novel chemistry.

A protein as chaperone

Darwin already pointed out the difficulty of repeating in our present world reactions from the origin of life. Viruses have developed ways to protect their precious genes against degradation today. One such way is described next.

A co-worker of mine wanted to analyze a viral protein from HIV, which covers the viral RNA and protects it against nucleases, typical of all RNA viruses. (*In HIV it is called nucleocapsid protein, NCp7, and in Influenza virus it is called nucleoprotein NP. Such RNA-binding proteins are positively charged and can neutralize the negative charge of the RNA, where they line up co-operatively, next to each other.*) Such proteins are even found in retrotransposons, for movement of nucleic acids.

They have been given the name "chaperones". Our grandmothers had chaperones — older ladies, commissioned by concerned mothers to accompany their unmarried young daughters. All RNA viruses use chaperones to protect their RNA. They are multi-tasking proteins. One property they have is that of keeping RNAs at a distance — useful for the young girls? Then chaperones can melt RNA folds and increase RNA–RNA interactions (which our grandmothers probably did not know about) and iron out wrinkles, secondary structures, in RNA. They can unwind tRNA cloverleaf structures to target one of its arms as a primer for RNA replication. Also, viral replication is much more efficient in the presence of the chaperones. The reverse transcriptase slides across the RNA more efficiently if it is disentangled by the chaperones. One might think intuitively that RNA-binding proteins are roadblocks that prevent the enzyme from sliding across the RNA; however, the opposite is true, as they in fact increase transcription rates. The proteins are very sticky, because of their positive charge, and in the laboratory they sometimes refuse to leave the pipette, but hang onto the inner glass wall forever. I did not want to believe this when my co-worker once told me about this problem with his project. But he was right. We once added chaperones to a ribozyme and RNA in a test-tube, and the cleavage reaction was accelerated crazily, becoming a thousand times faster. Then the PhD student went on a world trip, and the project got stuck. When he came back, I was furious — because "his" results had meanwhile been published in the journal *Science* by Peter Dervan. He had observed the same phenomenon — but published faster. Time matters! We made an emergency plan and got a publication in a reasonable journal, too. I mention this here, because I never thought of evolution while we were performing these experiments — not at all. Yet, for the beginning of life ribozymes might have received exactly this type of support from positively charged amino acids, not necessarily whole proteins, and the reaction was speeded up — catalyzed — a thousandfold. This may have accelerated evolution! Amino acids are assumed to be easily synthesized in the prebiotic world and might have been around early. Alternatively, some meteorites could have supplied them — some people say but not me. These proteins belong to the most ancient and conserved ones on our planet, and they often contain zinc ions, which bind to special structures, called "zinc fingers", for binding to nucleic acids and transposable elements. Chaperones must have helped at the beginning of protein synthesis and replication of retroviruses. Who learnt from whom? The viruses from the cell or vice versa? I think the viruses were the inventors of protein synthesis, which the cells then learnt from them. That is what I believe — with a little evidence as well.

From potato to the liver

After so much history it may be relaxing to talk about food. As mentioned earlier, there are viroids in many plants, and these can cause plant diseases. They are present and harmful in food. But they do not occur in humans — so humans thought. One single viroid, however, escaped. Nobody knows where it came from. There is the potato spindle tuber viroid (PSTVd), perhaps it was present in a potato dish or viroids in cucumber or tomatoes.

It ended up in the liver instead of moving through the gatrointestinal tract and getting secreted in the feces. Viroids are thought to be totally harmless for us — with this one exception. The viroid RNA must have found a surprising access to the liver instead of to the guts for disposal. It became the hepatitis delta virus, HDV. John Taylor from the Fox Chase Cancer Center in Philadelphia, USA, investigated the origin of this virus and proposed that route. When did that happen — hundreds of years ago? This is uncertain. HDV is a catalytically active hammerhead ribozyme, where the RNA can be aligned — at least on paper — as a hammerhead. The RNA is packaged in a protein coat. This coat originates from another liver virus, the hepatitis B virus, HBV. Thus the coat is borrowed and not self-coded, as for all other viroids, which are all non-coding ncRNAs. The two "viruses" replicate in the liver together after they met there. Then HBV supplies some of its own protein coat molecules to HDV — like St. Martin in the Christian legend, who cut his coat in half to share it with a poor man, as can be seen illustrated in many Gothic churches such as Basle Cathedral. This has an important consequence, because with the coat virus can form particles and leave the liver cell to infect other cells or even other people, not just move inside plants any more! If people are infected by the two liver viruses, HDV and HBV at the same time, a severe liver disease and cancer are the consequence, with a 20% mortality rate. Often this combination is associated with HIV, and then the patient is at even higher risk. In Germany there are 18 such cases per year, which fortunately is a low rate and may be the result of broad acceptance of anti-hepatitis B virus vaccination by the public. Studies on hepatitis D-like ribozymes indicated relatives in bacteria, insect viruses, and yeast besides plants. A second HDV-like ribozyme in humans, however, has never been discovered. What do these ribozymes do? At least such a viroid in humans is a singularity, and this is good news, because it means that we have no reason to refuse to eat "healthy" vegetables.

There are a few very rare other types of viroids around, for example a retroviroid found not in the liver, but in carnation flowers. It is called *carnation small viroid*, CarSV. The reverse transcriptase comes from the carnation pararetrovirus, a plant virus, which copies the ribozyme into a DNA. It does harm to the flowers — remember, diseases are normally the way in which viruses are discovered. Perhaps there was some commercial interest by the carnation industry, which wanted to get rid of diseased carnations — and detected a strange virus. The flower viroid has lost its catalytic activity and can no longer cleave RNA. This is a not even a rare degeneration of viroids. Depending on the environment, they become "lazy" (lose their catalytic activity) but need help. In the flower, as well as in the liver, there occurs some kind of synergy between two viruses, a viroid and a retrovirus (sharing the coat in one case and the reverse transcriptase in the other). Why? In the virus world, as in biology in general, everything has developed by trial and error. Synergy or mutualism or social behavior or symbioses are very successful strategies for all living entities, including viruses.

I have asked around at meetings whether anybody had heard of a retrophage. I think that they must have existed — even in large amounts before all the DNA phages developed presumably from RNA and retroviral intermediates during evolution. If there are retroviroids, why should there not also be retrophages? I indeed found one (!) single retrophage — known to young mothers. It accompanies the bacteria that cause a coughing disease, *Bordetella pertussis*. Small children get infected and frighten their mothers. The retrophage has not even got a name! The reverse transcriptase which is known to be error-prone, allows the phage to find new hosts. (*It mutates the gene responsible for selecting host bacteria, called tropism or host range, the gene is the MTD gene, major-tropism-determinant.*)

This example is worth mentioning because it shows how viruses find new hosts: by mistakes or infidelity or sloppiness on the part of the reverse transcriptase. This is well known for HIV — escaping from one major cell type early during infection to a new one later (called CCR5 and CXCR4) and mutating faster than the immune system can respond.

Viruses are the drivers of evolution — as are the phages of bacteria. Viruses and phages are related. Would it be a risky conclusion if I were to suggest that the DNA phages once started out as RNA phages? Have a look at Chap. 12, where this is discussed on the basis of replication rates. Then RNA phages or retrophages became outdated a long time ago! All this story-telling about retrophages is unfortunately of little help to the mothers of coughing children.

Tobacco mosaic viruses

When 30 years ago the Society of Virology was founded, there was not a single plant virologist in Germany who could have become a member. Viroids were not considered to be viruses! Therefore, viroid researchers were excluded. This is all the more surprising as the very first virus detected was from plants, the tobacco mosaic virus (TMV). Martinus Beijerinck showed in 1898 that the flow-through from diseased tobacco leaves contained the agent that infected healthy leaves. He coined the word "virus", to distinguish it from the larger bacteria. He always referred to his teacher Dmitri Ivanovsky in Russia, who had started these studies a few years earlier, but had always believed in bacteria, not a new agent. TMV inside the cells looks like a pile of parallel naked trees. The viruses form paracrystalline aggregates and look a bit like crystals. The lengthy crystalline structures were first observed by Helmut Ruska in 1930 using an electron microscope. Not he, but his brother Ernst received the Nobel Prize many years later for improving electron microscopes by using magnetic instead of glass lenses. His Institute in Berlin-Dahlem was built on thick layers of concrete to keep it free from vibrations from the nearby underground railway. Tobacco mosaic virus crystallizes almost by itself and presents an example of the efficient self organization of viruses. For the structural analysis, Wendell M. Stanley of New York was honored with the Nobel Prize in 1946.

He also showed that the virus crystals are still infectious even after 40 years of dryness on a suface — an interesting example of "almost dead" crystalline matter with biological activity. The infectious RNA is located inside the rigid protein structure, where it is well protected. One of Stanley's students was Rosalind Franklin. She speculated that the RNA must be inside the virus in a channel within the helical protein rod. She was right: the single-stranded RNA is surrounded in spirals by the proteins. She later on produced the "Photograph 51", the X-ray structure picture of the DNA helix, without ever knowing that her picture provided the basis for the DNA double-helix model.

Stanley's studies were a wake-up call for Adolf Butenandt, then President of the Max Planck Society, to establish virology as a new field of research in Germany. There were Max Planck Institutes dedicated to topics that did not seem appropriate to be continued after World War II. One of them was the former "Kaiser Wilhelm Institute of Anthropology, Human Inheritance and Eugenics" in Berlin. I still own a microscope with this engraved on it! The Max Planck Institute for Molecular Genetics was a follow-up. Then a new topic "molecular genetics" was established. Heinz Schuster, coming back from Pasadena to Germany, and Heinz-Günther Wittmann, from Tübingen, were appointed as directors. They had both carried out research on the tobacco mosaic virus and helped to identify the genetic material — not proteins, as was assumed at the time, but the nucleic acid, the RNA, which they had proved by mutagenesis studies using nitrosamine. Did these experiments possibly mutate the genetic material of the researchers also? Perhaps they did not use, or even have disposable gloves. Both of them died of cancer at rather early age. Schuster was diagnosed as having a work-related cancer.

The tobacco virus TMV was discovered as first virus, because it is so widespread and extremely stable. It can last for 50 years in a dry state and then be reactivated and become infectious again. In Berkeley, Heinz Fraenkel-Conrat could dismantle the virus and reconstitute it as infectious virus — an incredible experiment! Everything is so optimized that it can reorganize its fragments automatically — by self-organization! This is an important feature of the virus world. TMV was found in tobacco plants and named after them, but it can infect about 350 different kinds of plants, so that the viruses are collectively called "tobamoviruses". The viruses affect mainly cultured plants where they can do serious harm. It was assumed for some time that the viruses can only affect cultured and not wild plants. What an absurd idea. Viruses are everywhere.

Plant viruses have somehow "progressed" in evolution compared with viroids. They are no longer non-coding and naked, but produce their own coats from their own genes. This is what I call progress! Where the genetic code comes from we do not really know. Perhaps the viruses developed it by trying everything, and at some point it "just happened". Coats protect and increase mobility. While viroids with their non-coding ncRNA are illiterate, tobamoviruses use the triplet code for proteins. Indeed, the plant virus TMV studies helped researchers to understand triplets, the code and the corresponding amino acids. TMV codes for four archaetypical proteins, a minimal toolbox of a virus: one for replication, the RNA-dependent RNA polymerase, the "mother" of all replication enzymes. Another protein is the coat and a third one promotes movement of the virus from one cell to another one, the movement protein. A fourth protein is small and supports the polymerase. Remember, all this can be performed by the incredibly potent viroids without any information for proteins — all functions just secretly hidden in the non-coding, ncRNA.

A cloverleaf tRNA forms protective ends for the single-stranded tobacco viral TMV RNA against nucleases. For more information, the tobacco virus uses a trick known to many viruses. The motto is: make two proteins out of one by making a mistake! The polymerase performs an "illegitimate" elongation of the small protein from amino acid no. 136 to no. 186. The trick is based on "overlooking" a stop signal. A single nucleotide is ignored, resulting in a shift in the triplet code, a so-called frameshift. Two proteins arise with the identical beginnings but two different endings. Viruses are the world champions in finding such minimalistic solutions, and this one is extremely useful: HIV uses the same principle for elongation of the structural protein Gag and a frameshift leading to the Gag-Pol read-through protein. Overlooking the stop codon happens in one per cent of cases, and it not only changes the length of the protein but it also automatically reduces the amount of the longer one by a factor of one hundred. How elegant!

Another specialty of the tobamoviruses is their genetic stability. They are single-stranded RNA viruses and should be extremely variable. No, their low variability is in sharp contrast to the typical highly variable RNA viruses such as influenza or HIV. The reason could be that the plant viruses are not autonomous, but are more dependent on the plant as host. They adjust to the cell instead of escaping by variation as HIV does. Plant viruses are persistent; they survive in plants without forming and releasing particles. They are very resistant, and can tolerate temperatures up to 90°C; they can survive in saps or on dry surfaces as well as in the soil, in water and even in the clouds! They also tolerate UV light, which normally inactivates nucleic acids — but not so with the tobacco virus.

Even if the plant host dies, TMV can exist in a long-lived form for many years, more than decades. They "live" on something dead! Then, when conditions improve, they can be reactivated and become infectious again. The high stability of tobamoviruses is a sign of the robustness of evolutionarily very old viruses in a hostile world. Presumably they represent, together with the viroids, the evolutionarily oldest viruses there are. A long co-evolution between viruses and plants may have kept them away from humans, because they cannot function in mammalian cells and therefore do not cause diseases in humans. The tobacco virus can be produced in gram amounts and has become almost a kind of chemical; it is therefore well suited for research purposes. The TMV rods are currently being tested for possible application as cables in nanoelectronics, or as contacts in batteries.

Tobamoviruses were noticed in agriculture as pathogens and have been studied as such — and far too little — for their contribution to healthy ecosystems. Insects such as grasshoppers, or cigar-smoking humans, can spread the virus. Humans unintentionally help to spread the virus on their hands and instruments during seeding and gardening. Tobamoviruses cause mottles on the leaves. Plant viruses are slow and have long replication periods. The virologist Eckard Wimmer, from Stony Brook in the U.S., mentioned, when he was awarded the Robert Koch Medal in Berlin, that he preferred studying polioviruses, because plant viruses were too slow for him. A friend in Switzerland, 80 years of age, remembered the production of cigars by hand, called "Baeumli-Stumpen" (thick short cigars). They were not allowed to show any color spots indicative of a tobacco virus infection. They even used powder as "make-up" for the outside of the cigars. About TMV, a tobacco virus causing "mosaics" he knew nothing — it just didn't look nice on the cigars!

Viruses in "chili sauce" and my apple tree

Plant viruses pass through our stomach and intestines when we have eaten salad and are excreted in the feces. Plant viruses such as TMV are not as naked as the viroids, but protect their RNA in strong protein rods. These viruses are "real" viruses, even in the most conservative virology textbooks. The number of plant viruses that we eat in a salad is about 10^9 per gram — a huge number! We consume them and excrete almost the same number again. Most surprisingly, they are still infectious after passing through our intestines and upon excretion. They can infect other plants again. Does that mean that *we* can infect plants? Yes, indeed, in principle, except that the plants have developed some safety measures, such as trees with their thick bark. Viruses need insects, nematodes (worms), fungi or bacteria as vectors (carriers), which can then penetrate sores or wounds or enter through the roots. In California as well as in the Philippines plant viruses have been identified in human stool: mainly a pepper virus, the *pepper mild mottle virus,* PMMV. This persistent virus remains in the plants and never leaves them. Are all these people pepper fanatics? No! It is sufficient to eat chili sauce as a dressing, as even there, pepper virus PMMV is present in its infectious form. Viruses from the Philippines are present in salad dressing in California! Nobody expected that. The viruses are harmless, no matter where the sauce was produced, even if they came with chili from the Philippines. In any case, the virus makes it safely through our digestive tract and can go on to infect pepper plants. It specializes in pepper plants and does not adapt easily. That is good for us!

Rather exceptional viruses are the geminiviruses. The name is reminiscent of the Gemini space shuttles, two aircrafts coupled together, in this case both parts are identical. These viruses are a kind of Siamese twin, consisting of two fused icosahedra that share one side. Each of the icosahedra is the home of a circular single-stranded DNA molecule. The strands are not identical, but they are complementary. If one puts both strands together one would get a double strand. The strands seem to have separated, and each one occupies its own house. It looks as though packaging once failed, or that division caused problems. The result is a geminivirus. It certainly belongs into my collection of virus curiosities!

Its family name is "Begomo" virus, for *bean golden mosaic virus,* of which there are about 200 different kinds. They are not rare, but frequent! Again, they are only known because they cause trouble. They do damage to vegetables, aubergines, beans, tomatoes, cassava, and cotton. A virologist from the Philippines pointed out the problems caused by these viruses for farmers. In South Africa the geminiviruses are a focus of research because of their detrimental effect on maize. The geminiviruses are unique. First of all, single-stranded DNA viruses are rare in plants, and then this strange intermediate — half-way between single-stranded and double-stranded DNA — is curious. Double-stranded DNA viruses are much safer, more stable — but strangely enough they almost never occur in plants. There, RNA viruses prevail.

Geminiviruses represent a kind of transition or intermediate between naked double-stranded DNA and single-stranded DNA, like arrested intermediates in fused coats. That makes them very informative, and interesting for research.

It is the first virus whose effect on plants has ever been mentioned: in a poem by a Japanese Queen Koeken in 752, she noticed that the plant looked dry in the middle of summer in spite of sufficient water, like a premature fall effect — geminiviruses as cause!

One strange virus I did not even know myself until recently, is worth mentioning. This is a naked circular double-stranded DNA (plasmid), which can infect plants. The DNA rings are infectious agents designated as *phytoplasms* (*plant plasms*). Plasmid DNA is known to us from the world of phages in bacteria, yet in plants it is almost unknown. But everybody knows the consequences: They lead to yellow-colored leaves, and they infect fruit trees, vineyards, rice and palm trees. What do the yellow leaves on the apple tree in my garden indicate? I planted it thinking of Martin Luther, who recommended that everybody should, like him even the last day before the end of the world, plant an apple tree. Mine is now sick. Should I order a highly sensitive diagnostic test, a polymerase chain reaction, a PCR analysis of the DNA? And if I do, what then? With or without knowledge of the infectious agent, spraying against insects or cutting down the tree to start again are in any case the only alternatives. I do not need a PCR test: there is nothing one can do against a phytoplasm. Surprisingly, the virus-like naked DNA phytoplasms were totally unknown to me as a virologist. But they are viruses! I learnt about them from a plant bioinformatics specialist Michael Kube from the Max Planck Institute in Berlin and got interested. We started collaborating on a genome analysis and even published our study. In brief, we learnt that phytoplasm and geminiviruses are related, a "homeless variant" derived from "inhabitants of semi-detached houses". It may be worth mentioning that here the borderline between bacteria, parasites, genetic elements, plasmids and viruses has disappeared — and all that is going on in the apple tree in my garden.

Tulipomania: the first financial crisis caused by a virus

The striped tulips in my garden also fit into a chapter on plant viruses. They are imitations of old-fashioned tulips and are called "Rembrandt-like", since the real Rembrandt tulips became extinct. They were reproduced by directed mutations in the promoter of color genes to create the stripes, and are genetically stable. This is not what happened in the past.

This is what did happen: A colleague from the former East Germany mentioned to me about striped tulips, that his mother was enthusiastic about them, and bought the bulbs when they were on offer at the market before the Berlin Wall came down. Yet a year or two later no stripes ever showed up again on the tulips. She felt tricked, and blamed it on the dishonest Communist government — though she was wrong to do so. The same disappointment had already been brought about by striped tulips in the 17th century.

Striped tulips originally arose through infection by the tulip breaking virus (TBV), which "breaks" the colors. How? Even in todays' virology textbooks there are no real explanations. The gardeners were blamed, the greenhouses, the temperatures, the tools, the soil, even the communist government, none of this is quite wrong, but the answer is simply that environmental factors caused the stripes — and that means epigenetic effects. Back in 1576 the botanist Carolus Clusius had already recommended stable environmental factors to maintain the striped colors for the progeny.

Tulips came to Europe from Persia, through Turkey, from Suleyman the Magnificent (whose "turban" is the basis of the name "tulip"), via Vienna to Holland, where they were cultivated by C. Clusius, who established a botanical garden in Leiden. The tulip with his name *Clusius* is now being sequenced, and we want to learn with it how the tulip virus TBV can break the colors.

The tulips became a passion for the rich. They were novel, unique, a symbol of vanity but also of strength and male courage, and as a royal flower they conveyed social status. This went so far that the anatomist, the man with a hat in a famous painting by Rembrandt, adopted the name "Tulip", to mark himself out as a connoisseur of tulips: "Anatomy Lesson of Dr. Tulp" (1632). The tulips caused an unprecedented wave of hysteria and "tulipomania". This was the first financial crisis ever reported. Demand exceeded supply. Dutch people had become very rich by the East India

Company. The bulbs became so precious that they had to be protected against theft and were securely guarded in the gardens of rich people. Not only the aristocrats wanted to own them; also hazardeurs jumped onto the bandwagon, buying the bulbs in the Fall, without knowing what the tulips would look like in the spring but still hoping for big profits. Those who could not afford any tulips got them painted — which we can still admire today in Dutch paintings. Indeed, prices at flower auctions sky-rocketed, going up a thousandfold because of speculation. The unpredictable colors and patterns led potentially to incredible benefits — or disasters if single-colored tulips emerged from the bulbs. The virus was indeed the cause of the tulip bubble, because it was the virus that caused the unpredictable patterns. But nobody knew this, of course.

The most beautiful tulip, *Semper Augustus* — white with blood-red flares and stripes on the outside of the petals and blue shimmer from the outside, has disappeared. If one starts growing it from bulbs (which are always virus-free), it is single-colored — and viral infections have never since generated the desired stripes. Only a dozen specimens of this highly coveted tulip ever existed. The virus infection by TBV was normally not lethal and resulted in more than 1500 different types of tulips, a wealth of variety unheard of for any other flower. A double infection by a second virus, a lily mottle virus, causes death. All known experience with crossing of plants failed, and the outcome was a gamble. Tulips at the peak of the mania could fetch the price of a house. *Semper Augustus* cost 10,000 Dutch guilders; by comparison, Rembrandt got a mere 1600 guilders for his most famous paintings.

One day, on February 9, 1637, not a single tulip was sold at an auction, and that initiated the financial crash. A house for a tulip — that can well be called the first financial bubble, followed as it was by a collapse of the stock market. Prices plummeted and left many investors penniless, with huge debts. The government requested that the parties should settle by compromise without going to court, and that in future only partial down-payments should be made beforehand, not the whole price, to reduce risk and prevent bankruptcy. The reason for the breakdown of the market is not exactly known; the bubonic plague may have contributed to it, as people stayed away from an outbreak. The plague ravaged Europe during this period, coinciding with the Thirty Years' War, and it killed millions of people. Or the outbreak of a Swedish-Prussian War, which destroyed the hopes to export the tulips there.

Striped tulips are used as front covers of virology books, as attractive motifs on title pages. The stripe-causing virus is related to the tobacco mosaic virus discussed above, but the unpredictable appearance of the petals, which did not follow any known rules of genetics or breeding, resembled the results of jumping genes, which were later discovered in maize as a hallmark of epigenetics. Its connection with the stripes on tulips is not proven yet. However, already in 1576, Carolus Clusius described environmental effects on the changes of the phenotype! Here we are, 350 years later, studying epigenetics caused by environmental factors. Seen in that way, epigenetics was not discovered by McClintock around 1950 but by Clusius — around 1576! That is my conclusion and speculation and, to find out, I recently started a research project on epigenetics in striped tulips.

Tulips are difficult to analyze by genome sequencing because they have the almost biggest genomes known, 10 times bigger than ours: 30 billion base pairs (we have a mere 3.2 billion). However, the total genetic information is less than in our genomes (so tulips are not cleverer than we are!) and is full of repetitions. Tulips have something like three chromosomes, three similar genomes. In my forthcoming new research project, I am looking for sequence homology between the tulip virus and the tulip color genes — at least 12 genes are involved in color synthesis of the color anthocyanin. A sequence homology may result in "silencing" of the color genes at certain stages of tulip development, and dose effects may vary the color intensity. Epigenetic studies of striped tulips resemble Barbara McClintock's epigenetic studies on colored maize. It is going to be difficult, with all the genetic repetitions. Tulips can easily lose half of their genomes, without any problems! Many cultivated plants have genomes that are huge and repetitive — maize, wheat, potatoes — this is probably a consequence of breeding at least in some cases.

Tulips are present on most of the magnificent Dutch flower paintings from the 17th century. During a guided tour through the Zurich Art Museum the curator mentioned that the number of tulips present in a bouquet of flowers can be used in dating the paintings with respect to the tulip crisis: many more striped tulips were painted after the crisis than before.

The famous British mathematician Alan Turing who deciphered the ENIGMA code of the German submarines during World War Two, described the mathematics leading to stripes. It was only allowed to be published long after his death and is known as the "Turing Mechanism" valid for stripes in zebras, fish, giraffe patterns, tigers — and certainly for tulips. Two "morphogens" or chemicals, an activator and a long range inhibitor are interacting — but I do not know which of the two is the virus!

The 500 Deutschmark note with MS Merian

One of the first artists who painted the striped tulips was Maria Sibylla Merian (1647–1717), who has my highest admiration and respect. A tulip almost got her into prison; she had stolen several tulips from Duke Ruitmer, who lived next to the workshop of her older brothers and her home in Frankfurt. The Merian family was famous for the etchings of thousands of European cities, *Topographia Germaniae* (1642), very important documents of medieval city structures. She was so enthusiastic about the colors of the tulips that she stole the flowers to draw them in secret. Duke Ruitmer was not only impressed by the fourteen-year-old girl (who felt very guilty), but also by her superb paintings, so he forgave her — she could have gone to prison for the theft. Prison as a consequence of a virus-induced flower beauty!

Moreover, he also took over her education and paid for her to become a painter — a privilege of men in those days, and a project not greeted with enthusiasm by her elder brothers. Her passion for caterpillars and butterflies made her an object of suspicion for her contemporaries, and she almost ended her life as a witch on the stake — her brother prevented this by bribery, but she was still forced to leave the city.

Striped tulip caused by tulip breaking virus and 500 Deutschmark bill with plant, caterpillar and butterfly and some kind of a "virus" by Maria Sibylla Merian.

On one side of the old 500 Deutschmark notes, one of which I have kept, M.S. Merian is depicted with curly hair, and on the other side some sort of lionstooth wheat is shown along with a butterfly and a caterpillar. To combine these three elements, which are connected by metamorphosis and food, was her scientific achievement. During her lifetime her contemporaries believed in gases and miasmas. And her mother worried about how ever to get this daughter married, with her predilection for worms in the kitchen. She was also a master artist with her engravings. She learnt this technique from her elder, famous brothers. It was forbidden for females to use oil and canvas. If one looks carefully at the 500 Deutschmark note one can see wrinkled leaves — perhaps they are sick with viruses! So precise was her observation. What, however I am to make of the icosahedron-like structure at the top right corner, I do not know. It looks to me like a section through an icosahedral virus! Or is it perhaps a watermark, to protect the bearer against forged notes? Indeed, the butterflies and caterpillars she painted are full of viruses such as the poly-DNA-virus PDV (as discussed in the wasp, Chap. 5). With scientific meticulousness and correctness, she produced pictures of tulip diseases with stripes 370 years ago. She survived an expedition to the Dutch colony Surinam in South America, where she ran high risks to catch exotic butterflies, and where she caught malaria. She paid the natives to bring her rare butterflies and beetles — and failed to notice that they cheated her by attaching a wrong head to a body. She drew it, and it was printed — to her dismay, because she was too late to correct the error. With her drawings and etchings, she brought the South American plant and animal world to the attention of Europeans. Peter the Great, in St. Petersburg, bought some of her originals and paid on the day she died. A German research vessel has now her name. For a long time, nobody remembered her. She was an excellent scientist and an exceptional artist, and I do not know which I admire more.

Those who want to amuse themselves about stripes may read: "How the leopard gets its spots". No, not by the ancient Greek story-teller Aesop but J.D. Murray (see References).

9 Viruses and antiviral defense

Fast and slow defense

Almost every Saturday Jean Lindenmann, my predecessor, came to the Institute of Medical Virology in Zurich for breakfast. More than 80 years old, he would meet his former co-workers and his technician — who was then more than 90 years old, but still brought home-made cookies and reported about her recent skiing trips to Norway! Swiss women are among the oldest worldwide! Lindenmann never talked about the past, but he was curious about ongoing research and new results. He was a gifted writer and often published articles in the NZZ (*Neue Zürcher Zeitung*). Also, when he was honored by the Society for Virology with a prize for his life achievement, he would propose open questions and research topics to the audience, mostly young students or postdocs. Many questions about interferon are still unsolved — he discovered it. He looked ahead, not back. His modesty distinguished him from many contemporary colleagues.

Jean Lindenmann discovered interferon together with his then supervisor Alick Isaacs in 1957 in England with a simple experiment: cells that were infected with influenza viruses secreted "something" into the supernatant which protected healthy cells from viral infection. In the medium a factor was present, a transmitter or cytokine, which interfered with a virus infection on naïve cells and was therefore called "interferon". It is produced in only very small amounts, so that it was difficult to isolate it from culture medium and characterize it. Many decades later interferon was, on the basis of a proposal by Lindenmann, sequenced by Charles Weissmann, then at the biotech company Biogen in Geneva, the first company of its kind, of which Weissmann was a co-founder. A competition started with other groups as to who would determine the sequence first. Weissmann with his co-workers won the "race" on Christmas eve of 1980.

The sequence was the basis for the production of a recombinant interferon, rIFN, which could later be produced in large amounts in bacteria or cells. This was a new therapeutic, but nobody knew a disease to treat it with. I remember the headline in a newspaper: "In Search of a Disease". Only against influenza viruses? Today the drug is used to treat chronic hepatitis B- and C-virus infections and cancer of the liver, multiple sclerosis, lymphomas, leukemias, malignant melanoma, chronic arthritis — a standard treatment, available in every hospital. Surprisingly, interferon is effective not only against viruses, but also against cancer. Side by side with his Swiss colleague Rolf Zinkernagel and the Australian immunologist Peter C. Doherty, Lindenmann could have been included in the Nobel Prize as number three in the year 1996. Zinkernagel gave public lectures in 2015 about "How to win a Nobel Prize"— wait for a phone call! I am sure this was not Lindenmann's style. I do not know any person as shy and modest as he was. Modesty is not the criterion for survival of the fittest in publicity, not even in research.

Interferons, there are several of them, are guardians of the cell against viral and also bacterial infections, and they also have an antiproliferative effect that can inhibit the division of tumor cells. The antiviral effect is not limited to one virus only. It is more generally effective. A virus that contains RNA has to undergo transiently the formation of a double-stranded RNA as intermediate. This induces the interferon system. The infected cell switches on an alarm system, secretes interferon, IFN as a signal molecule and warns neighboring cells. (*There, a signal cascade, called the "JAK-STAT" pathway, activates interferon-response genes. A pair of molecular scissors — a hallmark of all immune systems — RNase L, destroys the viral RNA and terminates the replication of the viruses.*) This saves the cell population, and only the first cell is lost, while the others get ready for defense, great timing! This defense belongs to the innate immune system of an organism. However, an antiviral defense would not help against virus-free cancer. There, the interferon system follows another principle, by activating "natural killer cells" (NKC), which recognize specific surface structures on the tumor cells and are immediately ready to eliminate them. A single compound against diverse viruses, as well as against different cancers — that is unique in medicine. (*The hepatitis virus HBV vaccine designed by Maurice Hilleman at Merck Co. protects against the virus and liver cancer, but is specific to the same virus in both cases.*)

In addition to our innate immune system we have a second one, the adaptive or acquired immune system, which leads to the production of specific antibodies, the immunoglobulins. Antibodies are highly specific and recognize proteins (antigens), which are exposed by antibody-presenting cells, irrespective of whether the antigens originate from incoming viruses or from cells. The binding between antigen and antibody belongs to the strongest interactions known in biology.

There are about 1000 genes available for antibody production. Yet by combinatorial effects they can reach 2.6 million different antibodies and by further mutations up to billions. Such an efficient amplification system is unbelievable. The diversity of the DNA segments is based on the "cut-and-paste" system of McClintock's transposons and the colored maize kernels, which we have already mentioned (Chap. 7). The antibody diversity is indeed related by evolution to the jumping genes and arose from transposable elements. Transposons are distant relatives of viruses, which supply the enzymes required for "cut-and-paste": molecular scissors, enzymes for cleavage, the transposases. This must ring a bell for the attentive reader: our immune system is generated by viruses!

Indeed, in the immune system the same enzymes are present as in viruses; they merely have different names, because initially nobody recognized the relationship between transposons, retroviruses and the immune system.

The immune system is directed not only against viruses, but also against any other "foreign" antigen. This development took millions of years to evolve. Our uniquely complicated immune system is ultimately derived from viral elements. In principle it is very easy to understand that viruses themselves contributed to the antiviral defense. After all, they enter a cell with their own set of genes, supply them to the cell and normally prevent other viruses from entering the cell at the same time, until they have finished their replication. There are not enough cellular resources for production of a large number of progeny for more than one incoming virus. Retroviral elements are, furthermore, already integrated into the host genome; they are very versatile and can change quickly. They are present as additional genes. The infecting virus is immediately the immune defense of the cell after entry of the virus. In the case of retroviruses, the step from infection to antiviral defense is minimal, because the viruses integrate and become part of the cell. This is a simple negative feedback mechanism. We owe the viruses the credit for our antiviral defense, an impressive example. Since when? 800 million years ago? (*Antibodies arise by recombination of the immunoglobulin genes described as V(D)J-recombination, with V for variable, D for diversifying, and J for joining. The molecular scissors required for the recombinations are molecules (RAG-1 and -2, recombination-activating genes), which evolved from transposons and the scissors are related to the ones mentioned repeatedly, with a typical triad of three amino acids, DDE, in their active center as in retrovirial molecular scissors the RNase H.*)

In the best case, the antibodies bind to the viral surface and neutralize the virus, so that it cannot infect the host any more. Neutralizing vaccines exist against smallpox, polio, measles, rabies and mumps viruses, and to some extent against influenza viruses. Apparently a vaccine against one of the four Ebola strains has also recently shown some effects. Most vaccines were first produced without any knowledge of the molecular biology of the viruses. Killed viruses were injected into people to activate an immune response. This was often successful — but has not been so in the case of HIV. It is just too variable. Expert molecular-biological research is now in full swing — after decades there may be some hope coming up now. We want a vaccine against HIV, because history has taught us that a vaccine is the best means to fight a viral infection.

Immune responses can last for a lifetime, if memory cells are produced. They are then already programmed and do not need any time-consuming gene rearrangements (cut-and-paste) but can multiply immediately and therefore react fast when a known antigen shows up. Our immune system has an innate, fast, but unspecific response on one hand, and a second, slow, highly specific, adaptive immunological response on the other. The fast, imprecise reaction is probably the older one. We do not notice anything as long as we are healthy. However, if the immune system is defective as in the case of immune-deficient "ADA children", they have to live in a germ-free tent so as not to get infected by microbes or viruses. Also, organ transplantations can fail because of rejection of the foreign tissue or cells by the immune defense, so that the recipient requires immune suppression for acceptance. In the case of HIV, the virus infects the immune cells, essential for our immune defense. T-lymphocytes play the most important part in adaptive immune defense, and it is precisely these cells that get destroyed by HIV. This leads to immune suppression and AIDS.

A comment may be allowed: Our immune system has similarities to our decision-making processes, I think. We have our guts for what is called "gut feeling": fast, almost non-reflected decisions originally protecting against life-threatening dangers and triggering flight reflexes, in contrast to the brain's decisions, which are based on facts and evaluation of the "pros and cons" — these are more specific and take more time. And we have memories in our brains, just like the immune system. These analogies are somehow surprising to me. A book title says *Thinking, Fast and Slow* (by Daniel Kahneman) — nothing on immune responses, of course. However, I have never heard anything about the input of transposons to our decision-making. It is known that the LINEs jump during embryogenesis in our brain cells and may lead to diversity. Do jumping genes influence our decisions?

No color by silent genes

Colors allow us to "see" genetic effects. This started with Gregor Mendel, who detected a genetic principle by looking at the colors of blooming peas, from which he derived his laws. Barbara McClintock detected an epigenetic principle by analyzing the colors of maize. The stripes in tulips indicate viruses and epigenetic effects. Colors are wonderful, because they allow us to *see* with the naked eye the course of very complicated genetics. There is a follow-up: a virus changes the color of its hosts, and the host's antiviral defense also leads to changes of the color of the host. A good example for such a defense are the violet petunias decorating our balconies in summer, which often exhibit symmetrical white patterns. Where the flower is white, the color has disappeared. The symmetrical white patterns indicate that the loss of colors must have occurred early during development of the flower head. The patterns arise from the plant's defense against viruses. "No color" means "no color gene expression", which indicates that a color gene has been shut off; scientists call this "silencing". This word does not fit well, because not a noise, but rather a color, is turned off. (*Silencing occurs by two mechanisms: transcriptional gene silencing TGS, and post transcriptional gene silencing PTGS — two ways how silencing can occur, but the net result is the same, no color. Transcription of RNA is blocked or the RNA cleaved.*) The silencer is "siRNA", which stands for silencer RNA or "small interfering" RNA. The name is reminiscent of the abovementioned interferon, and correctly

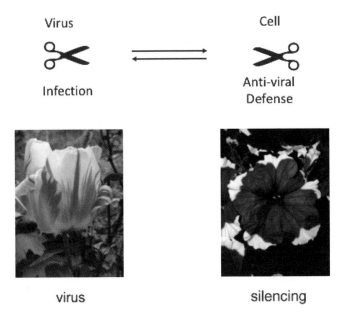

Virus and cell use the same toolbox such as molecular scissors for viral infection (striped tulip) and for antiviral defense of the cell (white pattern by silencing).

so, because the mechanisms are related. If the siRNA hits a color gene or the promoter of a color gene, one can see the silencing effect, the loss of color. A simple result; yet the molecular mechanism behind it, is rather complicated. And the silencing effect can hit any gene, whereby a color gene is just one out of many that can be affected. Incoming viruses can be silenced, too. It was David Baulcombe in England who described gene-silencing in plants, but plant genetics is a very difficult subject, many did not understand. McClintock, too, experienced this with her studies on the colors of maize, which nobody understood. Gene-silencing as immune defense was thoroughly studied in the worm *C. elegans*; this led to the award of the Nobel Prize to Andrew Z. Fire and Craig C. Mellow in 2006 without Baulcombe. Since then, silencing of genes has become one of the favorite and most informative experiments one can perform in the laboratory. Whenever one wants to prove the function of a gene, there is nothing better than knocking it out — silencing it — and seeing what happens without the gene: "loss of function" is the technical term. This has now become a standard procedure in every laboratory. Many universities even offer service units for knocking out genes by contract. No thinking about immune systems and antiviral defense is required — just order from a company a retrovirus with a given sequence, to produce the desired silencing siRNA. Then one can use a virus vector to inactivate a given gene and see what happens. It is safer to speak of knock-down, because genes are often not completely shut off. Sometimes a false gene is inactivated by a so-called "off-target" effect, which is another pitfall. Two groups once performed knock-out experiments using HIV-infected cells and obtained very different results. So the method still needs standardization and improvement.

We performed a silencing experiment after a Christmas party at the Zürich home of Tobi Delbrück (the son of Max). Colleagues wanted to find out what makes birds sing, and in our safety lab we helped to prepare the virus for siRNAs of various genes. We tested which gene resulted in silencing of the birds. The "FoxP2-like transcription factor", which influences the ability of humans to speak, was among them. We will find out which gene prevents or allows birdsong.

Silencing of genes in the worm *C. elegans* can be performed by feeding them. Worms can eat silencing siRNA like food. We used synthetic siRNA and dripped it onto the culture dish. In Nature, worms secrete silencer siRNAs as antiviral defense signals, to warn other worms in the population. Many worms populate the soil, and together they behave almost like one organism, so diffusion of the silencing siRNA is enough. This secreted signal resembles interferon, which also leaves an endangered cell and warns neighboring cells of invaders. Indeed the interferon system and the siRNA silencing system have some similarities, similar mechanisms in the RNA world and the protein world!

Secreted siRNAs exist in filaria (worms), fungi, plants, and even in mammals, where we found it (*to our surprise, but it is weak and dominated by other immune systems*). Worms are surprising. They led to the discovery of the antiviral silencer siRNA, but no viruses were detected in these worms. Are all their viruses "silenced" for ever? The worms then would be the only biological system on our planet without a virus. Recently, in France, a virus of *C. elegans* was discovered, the *Orsay virus*, in orchards close to Paris. However, as it turned out, the worms were a laboratory species and had a defective immune system, so they had no silencing siRNA defense system. Only then could the virus enter the worm and replicate. All viruses are normally kept out today. However, in the germ cells of these worms there are transposable virus-like elements, about 12 per cent, mainly DNA transposons. They are assumed to originate from former infections, then the worms must have encountered and integrated viruses before. They may keep other viruses out now. I do not quite believe in this claimed "single exception": worms as the only species on earth without viruses? There may be no more viruses today or viruses that we do not yet know. Some research is required here. The orchard in Paris is an interesting home of *C. elegans* and made them alcohol addicts via eating rotten apples. This is worth remembering because it will lead to a big surprise in the last chapter (Chap. 12).

One of my co-workers, Alex Matskevich from Moscow, found that we humans still have a weak silencing siRNA system in our cells. That is quite unexpected, because we have "better", well-developed immune systems, as mentioned just above: interferons and immunoglobulins. The small "interfering" RNA, siRNA carries almost the same name as "interferon" — correctly so. It exists but it is weak, overshadowed by the stonger ones. (*To demonstrate a silencing siRNA system in humans, Alex had to "silence" the interferon system first. Then influenza viruses replicated four times better.*)

We published this surprise. Eight years later a colleague published the same result, "absentmindedly" ignoring our results, hoping the reviewers would not notice. They indeed did not. How often one can get away with this the future will show. (Indeed he is in trouble now for other "forgotten" facts.)

I once noticed that not only the scissors, but also a whole toolkit — consisting of 12 components in total — was identical between the virus and the silencing system for antiviral defense. I mentioned the striking similarities to J.D. Watson during a Symposium in Cold Spring Harbor in 2007. They had there just published the structure of the defense scissors on the front cover of *Nature* which demonstrated the similarities between the viral and the antiviral scissors (*the RNase H and PIWI*). He was immediately convinced of my observation on the similarities between viruses and antiviral defense. He saw the implications and said spontaneously, "Submit a manuscript, I will talk to Bruce." He meant Bruce Stillman, his successor as head of the laboratory. Indeed, Watson had talked to him the same day, and early the next morning Bruce accepted my manuscript for the prestigious Symposium volume. How nice! Never ever had a manuscript of mine been accepted so quickly — normally a personal invitation would be required long beforehand. Best of all, I did not perform a single experiment, but made just a comparison. And I had not even written up the manuscript yet.

How nice also for the reader — this is so easy: viruses and antiviral defense are closely related. Who learnt from whom? (I believe the viruses are the inventors.) Interestingly, one cannot easily deduce the relationship of two pairs of scissors from their genetic sequences, but rather from their structures. This is not *quite* true, because there are a few structure-forming bonds formed by three typical conserved amino acids (D, D, E), which are, however, difficult to find among the other 120 amino acids. At least, we could not find them, even though we searched for them. Thus, structure is more important than sequence — something worth remembering! The similar functions are called "orthologous". The RNase H scissors have by now been identified as the most frequent protein folds and domains in the whole of biology — a recent discovery by the evolutionary and bioinformatics specialist Gustavo Caetano-Anolles from Illinois, and one that surprised me. RNase H is the oldest pair of scissors in the world! In the most recent analysis of data from numerous species from the "Tara Oceans" experiment in 2015, where millions of sequences taken from plankton were analyzed, RNase H and reverse transcriptase turned out to be the most abundant proteins — presumably because retroviruses, retrotransposons and antiviral defense are so ubiquitous, even in the plankton of the oceans. Everybody was surprised. (*How does silencing work? For those who want to know: a newly invading RNA, if it meets a similar one from earlier infections, is recognized and cleaved by molecular scissors. The toolbox for infection and defense are alike. The structures of the scissors are identical, but have different names, RT-RNase H in the virus and Argonaute (PAZ and PIWI) for defense. The scissors are present in many biological systems: not only in viruses, but also in bacteria, in mitochondria, during human embryonic developments and in human genetic disorders (AGS was mentioned above). Very recently, the antiviral defense system of bacteria CRISPR/Cas9 was discovered, where Cas9 is a scissor defending bacteria against new phages. Newly discovered is the PIWI-interacting RNA or in short "piRNA" with consequences for transgenerational inheritance — see below.*)

In summary, there are many names for scissors. Where do the scissors come from? The cut is normally unspecific, and RNases H need to be directed to specific sites and then cleave with high precision. They are almost real scissors, because they cleave real chemical bonds. The RNases

H initially recognized and cleaved only related invaders, but they became more universal with time. The scissors are active inside the cells — and once came in with the first invading virus. The protein scissors may originally have evolved from cleaving RNA, the trend is known from RNA to protein, from ribozymes to RNases H, from RNA scissors to protein scissors. To think about this scenario I had to go into retirement. There are about a dozen molecular scissors, RNases H, and for historical reasons they have different names. I discovered the retroviral RNase H and studied it for decades, it occupied my thinking for years, but I missed all the others.

What an embarrassing and also amusing conclusion. This is the result of specialization, and the dominance of the daily needs dictated by a full-time job and — somewhat in my defense — almost all of that is new. Nevertheless, how could I have failed to notice all this? I could have discovered it — but I did not!

Inheritable immune system in bacteria — and what about us?

Do bacteria need immune systems? Can bacteria get ill? How can one recognize sick bacteria? Do they need antiviral defense? Can they protect themselves? Indeed, there are no really ill bacteria. But they harbor viruses with a special name, the phages. Can phages induce diseases? I do not know, but phages can do much more. They can lyse bacteria and completely destroy them. This happens under conditions such as environmental stress, high temperature, lack of food or lack of space. Would bacteria need an immune system against their viruses, the phages? Yes, indeed: bacteria need an antiphage defense system, not against diseases but against death! Since most phages contain DNA, bacteria need an anti-viral DNA defense, just as RNA viruses need anti-viral RNA defense.

DNA phages are rejected by DNA mechanisms, whereby the defense of one phage is directed against a second similar phage. It is not the cell, but the virus, which fights back against other competing viruses; the cell profits from it. Each phage inside a cell will defend the cell by its integrated DNA. There is probably a simple reason for this: the resources of the cell are limited. One virus producing hundreds of progeny per cell is enough. The same mechanism applies for all virus systems. This is also related to the anti-RNA mechanism, the silencing or siRNA against other RNA viruses. The viruses always regulate this, the cell supplies only the milieu and profits from it. This is a basic general mechanism of all viruses and all immune systems. All the various cellular immune systems started by viral invasion of a cell and rejection of other viruses. That happened all the way to us, so it is very important!

The DNA-based immune system of bacteria, which destroys the newly infecting DNA phages, is called CRISPR/Cas9; CRISPR is an acronym for Clustered Regularly Interspaced Short Palindromic Repeats and Cas9 for CRISPR-associated enzyme number 9, there are many! Just remember "Crisp" (for new or fresh or crunchy) — and since even the Nobel Prize laureate C. Nüsslein-Volhard writes on her slides "CRISPER" this is no longer a spelling mistake but legitimate!

The CRISPR system resembles the human adaptive immune system, the specific one recognizing the invader. But it is inherited and could therefore also be called an innate immune system! Both! The bacterial DNA of the incoming phage is scrambled, spacers are added and the fragments are integrated into the bacterial genome. If a second phage comes in, it will be confronted with the RNA transcript of the integrated previous phage. If it is identical, it will be cut to pieces by the Cas molecular scissors. (*Cas is a slightly modified RNase H — with a similar structure, but specialized to cleave the DNA in a hybrid, not the RNA, and it therefore functions as a hybrid-specific DNase H! H indicates the RNA-DNA hybrid. After all, the invading virus has a DNA genome, which needs to be destroyed, whereas in the case of retroviruses the RNA has to be destroyed in the hybrid by its RNase H.*)

The aligned DNA phage fragments are inserted into the bacterial genome, waiting to be activated as antiviral defense. Again, one virus forces another one to stay outside. Thus virus and antiviral defense are very similar — all this has just been described above for retroviruses. The protection of bacteria against the phages is called "superinfection exclusion", something like "no entry".

The surprising immune system of bacteria has another unexpected property: it is inherited from one generation to the next one and protects all bacterial progeny. This is due to the integration of the fragmented phage DNA of the first phage. Even many generations later, the mRNA transcript will help to eliminate newly infecting related DNA phages.

What about humans? In public lectures during a World Congress on RNA in Davos in 2013 the plenary speakers praised the CIRSPR/Cas9 system as the *only* inheritable immune system. In humans, the mother gives some of her antibodies to newborns, which protect them against infectious agents for a limited period of time, as a start-up help, but they are short-lived. We develop our immune systems and an immunological memory later in life. Cells that save our body from invaders can stay as memory cells much longer, and they very quickly become mobilized and expanded in the case of a repeated infection — a life-long immunity. Yes, but all this is only for an individual's lifetime and not for the next generation.

Why are bacteria so much better equipped, with their "inheritable" immune system, than humans are? This is very surprising. Perhaps it is not true! This occurred to me only when I wrote this chapter. We also have an inheritable immune system: the endogenous retroviruses in our genome, which can act as antiviral defense, because they force other retroviruses to stay outside throughout evolution. We do not notice this any more. The endogenous viruses protect us against the exogenous related ones — many, or even all, of which have disappeared with time; they may not exist any more. Think of Phoenix, the retrovirus reconstructed from a dozen endogenous crippled Phoenix-like structures. An infectious virus was reconstructed from fossil ancestors. What an advantage for both the viruses and the cell! A virus cannot enter an infected cell, a cell that is already occupied, unless it undergoes changes — it is forced to be inventive and develop something new, evading the previous occupant, mutating for example its surface receptors for a new virus–host interaction, called changing their tropism. Something new is required — and viruses are the most inventive entities that this task could go to!

Perhaps that is the reason why, throughout more than hundreds of millions of years, so many residual viruses have accumulated and today make up about 50% of our genome. They built our genome and defended our cells against other viruses. Viruses supplied the antiviral defense to the cells: that is one of my most important messages — my *ceterum censeo* — again. We do indeed have an inheritable immune system as good as the one from bacteria.

The retroviruses must have been very useful and may still be. If they do not serve any purpose any more, they deteriorate. In conclusion, everything in our genome closely resembles that of bacterial genomes; phage

DNA-proviruses in bacteria resemble the retroviral DNA-proviruses in eukaryotes, and thus in us. Both of them integrate and get inherited. They even have the same names, DNA proviruses. So everything fits together well. Humans also have an inheritable immune system; it is not a privilege of bacteria.

There are cases where this can be clearly demonstrated and indeed proven: an endogenous retrovirus of mice forces other viruses to stay outside, a phenomenon termed "restriction" rather than "interference" (*Friend virus-1-restriction, abbreviated as FV-1*). The defense gene originates from an endogenous retrovirus ERV, its structural protein Gag. These discoveries were published about 40 years ago and were subsequently almost forgotten about. A viral protein (instead of RNA) prevents entry of a similar virus into the cell. The koalas from Australia have already been mentioned: they also endogenized a dangerous monkey retrovirus, which infected their germ cells and thereby protected the next generations from the same virus infections. Immunity of koalas was thus inherited by endogenization of retroviruses. This is one proof. There are more, the monkey virus SIV protects monkeys against a simian version of AIDS. The paleovirologists mentioned above also detected bornavirus sequences in our genome, dating from 50 million years ago. Endogenous bornavirus expression protects us against new bornavirus infections. Horses, which do not have endogenous bornaviruses, accordingly fall ill with these viruses and get depression as a disease. Bats, too, carry many endogenous viruses and do not get ill — but they transmit many of these viruses to humans, and we get ill, because we lack the endogenous viruses and die of Ebola or HIV.

Thus, we also have an inheritable immune system, our endogenous viruses, called HERVs, LINEs, or SINEs. Our inheritable immune system follows the same principle as in bacteria. CRISPR phage DNA proviruses accumulate in the genomes of bacteria and also of archaea. These protective sequences can vary in numbers depending on the need for defense against phages and can be lost again. For comparison, 50% of our genomes is related to retroviral sequences. Probably in both cases we do not recognize older viral sequences any more after some time. The exogenous viruses may have disappeared as exogenous viruses. The resurrected Phoenix virus, obtained under laboratory conditions, does not exist in Nature any more. Therefore, the Phoenix sequences became crippled and useless as defense. Now we understand why we have so many endogenous retroviruses in our genomes — they once protected us. They were our anti-viral defense and kept related viruses out. What a simple answer to such an enigma! Perhaps they will get lost after yet more time has passed.

Does this prove that the viruses "invented" the cellular immune system by supplying their genes to the cell? Or could it have been the other way round: the viruses "stole" their components from the cell? Many textbooks of virology suggest the latter idea, and I do not believe it, though some colleagues may.

It is striking how similar all the immune systems are, even though at first glance they seem to be so different. In all cases the same principle operates: viruses cause antiviral defense with related molecular scissors. This observation summarizes one of the most important principles for understanding the prominent role of viruses as drivers of evolution. Only that sentence is worth remembering — forget all the details! The particulars are more for specialists, to persuade them of the rightness of my argument — and supply facts. One can even read this in a peer reviewed scientific paper from 2015.

Therapies imitating antiviral defense — CRISPR/Cas9

How can one find new drugs against a disease? One could try to imitate Nature, and fight against nucleic acids by using other nucleic acids and molecular scissors. Nature exploits such possibilities. Three possible mechanisms for therapeutic nucleic acids have been put forward: antisense DNA, ribozymes, and silencer RNA (siRNA). In all three cases, naturally occurring antiviral mechanisms are imitated by therapeutic nucleic acids. This initially appeared very encouraging to drug designers: what Nature can do, we can ourselves copy and try out. If we want to fly, let us imitate the birds. Every time a new therapy was announced, it was celebrated as a revolution, and was accompanied by enthusiastic press releases, company start-ups, and mobilized business angels.

The drug development of therapeutic nucleic acids does not require crystal structures or high-throughput drug screening: it can simply be done on paper by sequence comparison. The only thing one has to know is the pairing rule of the double helix: A–T and G–C, or in the case of RNA A–U and G–C. For antisense therapy a piece of single-stranded DNA or RNA is designed against a dangerous gene that needs to be inactivated. The mRNA of that gene forms a local hybrid with the designer DNA, which activates the scissors RNase H and then the RNA is cleaved. That terminates the production of the respective protein. Alternatively, a blockade of mRNA can terminate protein synthesis and stop the ribosomes. That also leads to the inhibition of the production of proteins — including viruses.

Instead of antisense therapy, another approach for gene inactivation is based on ribozymes against mRNA. In that case the RNA itself is the enzyme that performs the cleavage, and no scissor RNase H is needed. The ribozymes bind to selected sequences and cleave with their endogenous scissors. They can also be designed at the desk against the RNA to be eliminated. One adds about seven nucleotides on either side as little legs, flanking the critical cleavage site, which carries the meaningful name "GUN", the nucleotides G, U and N (N stands for any of the four nucleotides), as the cleaving triplet. This approach was used against the oncogene "Bcr-Abl RNA", which is typical of chronic myeloid leukemia. A ribozyme of this kind against HIV, in combination with other factors, is also undergoing clinical trials in California, in City of Hope, a suggestive place-name.

The third general approach to silencing genes is the use of siRNA. This consists of short double-stranded RNA, about 20 nucleotides in length, separated into single strands. One of them binds to the selected target RNA and then activates the scissors (Argonaute with its PIWI-RNase H) to cut the undesired RNA.

All three approaches originally raised great hopes for therapies. What is left of them today? A lot of research data on mechanisms of antisense, ribozymes and siRNA, but not a single drug, and certainly no blockbuster. They all suffer from the same problem: how to direct the therapeutic nucleic acids to the target site for cleavage — in tumor cells, in the brain, in organs. Nature has developed a splendid procedure to achieve this: viruses! Right, was the conclusion, let's try synthetic or modified viruses, liposomes, nanoparticles, or "degutted" viruses. A whole research field, bionics, is based on copying Nature. One can exploit the modular character of viruses and assemble the most valuable viral components, "the best of all virus worlds" as the authors called it, combining genes from a dozen of different viruses; great fun in the laboratory but, again, little therapeutic success emerged from it. The only such drug I am aware of is an antisense therapeutic against a herpes virus, the cytomegalovirus infection of eyes. Eye drops have to be applied by the patient — not by a virus or a nanoparticle!

Antisense DNA was first described by Paul Zamecnik, the "father of antisense". Only later was it recognized that viruses and bacteria also use it as a regulatory principle. That made the approach even more attractive for drug design: what Nature has designed, we can develop accordingly. Herpes viruses use antisense nucleic acids to keep the virus in the latent state. Also, maintenance of immunity of the phage P1 against superinfections is based on this principle. That was pioneering work performed at the Max Planck Institute in Berlin in Heinz Schuster's laboratory. I remember well that in 1994 a repressor protein was searched for in phage P1 immunity — and a nucleic acid was discovered instead. This was published in the journal *Cell*.

Unexpected new possibilities have emerged to exploit these technologies for the knock-out of genes. Complete inactivation of a gene, "loss of function", allows us to find out what the gene's task is. There is no laboratory in the world that does not use this antisense technology to understand metabolic pathways, diseases or cancer. It can be performed in cell culture, and also with animals, although not in humans. What is left of this? Good research and the identification and "validation" — confirmation of the essential role — of target genes.

We also used a piece of DNA to inactivate HIV, a "silencer DNA" instead of a silencer RNA. A hairpin-looped DNA is targeted to a conserved region of the viral RNA in the particle and activates the molecular scissors, the viral RNase H, driving the HIV to "suicide". We designated it as silencer DNA, "siDNA" in the Cold Spring Harbor CSH-Symposium volume of 2006. Perhaps it can help women to protect themselves, as a vaginal microbicide against HIV infections — a long expensive way.

There is now a new technology, the fourth therapeutic nucleic acid, derived from the immune systems of bacteria. The whole world is fanatic about it!

Bacteria save the genes of an infecting phage by integrating its DNA into their genome. If a similar phage enters the bacterium, it is recognized by the RNA transcribed from the first one and a cut destroys the new phage. The system is called CRISPR/Cas9, whereby Cas9 is the molecular scissors, the nuclease that performs the cutting. Researchers imitate this immune response by combining a nuclease with RNA. The RNA is called "guide RNA" and has two functions: one is to find the cleavage site by sequence homology, and the other is to bring in the information for genetic changes. Furthermore, the RNA has a hook, to hold onto the nuclease with. The genetic changes are described as "editing". The *New York Times* published the headline "edit DNA". Remember that "edit" is a command on your computer and means "change". The target DNA, together with the guide RNA, form an RNA–DNA hybrid, whereby the DNA is cut by molecular scissors Cas9. It can cut out regions of the DNA, and the ends will be joined by the cell. Or the cell copies the guide RNA for editing. In contrast to other DNA-modifying technologies, CRISPR can be used to alter specific genes without introduction of foreign DNA, without "recombination". This is a big advantage. Also, there are no laws or restrictions yet. It was one of the "top 10" breakthroughs celebrated in *Nature* and *Science* at the end of 2013. Journalists immediately described their visions: deleting genes, editing (changing) viruses, bacteria, plants, genetic defects, curing diseases — modifying all imaginable systems on our planet, including humans. Any desired modification of the DNA in a living cell or whole animal is possible; in fact, many changes can be performed at the same time: just combine the desired sequence changes on a long guide RNA. The cell will copy it. Commercially available kits pop up almost every day to make the procedure as simple as possible. Companies offer viruses to transport the complex into cells. At present I am getting three e-mails with offers every day! I have seen up to 62 edits carried out at a single stroke, to "humanize" a whole pig so that spare parts of the heart or other organs can be transplanted into patients without immune rejection. Any student can use this method, and alter genomes of cells and embryos to produce transgenic animals — in almost no time, which used to require up to three years of hard work by a PhD student, and even then not always with success guaranteed. Now it takes three weeks.

Editing is a general phenomenon in biology, not invented by Google. The same mechanism is present in archaea as well. African trypanosomes also show extensive editing and genetic variation, and humans have 35 times more editing in the embryonic brain than monkeys do. So that is a natural mechanism. Can one make multi-drug-resistant bacteria sensitive to antibiotics again? This is one of the most urgent needs worldwide. The synthetic biologist Timothy Lu, at the Massachusetts Institute of Technology, is trying this by infecting bacteria with phages for editing to make them sensitive to antibiotics again. Especially biofilms, consisting of layers of bacteria not very actively metabolizing, are targeted by the editing phages. Editing by CRISPR/Cas9 technology is the newest hype in nucleic acids research. This enthusiasm has infected me too. The novelty is the speed and the simplicity of the approach. There are 20 kinds of Cas-type scissors, with different specificities — some even cut single-stranded RNA. How will they be used?

Some scientists are worried about potential abuse of this method. Designer babies have already been announced in the press! Indeed, Chinese researchers modified genes of human embryos early in 2015 — even though the embryos did not survive. However, one could imagine experiments with real human embryos. "Engineering the Perfect Baby" was the headline of an article published at the MIT early in 2015. A one-year-old child has been treated to accept a transplant it would have otherwise rejected. The method is not 100 percent precise. There are so-called "off-target" effects, which may lead to side effects and editing of wrong genes. This problem has to be overcome — it is typical for all nucleic acids and not trivial to overcome. Early in 2016 the UK authorities approved the use of human embryos in the laboratory for about a week — from the four-cell to the 256-cell stage — to apply the CRISPR technology and to test for the role of certain genes during early embryonic development. The cells then have to be discarded. In mice, Duchenne Muscular Dystrophy (a genetic disorder of muscles in boys) was cured by cutting out a region with a genetic defect. Ethics committees are getting ready to evaluate the technique. Self-restriction by scientists is something that we have had before — in connection with recombinant DNA technologies. It was incepted at the famous Asilomar Conference in 1975, out of fear of biohazards. At the end of 2015 a committee from the International Summit on Human Gene Editing declared that the new editing technology should not be used to modify human embryos in pregnancy. David Baltimore, from Caltech, chaired the Summit and said: "Scientists must be cautious, because once you have made a genomic alteration there is no taking it back" — which some people find too soft a statement. Every year, eight million children are born with genetic defects. Do they need an edit, a change? Where and when?

It is very surprising to me that, in both cases where scientists proclaimed self-restrictions, the methods were based on bacterial immune systems: a bacterial defense system was first, the use of "restriction endonucleases" discovered by Werner Arber in Switzerland, the basis for recombinant DNA. Now it is the immune system CRISPR/Cas 9. Bacteria employ both of these defense systems against phages, in about equal measure. We are thrilled, but also troubled. Arber received the Nobel Prize in 1978. Who will be next?

Two young ladies are driving this technology with breathtaking speed, and both have already made it to the short list of candidates for the next Nobel Prize. One of them is Emmanuelle Charpentier, who has been appointed as a new Max Planck Institute Director in Berlin. Jim Watson wanted to know more about her; he seemed impressed and a bit curious, too! The other is Jennifer Doudna in Berkeley, California. A third young person, Martin Jinek from Zurich solved the Cas9 crystal structure. The enzyme is a variant of RNase H — except that it cleaves the DNA in hybrids, so it should be called DNase H. Moreover, Feng Zhang at MIT is driving the various applications and has facilitated the techniques. Behind the scenes a patent war is being waged between Berkeley, which filed earlier, and MIT, which filed later but for more applications. Unfortunately, there is a lot of money involved. The Bayer company invested €300 million for three diseases, blood disorders, blindness and congenital heart diseases. Three more big companies got involved and many more investments will follow. 600,000 research groups got going — this technique is applicable by everybody. It is important to note, that this method can be used to mutate (edit) or inactivate specific genes without introduction of DNA of a foreign organism. Such edited organisms are presently not covered by the restrictive laws imposed on GMOs (genetically modified organisms). Genes which are copied but not recombined would be less dangerous. But discussions by regulatory authorities have already started. This story will continue!

From horseshoe crabs and worms for immunization

There is an ancient-looking animal, hundreds of which can be found on the beach of Long Island and in the Mississippi delta. This is the time for the horseshoe crabs to mate. They have three eyes on a horseshoe-shaped shell that gives them their name. They also have a long, thin tail like an arrow — used in their other name, arrowtail crabs. They do not have an immune system. How have they managed to survive for so long? The animals have blue blood instead of red: they contain hemocyanin instead of hemoglobin, and their blood contains copper instead of iron. Copper is toxic for microorganisms. This is why in earlier times the incubators for growing cell cultures were made of copper, as it has an antimicrobial effect — a leftover from Robert Koch's days. I can still remember working with them. If the immune system is missing in humans, as is the case with ADA children, they are at risk of life-threatening diseases and must be kept isolated in sterile tents. Invertebrates and insects can also get away with blue blood for survival. I once took a crab shell home, dried it and kept it as living fossil until one day it started to decompose and had to be disposed of. Do they really have no reverse transcriptase, no siRNA, no CRISPR, no ERVs, no IgGs? Can all these immune systems be replaced by some copper in the blood? For the past 500 million years? — surprising!

What can people do to improve their immune systems? Eat worms! That does not sound very appealing. One can phrase it more scientifically and describe the worms as helminths, though worm-eating is what "helminth therapy" means. Allergies are on the rise because our immune systems are not busy enough with foreign threats. They are bored and search for other alternatives — including, unfortunately, the body's own antigens. This leads to allergies and autoimmune diseases. These are frequent in modern industrialized habitats, where people do not come into contact with parasites or other infectious germs often enough, so that they lack sufficient stimulation of their immune systems. The children of farmers have become city inhabitants, with vacuum cleaners instead of cowdung. In autoimmune diseases, the immune response is directed against "self" instead of "foreign", against parts of the organism's own body instead of foreign invaders. Children in Switzerland are kept under the cleanest conditions — perhaps the cleanest worldwide, according to one estimate. So the Swiss immunologist Rolf Zinkernagel announced in an article in the popular newspaper *Blick* that worm therapy was an option against autoimmune diseases. He was not the first to propose this. Years ago, an American helminth "advocate" offered worm eggs on the Internet as a treatment against allergies, but the US Food and Drug Administration put a stop to this. So he migrated to the UK and continued offering his therapy. Even *Nature* devoted an article to this topic: patients with inflammatory bowel disease (IBD) are offered a drink containing thousands of worm eggs. The eggs are transported to the intestines, where small larvae are released and develop into veritable worms. They are fortunately so small that one does not recognize them. Nobody wants to be reminded of the small white worms in children's feces. The therapeutic worms are derived from a swine worm, so they die out within eight weeks in the human gut. A physician, Joel V. Weinstock, working at Tufts University in the US, has described our immune system as "unemployed", jobless. If it now starts to direct itself against the worms, it is distracted from the autoimmune reaction. Weinstock treats patients with chronic diseases such as Crohn's disease, ulcerative colitis, type I diabetes, psoriasis, multiple sclerosis and food allergies with worms. In Freiburg in Germany, a clinical trial of this worm therapy with 330 patients is under way. The interest is huge. Nobody has to be talked into joining the project.

The name of the worm is *Trichuris suis*. Perhaps one should swallow the eggs repeatedly, following a certain schedule, to suppress autoimmunity more permanently. Such trials have also been initiated at the Mount Sinai School of Medicine in New York. The "hygiene hypothesis" posits that multiple sclerosis patients who receive parasites as therapy show a slower progression of their disease than patients receiving a control treatment. Today, sticking plaster with larvae of nematodes (worms) are stuck onto the skin; the larvae drill into the skin and end up in the intestines. This therapy is cheap and simple. There is little experience as yet regarding the possible consequences of this treatment on other infections. What will happen to the big anti-worm campaigns in Africa? There, schoolchildren have been treated to get rid of worms, which significantly improved their performance and increased their learning abilities at school. Children in Africa will probably not get autoimmune diseases, even without worms, since they are sufficiently exposed to parasites or "dirt", which keeps their immune system busy. Another aspect of worms is their use for treating deep sores that refuse to heal, such as *Ulcus cruris*, often located at the tibia and due to weak blood vessels. Or pressure ulcers in hospital and nursing-home patients (bedsores, or *decubitus*) caused by immobility and skin pressure. Worms will feed on the infecting microbes and "clean them up". This procedure is applied at University hospitals where patients require stationary treatment. Very recently phages have also been discussed as possible means to cure precisely those kinds of lesions, by destroying bacterial infections.

As well-educated a scientist as Freeman Dyson once told me that he had earlier suffered from chronic arthritis. Then he was bitten by a dog in Warsaw — and this made the arthritis disappear. One day we shall understand the mechanism of this remarkable cure, he predicted. The bite probably activated his immune system and redirected it to the dog's microbes in the wound, away from his autoimmune arthritis. Some people are propagating the idea that children should grow up with dogs, or spend holidays with animals in farmhouses — as training for their immune systems.

Dyson also mentioned his family doctor in La Jolla, California, to whom he took his six children when the need arose. The doctor had noticed that the children on the Mexican side of the border had hookworms but no allergies, in contrast to children on the Californian side, who had allergies but no hookworms. Small worms would do!

Even viruses have been used for years as "virotherapy". Vaccination with inactivated viruses has shown unspecific immunestimulatory effects. In one patient, a tumor disappeared after the patient had been vaccinated with a rabies vaccine. In another case, a boy's leukemia was reduced after a smallpox vaccination. These were case reports and were never followed up systematically. Whether worm therapies have any potential for the future, or will be of long-lasting interest, will have to be seen. There will be better therapies one day.

Can too much hygiene make us ill? In the train I once happened to meet a professor and his family from Bangladesh. It was the spring of 2013, and the river Elbe had reached unusually high water-levels, flooding the surrounding marshland. He mentioned that in Bangladesh the children swim in the rivers and swallow dirty water without getting sick. Clean water will make us ill, he explained — the opposite of what one might think. This can be observed with adults, who get sick because they no longer train up their immune system on dirty water. There are about 50,000 virus types (mainly phages) in sewage, and 250 other types of viruses, 17 of which are known to cause diseases such as polio. In the Indian holy city of Benares (or Varanasi) I once observed how the pilgrims used special little ritual cans made of copper, which they filled with water from the river Ganges, to wash their mouth. It looked terrifying to me, because the water seemed so dirty. I need not have worried — they were immunizing themselves! Some phages in the Ganges river were described decades ago as having antibacterial efficacy. The "holy" water of the Ganges was supposed to help against cholera and leprosy. Perhaps they used the copper cans for further antimicrobial efficiency. I brought one home as a souvenir. Today it is accepted that too much hygiene can be a cause of allergies and chronic intestinal diseases — but also against asthma and autism? Let them try worm eggs. However, children in Bangladesh get still poliovirus infections. So river water alone is not safe enough to protect against diseases.

Most recently, the interaction between pathogens and the host's defense have been looked at from a completely different perspective: if a mouse is free of microbes, including its gut microbes, then it develops a weak immune system. Intestinal microbes are a prerequisite for a robust immune system. This is one of the results of the Human Microbiome Project (HMP). A good microbiome and immune system are already established during birth. Early in 2016 it was shown that birth by Caesarian section, which is used in about 90% of deliveries in China, may cause allergies and arthritis later in life, around the age of 40. Newborns have to be treated with the regular gut microbes recovered by lavage from the mother's birth canal as a protective measure: a shower of microbes for a newborn! Reports about the first four children to be treated in this way appeared early in 2016. Immature, newborn babies are at the greatest danger of infection by hospital germs. We studied some of them who became infected with multidrug-resistant bacteria (MRSA). They will be given a treatment with their mothers' microbiome in the near future.

The newest therapies against cancer are increasingly based on immune therapies, whereby the tumor cells are first made immunogenic, to allow the immune system to recognize and eliminate them. The signal that cells use to escape from immune surveillance is called "do not touch me", or more scientifically PD-1 (programmed cell death protein 1), discovered by Irving Weissman from Stanford University decades ago. Only recently he was honored for it by a generous Swiss Award the "Brupbacher Prize". PD-1 needs to be blocked by an antibody — and then the immune system can destroy the tumor cell. Let us hope that this will work!

Viruses and psyche

More than once, I have mentioned stress as applied to bacteria and phages. For them, stress is induced by lack of space, scarce food supply, or extreme temperatures. For mice, stress is induced by changing their cages several times a day and by giving them new partners. This is almost the same as for people who lose their home, their job or their partner — the three strongest stress-inducers for humans. In bacteria, stress activates factors, repressors are displaced from their location on the DNA, phage DNA is released from the genome of the bacterial host, and the bacteria enter the lytic phase. In humans, stress activates signal kinase cascades of the type mentioned in connection with cancer cells. They activate stress-response genes or stop cell division. Otherwise, human stress is similar to bacterial stress, also activating herpes viruses. These are fantastic "psycho-sensors", and notice the trouble before we are even aware of it ourselves. They are activated by sorrow, by too many appointments when on the job, by too hard work, by examinations, by dentist's appointments, by lack of sleep, sometimes by hormones, and sometimes also by too much sunshine. Then herpes viruses "crawl" out of the niches that gave them their name. They persist chronically in the spinal cord and ganglia and are always there, even if we do not notice them. They can persist unnoticed for decades. During christening ceremonies, aunts and uncles often unknowingly bring herpes viruses as little presents for the baby, by kissing it. Ninety per cent of us have them. Under stress conditions the viruses leave their hideaway deep down in the spinal cord and migrate through the nerves into the periphery, such as the lips, and cause lesions with enormous numbers of viruses and high infectivity. These lesions are also called cold sores, because cold (or the fever accompanying a "cold") is a kind of stress for the immune system. They are often recurrent and return repeatedly and never go away, once herpes always herpes. Other diseases, such as shingles, are caused by the varicella zoster virus, again a herpes virus infection. It may be a serious warning signal pointing to other severe diseases, such as a tumor. If viruses reach the brain they may cause encephalitis and be life-threatening. A study has shown that females with breast cancer have a better chance of longer survival if they have a supportive family, friends, and a positive environment. Psychological effects can indeed go this far. The immune system is the most incredible sensor. Even if we just encounter something that disgusts us herpes viruses may get activated. This I find most surprising myself. The herpes viruses are the most sensitive detectors of our inmost, psychological problems. Perhaps this is also true of other viruses and diseases, but we have not found out this yet.

10 Viruses and Phages for survival?

The forgotten phages

The application of phages instead of antibiotics to kill bacteria was introduced by Félix d'Herelle in 1916 — as early as 100 years ago! However, he was unlucky, because only ten years later (1928) the detection of the antibiotic penicillin, by Alexander Fleming, put an end to his plans and his hope. Antibiotics are much easier to handle, they act unspecifically against a variety of bacteria, and they are more effective. As therapies, they completely out-competed the phages. So phage therapy, as a means to kill bacteria, was forgotten about before it had even got off the ground. This was a big disappointment for d'Herelle, who had immediately seen the implications of his phage discovery and also demonstrated its effectiveness. He travelled, quite frustrated, to various countries where he successfully treated epidemics such as cholera in India. Together with Georgi Eliava he founded a Phage Institute, which still exists today, in Tiflis (today Tbilisi), Georgia.

It is this tradition which still makes Eastern Europe and Russia the front-runners in phage therapies. In Russia one can go to a pharmacy and buy phages as OTC (over-the-counter) products. Admittedly I did not succeed in this, but only because I do not speak Russian and had no interpreter. By now, millions of people have received phage therapies. They do not cause harm; but do they really cure? Often only case reports have been published, without control groups. Consequently, the American health authorities (FDA) refused approval of such therapies because the criteria for quality control have not yet been fulfilled. An approved drug has to be tested for side effects, it must undergo toxicity studies in animals, and defined starting material has to be retained for repeated testing. Bacteria change quickly; they also form quasispecies, swarm populations. Thus, the issue of phage therapy is not a trivial one.

In 2010, during an international meeting about phages at the Institut Pasteur in Paris, a former patient in the audience got up and told the meeting participants about three shipments of phages from Georgia which, he said, had cured his skin disease. No doctor had been able to help him. The US health officials clearly opposed such a therapy in a subsequent session — but at least they were present, which after all meant something. Case reports are never easily accepted, and they sound unscientific. However, one should at least pay attention to them. During the Russian–Finnish War in 1939, wounded soldiers were successfully treated in field hospitals by phage mixtures against anthrax, a bacterial infection that otherwise would have ended in limb amputation — or death. The phage cocktail was dripped directly into the wounds — local treatment is still the simplest and most promising therapy. Such successes kept up the reputation of this kind of treatment. Which phages should be used, how and when? — That is the question. They need to fit the case to be treated. This problem was already known to d'Herelle. His antagonists could not repeat his experiments, because they used different or ill-suited phages, so that d'Herelle suggested early on that one should use phage mixtures, cocktails, to increase the chance of administering the right phage to destroy the bacteria.

In Poland, prostate infections are being treated with phages. The results are not overwhelming, and the details have not been clearly reported. In 2012 about 40% of 157 patients were described as having responded successfully. The Swiss company Nestlé has also started phage studies. Why Nestlé? Perhaps one can use phages to kill bacteria and produce clean water? Or cure diarrhea? Nestlé started with the well-known laboratory phage strain T4 in Bangladesh, using it to treat children for diarrhea. The control group of children in Switzerland showed a much better outcome, possibly because the children in Bangladesh had many more different bacterial strains than the Swiss controls, so that one phage for treatment was insufficient. It came as a surprise that the diarrhea in Bangladesh was not due to the typical *E. coli* strain T4, which had been presumed to be the cause of their diarrhea. Thus basic questions need to be answered first.

In the summer of 2014 headlines in the journal *Nature* attracted attention: "Phage therapy gets revitalized." And another title was downright provocative: "The age of the phage: It is time to use viruses that kill bacteria again." Again! We are getting up to speed. Wonderful! Where will phage therapies be useful? The best-suited applications are on surfaces. Burns are among the top indications for phage therapy. An international consortium was recently founded and is financially supported for treatment of burns by the European 7th Framework Programme, called Phagoburn, in which burn wounds infected with the bacteria *E. coli* and *Pseudomonas aeruginosa* will be treated with phage suspensions.

Furthermore, there have been impressive success reports for diabetic toes, which occur as a consequence of obesity and diabetes; this condition affects or threatens about 6 million people in America and 600 million adults worldwide. During a scientific conference in Zurich in 2015 diabetic toes before and after phage cocktail treatment were described; they healed within a few weeks, and the phage therapy made amputation unnecessary. Impressive photos were shown. No other therapies can replace this one, because depending on the depth of the sore (which can reach all the way to the bone) different bacterial strains are active and therefore require different phages to be present in the cocktail. What a great success story! Sores in the lower limbs, *Ulcus cruris*, which often never heal, have also been treated effectively with broad-spectrum phage cocktails. Such success justifies increased effort. Also, senior citizens in nursery homes, who are threatened by lack of mobility and pressure sores (*decubitus*), which do not heal easily, would benefit from the phage therapies. Chronic middle-ear infections and cystic fibrosis, with bacterial infections in the lungs, are important phage targets as well. Some surgeons treat the wounds or the surgery catgut (thread) with phages to keep bacteria out when sewing up operation wounds, to prevent hospital-acquired infections. Phages against methicillin-resistant *Staphylococcus aureus* (MRSA) are urgently needed. 600,000 infected people with 15,000 death cases have been reported for Germany. This is probably an underestimate and amounts to a death toll higher than 40,000, perhaps 60,000. And then there are other resistant microbes such as the bacteria ESBL (extended spectrum beta-lactamase) or VRE (vancomycin-resistant enterococci). Even tourists visiting India catch resistant ESBL, in more than 50 percent of the cases within two weeks, but the bacteria seem to disappear rapidly again at home — at least in healthy individuals. Millions of people worldwide will die from antibiotic-resistant bacteria. And the G7 Summit in Germany in 2015 had MRSA bacteria at the top of its priority list for new action, including the reduction of antibiotic use in humans and animals. Even the hip replacement of the King of Spain was complicated by MRSA infection. He had to undergo repeated surgery — and finally abdicated (also for other reasons!). Thus it is not just a matter of finding the best doctors, which he probably had done anyway.

Some countries, such as Holland or Norway, require certificates from patients, health-care workers, employees and students to prove that they test negative for MRSA before they enter the hospital. If they cannot produce the certificate they are sent back home with instructions on how to clean up. In Scandinavia, emergency cases are housed separately until they have been screened and have tested negative. This practice is unheard of in Germany.

A neighbor in my hometown by the Baltic Sea suffered from rupture of an abdominal aortic aneurism complicated by a bacterial infection, which is difficult to control or cure. Also Feodor Lynen, the German Nobel Laureate in medicine, died of complications of an aneurism by bacterial infection of his abdomen. Can phage therapy be of help there? If one cannot travel to the Eliava Institute in Tbilisi any more, one can still send a swab and order a phage cocktail, according to the information from the Institute's website — which I verified by a phone call. However, that would take a while, I was told, because they need to grow the bacteria and run test series for their sensitivity towards a number of phages to find the suitable ones. It took too long for this patient. The phages could reach niches and would find even hidden bacteria to destroy them. The phages grow better when more bacteria are around. Now even the Institute of Medical Microbiology of the University of Zurich is getting involved in some studies with phage therapies, and has started cooperating with the Eliava Institute. The EU has recently financed a safety laboratory in Tbilisi, to guarantee better quality control of the phage production. Things are starting to move. This is personalized medicine, and its is self-limiting when no bacteria are left. Some people pooh-pooh it as "alternative" medicine — but let us wait and see.

My vision is to use phages to kill resistant bacterial contaminations in hospitals — that would be a huge success. Cleaning a surgery room with a phage spray appears feasible to me. Doctors and colleagues are pessimistic, though, asking which phages should be used, and how can one know which ones fit the need. One would need a mixture of many phages, hoping that some would fit, as d'Herelle already knew a hundred years ago. Then there is the commercial argument that one cannot use phages in business, as they are well-known and cannot be patented. However, a mixture could be patented — why not? There are some companies, such as Intralytix in the US, that are developing phages against bacteria such as *Listeria* (a danger for a gourmet product, the Swiss cheese Vacherin, but for other food products too), and also against *Salmonella* and a dangerous *E. coli* strain (*E. coli 0157*). In India the company GangaGen tries to fight dangerous bacteria, destroying colonizing nasal *Staphylococccus aureus* by phages, "StaphTame." In 2012 a meeting supported by EMBO and the Belgian military took place in a baracks in Belgium. Such a combination, and location, had never been seen before. But both were appropriate: the military was the first customer and still is an important one. The organizers reserved a whole day out of five for presentations on phage therapy. Someone must have remembered the soldiers who were cured back in 1939. A dentist mentioned that he tried to use phages to prevent bacterial infections of implants, which can lead to dangerous complications such as infection by *Clostridium difficile* — a side effect of dental treatment with a potentially fatal outcome. The age of phages has come. (For a detailed description of one case, see below.)

Much speaks in favor of using phages: they are self-dosing (the more bacteria, the more phages will grow to kill them) and they are self-limiting (they die out when no bacteria are left). That is unique! Last, but not least: phage therapy is cheap — a great advantage for developing countries. Most suitable targets for treatment are humid open wounds or sores, which can be reached from the outside and allow the phages to multiply. Swallowed phages will die during passage through the stomach, so they have to be encapsulated into pills for treating intestinal infections.

What are the limits and the dangers of phage therapy? Phages transfer genes by horizontal gene transfer. No transfer is allowed for genes such as those conferring resistance against antibiotics. Furthermore, phages for human use are not allowed to integrate into the bacterial genome; they must be lytic, meaning that they multiply and then lyse and destroy the bacteria. This is reminiscent of the new anticancer therapy with "oncolytic" viruses, which also replicate and then lyse and destroy the tumor cells — the same principle.

Still, there are some important side effects that need to be ruled out. Do phages affect the bacterial genome and possibly induce new phage-resistant strains? Resistant bacteria, ones that cannot be killed by the phages, are a major concern, since bacteria can change their surface receptors and prevent entry of the phages. Therefore, multi-therapies are needed. We know, at least since the introduction of successful anti-HIV therapies, that multiple therapies are superior to monotherapies. D'Herelle also knew this already, 100 years ago. Thus, many control experiments need to be performed and are required by the health authorities. The regulations governing use in humans are enormous.

We know a lot about phages in the laboratory, because for 50 years, phages constituted a major field of research for molecular biologists. The contribution of the famous phage specialist Max Delbrück, who introduced the phages as a model system for life (displaying as they do the properties of replication and mutagenesis), deserves the credit for opening up the field of phages. Thus, we know a lot about the molecular mechanisms of phages, but only in the laboratory and not so much in real life or in natural habitats such as the gut. There, everything appears to be rather different. In the laboratory T4 (type 4) phages grow slowly and T7 (type 7) phages grow fast, yet in the gut there is no difference. That probably also depends on the individual gut. Before phages can save our lives, we shall have to learn a lot about their behavior in patients. What we need to learn about is the specificity of the phages for the respective bacteria. In this respect, antibiotics are far superior to phages: they are effective even if the bacteria they are to act against are not known. However, this is changing, not least because of the emerging antibiotic resistance of bacteria, which requires pre-testing of antibiotics to find out which ones are effective in a given situation. In the Spring of 2012, the WHO declared that the end of the era of antibiotics had arrived. Also, the Cold Spring Harbor Laboratory dedicated a whole conference to this topic: a Banbury Conference, a think-tank for the future. Time is running out. In 2016 the first World Phage Therapy Meeting is due to take place in Paris.

Bean sprouts poisoned by phages

I have to warn the reader, because my praise of the phages may be misleading. Like all viruses they can also do harm. But if they do, then we have to blame ourselves, because we are responsible for their pathogenicity and diseases. Here comes one more example: the EHEC food scare in 2011 in Hamburg, Germany, which was caused by bacterial viruses, that

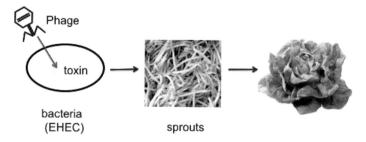

Phage with poisonous gene (toxin)-infected bacteria (EHEC) were used as fertilizer of sprouts, they got ingested with salad by people where it caused a food scandal.

is, by phages. EHEC is a bacterium, enterohaemorrhagic *Escherichia coli*, which is related to harmless gut bacteria. EHEC arose because phages had transferred genes for the Shiga toxin. This causes bloody diarrhea and other complications, some of which are deadly for humans. About 3900 were people were infected, almost one thousand became seriously ill and 35 died. Such bacteria should never end up in salad decoration. They originated from animal feces, and should not have entered the human food chain.

I was sitting at lunch in the Institute for Advanced Study, Princeton, and the colleagues were making fun of the German agencies, which were trying to find the cause of the EHEC epidemic. Cucumbers or tomatoes? No way, they said — bean sprouts! That was their first reaction. They had experienced more than one such outbreak in the US. Everybody knew about it — except me, I had never heard of it. The bean sprouts are grown in warm humid chambers, which are optimum growth conditions not only for the bean sprouts, but also for bacterial growth — especially if the bacteria are derived from the animal feces used as fertilizers. The bacteria were infected with poisonous, toxin-expressing phages. The animal fertilizers may possibly have originated in Egypt. This caused one of Germany's worst food scares in decades. "Call the Robert Koch Institute in Berlin and tell them," urged my colleagues in Princeton. In Berlin they spent weeks sorting out the delivery papers and food-suppliers' bills to restaurants, trying to identify possible causes, because no patient ever remembered having had weeks ago bean sprouts on a salad dish. So interviews with patients did not help to trace the cause of the disease.

Bean sprouts can be easily seen in a salad. But they are difficult to avoid in restaurants! I try not to eat bean sprouts any more.

Which is dirtier, the refrigerator or the toilet?

Never ever should we shake hands when we meet somebody. That should be banned, at least in clinical settings, when one is greeting doctors or medical personnel. However, shaking hands makes patients feel good, so we cannot do away with it — that was the answer in a meeting of the Medical Faculty in Zurich when I brought up this issue. Why not copy the Japanese and bow to your friends? Rubbing noses as a sign of welcome, as the Inuit do, may however be even worse! In almost every clinic there are now hygiene specialists training health-care workers in how to disinfect all possible surfaces — especially hands. They use artificially manipulated green-fluorescent bacteria and shine UV on them to demonstrate how heavily contaminated things are: stethoscopes, door handles, keyboards of cash dispensers, computers, telephones, buttons in elevators, remote controls of TVs. Yet most of the time we do NOT get sick. A normal immune system takes care of these bugs, and even needs them for perpetual training. It is a fact that not the toilet, but the refrigerator, is the place most highly contaminated by microorganisms in a household. Should we get some green-glowing bacteria to find out where cleaning is required? In McDonald's fast-food restaurants the toilets are cleaner than the trays, according to a TV report. The seats in the subway in New York City are among the most heavily contaminated sites, where microbes last for a week. What about airplane seats? I found this hard to believe and wondered whether this report had got it wrong.

We should not get frightened in our everyday life. The rule should be to wash your hands when coming home before cooking, and in hospitals there are special hand-washing rules: use soap, wash five different regions (mainly the fingertips and thumb) for 30 seconds in total. There is an international global hand-wash day, the 15th of October, to increase awareness of hygiene worldwide. Bacteria can be infected with phages carrying the gene for Green Fluorescent Protein (GFP), which can be detected by shining an ultraviolet lamp onto them. This test is superior to the PCR technique, because one can differentiate between living and dead bacteria, which PCR cannot. GFP is only synthesized by living bacteria, and is detected by fluorescence. A phage with the fluorescent protein can also be applied to control food, such as the Swiss speciality Vacherin cheese, to check for contamination. There is even a department at the ETH, Zurich, specializing in this technology. Fish, salmon, shrimps, chickens, hamburgers, or sausages have been tested this way and, if necessary, withdrawn from the market in Switzerland — this has already happened more than once.

Phages can be designed to kill bacteria. They are sprayed onto the chicken eggs that are used for production of vaccine against influenza viruses. Even chickens for food supply are sprayed with phages before being packaged. Phages are applied to potatoes to disinfect the potato bulbs before planting them, which according to Dutch scientists has led to a fivefold higher crop yield. Phage showers are also being applied to grapes for better harvest. In South Korea, even milk is sometimes treated with phages to sterilize it; there, phages are replacing the pasteurization procedure (which involves heating). I cannot read the product description on Korean milk packs, so I do not know whether people are informed about it. Perhaps many consumers would refuse to buy milk if they knew that it had been treated with viruses! Is such treatment good or bad? We should use the phages in animal husbandry to replace antibiotics, which produce too many multiresistant bacteria.

The ETH in Zurich offers programs for start-up companies, "Start your own company", which I would love to do. I have discussed phage applications as a company profile with several colleagues. However, they all argue against it: phages are too difficult to handle and, moreover, one cannot secure patent rights any more, as phages are well-known already — "prior art." However, I think that one can even make money with soap! Perhaps a reader can make use of a business idea of this kind. I envisage phages for eliminating bacterial contamination in intensive-care units in hospitals. To kill bacteria, we should enter into alliances with the phages. Furthermore, they may have a great future in the food industry. The use of phages is not limited by laws on GMO (gene modified organisms), which restrict other gene-modified products. Phages sit on the outside of a product. Eating them is normally not dangerous. And if you still don't want to — just rinse them off!

The "Zurich Case" on fecal transfer

One day a colleague from the University of Zurich came into my office. She wanted to know whether I could find out, during some international meeting, what kind of therapy options there might be against *Clostridium difficile* (*C. diff.*). She was just getting the third round of very high doses of an antibiotic (Vancomycin) against *C. diff.* Whenever the dosing ended, she again suffered from severe diarrhea. About two years earlier, she had been treated by a dentist and incurred a bacterial infection of the jawbone. Antibiotics do not penetrate there easily, and bones have a rather slow metabolism. Accordingly, the therapy was harsh. It required several rounds of various antibiotics at high doses for several months. The result was a ruined gut microflora, where only one bacterium was left and grew: *C. diff.* It sounds ironic, but the only way to get rid of these resistant bacteria — caused by antibiotics — is by treatment with a specific antibiotic again. Only a single effective one is left: Vancomycin.

With this problem in mind I went to Paris to the Institut Pasteur in 2008, where 100 years earlier the phages had been discovered, to attend an international congress on VOM, Viruses of Microbes. Yet there was nothing new, no speaker, no poster mentioned anything new about *C. diff.* — no hope in sight. Then I was also invited to a World Summit of Virology in Busan, Korea. Never had so few participants attended such an event, which was irritating; however, the few participants stuck together more than is usual during meetings, talking in depth with one another. I raised my perpetual question about *C. diff.* therapies during breakfast. A colleague immediately cited an article from *The New York Times* and promised to send me information. The answer was: fecal transfer (FT), that is, stool transplantation.

Back in Zurich I collected reports from the database "PubMed", the world's most important repository of scientific literature in the field of biology, medicine and the like. This therapy has been in use since 1950, and results have been published from more than 200 studies in about 10 countries. Hundreds of patients have received FT by now. Even sick cows were subjected to FT as long as 300 years ago, with feces from healthy cows, a method called "transfaunation". The cows had an opening on the backside for FT — horrifying pictures! And in China a "yellow soup" was once used as therapy — it contained fermented human feces — 3000 years ago. Bedouins eat the feces of their camels as therapy.

FT is performed by means of an enema, a kind of syringe from behind or through a nasal tube into the gut. Stool donors are normally relatives, who supply a spoonful of feces, which is sufficient. It is diluted, filtered and infused. Several precautionary diagnostic tests need to be performed beforehand on the donor's stool in order to avoid transmission of viruses such as hepatitis viruses or HIV. Toxins present in the recipient's stool prove the presence of a *C. diff.* infection. Some groups have developed patients' questionnaires, which can be downloaded from the Internet. The patient in Zurich was immediately willing and eager to get this treatment. She understood that the replacement of her ruined microbes by fresh microbes from healthy stool would at once supply her intestine with thousands of different bacterial types — phages, viruses, and archaea — which alone could never build up from scratch. The results from the Human Microbiome Project, had just appeared (2010), demonstrating that humans have about 1.5 kg bacteria amounting to a total of 10^{14} bacteria in their gut, and more than one thousand different species. We need them for digestion of some of our food. The malfunctioning of this bacterial community causes numerous diseases. The microbes, bacteria, archaea, fungi, and viruses lead a life of peaceful co-existence, not in a permanent "war". Where should the bacteria come from after a devastating antibiotic therapy? Probiotic medications or yoghurt contain 5 to 10 species, which is far from being enough. Colleagues from the Medical Faculty in Zurich were not willing to perform such a therapy. I was not taken seriously and was even told to keep my mouth shut as a non-physician — even though I was a member of the Medical Faculty. But the patient insisted. She would otherwise conduct the procedure on her own at home! That is what she had read. Happily, there are enough doctors in Basle, or Freiburg, or cities as far as in Australia, the US or Scandinavia. So we finally persuaded the Swiss colleagues. Only a week after the treatment the patient rang me and said that she felt healthy — after many years of suffering.

We collected stool samples from the donor and the recipient at the beginning and at several later points in time, after a few months up to five years, to keep a record of changes in the microbial composition by using up-to-date methodology: by sequencing the microbiome. This approach gathers information from all the bacteria and other microorganisms at the same time, without distinguishing between individual ones. The total nucleic acid needs to be extracted from the feces sample, which turned out to be more difficult than anticipated — it was not exactly a standard procedure in a biochemical laboratory! The sequences obtained from the samples can then be compared with known sequences of microorganisms deposited in comprehensive databases. We also looked at the virome — all the viruses at the same time. The amount of data obtained was indeed gigantic. A hard disk was transferred, well-protected in my handbag, between Zurich and Berlin, with hundreds of terabytes of data on it. The largest computers from the ETH and the MPG were occupied for days to process the data. The composition of the gut flora was a mixture of the patient's and the donor's, and the dangerous *C. diff.* had disappeared. Normal gut bacteria from the donor outgrew and displaced the dangerous ones. That is the normal protective behavior of bacteria: the healthy ones win out over the dangerous ones, where "healthy" means "our" bacteria, those that are part of an established ecosystem, though there is no general definition. Healthy bacteria keep us healthy!

We found about 100 bacterial types in each fecal sample, only a tenth as many as I had expected. Then we detected some archaea. The functions of archaea in our intestine are not well understood, but they may be important for digestion, and they have been found before in human feces, indicating that not all archaea are extremophiles. Furthermore, we also identified 22 viruses; 21 of them were phages and one was a kind of *Chlorella virus*, a giant virus from algae. Such a virus we had not expected, and neither did James Van Etten, who studies this type of virus, when we told him over the phone. This was the first time a giant virus had been found in a human fecal sample. This made the patient nervous, but there was no reason to be, as giant viruses are not known to cause diseases. We asked her whether she had any dietary preference, such as for sushi. No, she hadn't; but ordinary tap water in household pipes can also contain this giant virus. We expected about 100 times more phages, not so few. Was that due to sample preparation under non-optimum conditions — using methods that had been designed for larger bacterial genomes and not for small phage genomes? Later, others showed that a healthy microbiome harbors the phages in the integrated state in the bacterial genome and that they do not show up as free phages, except during diseases. Thus, this was perhaps another indicator of the normalized gut bacterial content in the patient. We have followed the patient up for five years now, and she has never experienced any problem, and the antibiotics that she twice had to take for other reasons did not cause her any harm. A systematic trial, including control groups, has been performed and its results were published by Dutch researchers in 2013. It showed that the fecal transfer was even better than in the previous case reports, which makes this therapeutic approach now legitimate.

What else one can expect in the stool may be deduced by studying bats. They are known to harbor many viruses — probably as a consequence of their lifestyle, hanging head-down closely packed from the ceiling of caves, with meter-high piles of feces below. Many of the diseased bats may die; the resistant survivors do not fall ill any more, but they can transmit the viruses. (*Some scientists suspect that bats have different immune systems, which needs to be confirmed.*) Some of the viruses transmitted by bats are pathogenic human viruses such as SARS corona, Ebola, Hanta viruses and 66 different paramyxoviruses such as measles virus and many more.

The colleagues in Zurich now see patients who travel from far off, and they have received a research grant of around 2.3 million Swiss francs to investigate patients with also for other gastrointestinal diseases such as Crohn's disease or inflammatory bowel disease (IBD). The procedure of fecal transfer can even be performed at home, as our patient explained to everybody's surprise in a television show, talking about a bathroom, a kitchen blender and diapers as a filter — tools available in every kitchen or household. However, I would strongly advise interested persons to consult a doctor. Most recently, one can even order pre-tested stool samples on the Internet. The method will become more practical if universal donors can be found — if "one size fits all" then we shall not need individual donors any more. Some start-up companies have also appeared that sell quality-controlled samples. I have not yet heard any reports about their use or success rate. Also, there are gelatin-coated stool pills on the way to the market; they can be swallowed without infusion. At the Sanger Institute in Hinxton, UK, Trevor Lawley is developing the best combination out of 18 bacteria in groups of six as a drug. First results in mice looked promising. The open questions are about the stability of the therapy, its sustainability, and its possible side effects. One case was reported, where a patient gained weight afterwards. Was that due to the FT, or to eating habits, which certainly need to change after such a treatment? One person received an FT but had lesions in the gut; he survived a subsequent sepsis, but such lesions should exclude a patient from treatment. Also a phage therapy has been reported; this contained only phages from a donor's stool, and no bacteria. Yoghurt producers are probably already arming themselves with more complex products as well.

Initially, there were no regulations regarding FT. However, in 2014 the NIH stopped the FT procedure. A patient organization in the US then submitted a petition for continuation, since there was no other cure for a life-threatening infection such as *C. diff*. Against this background the treatment was approved again, but only for patients in whom previous treatments had failed. An investigational new drug (IND) application is still required, and the Swiss regulatory authorities have followed this decision. Yet FT against obesity is not approved — and not sustainable, as it only has a transient effect. This is under investigation for the bowel diseases. The procedure is incredibly simple, and can be repeated if it fails, though failure is rare, with a success rate higher than 90%. It is also cheap, it does not hurt, it can be repeated, and it can be applied in all parts of the world.

In the meantime, I too have had a serious dental infection. Before getting an antibiotic therapy I collected a spoonful of a stool sample at home and put it in a sealed tube into my freezer. Addition of some glycerol would have probably been even better in protecting the sample from damage caused by freezing, but I did not have any at home. My dentist did not want to believe what I had done, but I am sure that one day this will be standard practice. After all, some people even store umbilical cord blood of newborns as a "life insurance" against diseases that they may acquire later in life — often a present given to the infant by far-sighted grandparents. Microbiomes of stool undergo changes with age, however, which has to be remembered.

Infections by *C. diff*. are increasing at a frighteningly fast rate. In Germany there are 150,000 hospital cases of it each year, and about 15,000 deaths, while in the US there are some 500,000 patients, for about 15% of whom (75,000) it proves fatal. What measures are taken in daily clinical routine? None in the case I know of. I happened to overhear an urgent phone call in a hospital late in the evening from their diagnostics department — a patient tested positive with *C. diff*. The patient was the friend I was visiting. Next day I came equipped with coat, gloves and a special soap (because the ethanol-based Sterillium is ineffective in this case). I was not allowed to use any of this, and neither had anybody — personnel or doctors — taken any protective measures. Nobody was even supposed to know: "Do not cause a panic, please! How did you find out, by the way?" That was their only concern! Did the patient get the infection in the hospital — I wonder? Hospitals are the most common site of infection with *C. diff*. I am still affected and depressed by the subsequent death of this close friend. We need better training of doctors and personnel.

What do we have to do? Wash our hands. That is not all, of course, but it is a good start. The hand-wash "rule of five" is: soap the palm, sides, back, fingers and thumb for 10 seconds each, then dry with a clean towel. In 2013 a special Robert Koch Prize for Hospital Hygiene was initiated in Germany. No nurse, however can wash her hands 100 times a day….

How to fight obesity

During a meeting to celebrate the 200th birthday of Charles Darwin I met a former friend, with whom I had studied molecular biology at the University of California in Berkeley, decades earlier. She was chairperson of an important session — a great honor — and she had also published several textbooks. She had been appointed Professor of Biochemistry at Berkeley 40 years ago. Already then in the US there was a gender "quota": a woman had to be appointed before a man could become a faculty member. In Europe nobody had ever heard of that yet. She was badly treated — and put on weight. Now it is difficult for her to leave her apartment with its spectacular view over Central Park in New York City. Boxes of instant food for diets are piled up in her bookshelves. Doctors are sending her home. "Exercise" is the only instruction she gets. She rides a bicycle through New York City. Even as Professor of Biochemistry she had little success in losing weight. But when I invited her out for dinner she ate much more than I had expected; she would feel cold otherwise, she said. Clearly her metabolism was out of control.

Another case concerns a former colleague, the institute's animal keeper in Berlin, who did not fit into airplane seats any more because of his extreme obesity; he could not do up his shoes any longer as they did not close; he had apnea attacks at night, a heart attack, arthritis in his knees because of his weight, and type II diabetes. Then he heard on TV a few years ago about gastric bypass surgery. He was very carefully tested, and was asked whether he understood the consequences, that there was no way back, that he would not be able to eat normally but only in small portions. He agreed and has now been well for several years — without any regrets. He lost 60 kg! He feels like a new person now, and most of his severe problems, including diabetes, disappeared within less than two years. His microbiome has changed, not only in the gut but also in the stomach and even in the esophagus. Are you hungry, I asked him? No. Many people are, however, too afraid to undergo such severe and crippling surgery. Often, more than one operation is required, because the excess skin has to be removed as well. Special surgery tables and lifts are required because of the body weight of the patients — as my sister, an anesthesiologist, told me.

A third case may be worth mentioning: Liping Zhao, from the University of Shanghai, who had studied at Cornell University in the US for several years. While in the US he ate American-style food and gained 30 kg of weight. When he returned home to China, he reverted to a Chinese diet. He then constantly tested the microbiome composition of his digestive tract. Two years were required for his microbiome to adjust back to traditional Chinese food. Within those two years he lost 20 kg and his microbiome changed, 80 types of microbes out of 1500 were different. Most striking was the increase in the bacterial species *Faecalibacterium prausnitzii*, which increased from zero to 14% of his total microbiome, indicating a healthy composition. Less than 5% of this species in the total intestinal bacteria indicates obesity and disease. In this case, there may still have been some residual bacteria left over from his former years with Chinese food, which just had to grow again. This might take longer in other cases, I suppose. Comprehensive data do not yet exist. How long can former bacterial populations survive and then regrow?

The microbiome of obese people does not change quickly; people try hard to change their eating habits, but in spite of these efforts they do not return to normal weight or, if so, then often only transiently. And afterwards a "rebound" effect frequently leads to an especially high increase in body weight. The ups and downs of weight are sometimes called a yo-yo effect.

Therefore, it is even being debated whether one could exchange the intestinal microbes of obese people with that of a normal person by FT. However, any such effect has only proved transient so far. At conferences it has been reported that after eight weeks the original microbiome catches up again. Especially if the gut microbiota is not removed by antibiotics before fecal transfer, it finally wins again. Yet nobody wants to ablate the microbiome completely before a stool replacement. That is only appropriate in life-threatening situations, not for merely losing weight. FT is therefore at present not an approved therapy for obesity.

Recent studies indicate frightening results. There is a point of no return: if the microbiome of an obese person has changed "too much", then reversal is impossible. Beyond this limit, dietary restrictions may not allow a cure or a return to a normal gut microbiome. As mentioned, we have not only our own, human genome, but also, supplementing it, a "second genome" from the bacteria that populate us, contributing millions of genes to our own. What roles do the additional genes play? The gene richness of the microbiome in healthy people is about four to five times higher than that of obese people. This can even be quantified by numbers, the so-called "gene counts". The gene counts of lean and obese people differ significantly. Lean people not only have more bacteria but also a much higher complexity or variety of bacteria in their gut, with gene counts of around 800,000. This can go down to about 200,000 in patients with high body mass. Low gene counts in the microbiome mean that the number of different microorganisms is reduced, that is, the bacterial diversity has declined. Gene richness has become a criterion of good health.

To revert to a normal microbiome is apparently very difficult and cannot be achieved simply by dietary restriction. The microbes become less complex and "lazy" under "luxury" conditions in the guts of obese people. The rich food supply reduces the metabolic rate of the microbes and their diversity. This means stress for the phages and the bacteria, they are lysed by the phages. The number of phages increases accordingly, an indicator of an unhealthy microbiome in parallel with decreased diversity of the bacteria.

Obese people often harbor microbes that are also very efficient at extracting calories from the diet — thereby increasing body weight. Thus the situation of a person who has reached an obese state is very complex and not easily reversible. There seems to be a point of no return! The balance between eubiotics and dysbiotics, good and bad microflora in the gastrointestinal tract, becomes disturbed irreversibly, and results in a sustained alteration of the gut. These new results are rather disillusioning for those who are trying hard to lose weight, they explain many of the failures that people experience and they may lead to new, microbiome-targeting therapies against obesity. At least that is the goal of many researchers, such as Joël Doré, who is one of the leading figures in France on Food and Gut Microbiology at the INRA (Institute for Agricultural Research).

Another surprise coming from the microbiome studies is the question: who dictates what? What controls the regulatory circuit? Does the microbiome impose the demand for food on the person or, conversely, does a person decide freely how much and what to eat? It is most probably an influence of both on each other. This is attributed to the "gut–brain axis", a path of communication between these two organs. The gut has 500 million neurons, compared with the brain which has 85 billion. So there is some relationship and communication. Signals from stimulating factors such as brain-derived neurotrophic factor (BDNF), dopamine, immune stimulators, cytokines and metabolites secreted by intestinal bacteria such as butyric acid, are transmitted between gut and brain. 50% of the pleasure hormone dopamine and 95% of serotonin involved in sleep, memory, depression, and social behavior is produced in the gut. Morphines given against pain are known to influence both the brain and the gut. The gut-brain axis is even reflected in our language: we make decisions based on "gut feeling" and have "butterflies in our stomach" when we fall in love!

The microbes become specialized by the dietary habits of the host, and they dictate what and how much to eat. Does the gut then dictate initially, and later even override the brain? Who is the chief regulator? Gene richness of the microbiome has become a testable predictor of responsiveness to dietary restrictions. Samples for counting microbes and phages can even be taken from saliva, because mouth and gut are connected. There will be consequences on preventive medicine. The bacterium named after the German bacteriologist Otto Prausnitz, the *Faecalibacterium prausnitzii*, mentioned already, is a predictor of health independently of gene richness and dysbiosis (bad microflora). It is a leading component. If it is underrepresented in the gut microbiome, together with low bacterial species richness, then this is a predictor of poor success in surgery of a person with Crohn's disease. Thus, certain bacterial strains are important for health, as noticed early on by Liping Zhao.

The greatest complexity of a microbiome was detected recently in an African tribe living, rather isolated, on whatever food and meat was available from their habitat. Microbiome complexity as a survival strategy — worth thinking about! Then the viruses come into the game too: the reduced complexity of the microbiome in obese people is accompanied by a significant increase in the number of free phages. Bacteria experiencing "stress-full" conditions change from the lysogenic (chronic or persistent) behavior, typical of integrated DNA prophages, to the lytic phase, in which bacteria are lysed and destroyed by replication of the phages. Indeed, obese people with reduced bacterial complexity show increased numbers of free phages in their intestines. Lysis is induced by stress, whatever that might be in the case of obesity; among other things, obesity leads to chronic inflammation in the gut. Then the bacteria and phage interaction system leads to bacterial lysis, to release of large numbers of phage progeny and to a reduction in bacterial diversity. Low phage counts correlate with a healthy microbiome. Today, this can be measured in the saliva, as a rapid diagnostic test. And we concluded that the low virus count we measured in the "Zurich patient" after successful FT was a good sign: her microbiome had become normal.

Smoking, even passive smoking, reduces the gene richness of the intestinal microbiome. Liver cirrhosis, no matter whether caused by a virus or by alcohol or by other factors, influences the complexity of the microbiome. This can even be used for diagnostic purposes, by determining the gene richness of the stool of a patient — a non-invasive procedure. Even ranking of the patient as mildly or severely ill can now be measured and defined by the microbiome. Nowadays, diagnoses even for disorders such as cardiac disease can be made on the basis of the oral microbiota.

I wonder what we exchange while we are kissing someone or, less dramatically, if we shake hands. The latter is well known as a route of infection — but kissing? Kisses indicate brotherhood. Remember the fraternal socialist kiss between Erich Honecker, former East German leader, and Leonid Brezhnev, then leader of the Soviet Union? — It was a real kiss, not only symbolic! Kissing is almost like an equilibration of microbiomes characterizing kinship, I think! Not a very romantic interpretation. How long-lasting that is, may be another question.

There are also genetic factors that contribute to obesity — we are just learning about them. There are genes that contribute to obesity, genes that encode satiety factors. Such a factor signals to the brain that food uptake was sufficient, leading to the feeling of satiety and termination of food uptake. This may take some minutes — so some people are not patient enough to wait for this signal and continue eating. There are leptins or the gene FTO (sounds like fat — easy to remember), or the brain-derived neurotrophic factor, BDNF or transcription factors such as IRX3 (defects in it lead to 30% loss of weight in mice). Up to 20 genes are already on the list, some of which are, however, controversial. They are probably more indicative of a predisposition toward obesity. Thus, for obesity one has to take into account the fact that genetic factors play a role alongside many other factors. This field of research is still in its infancy. It is generally assumed that eating habits need to be changed, as a matter of self-control and discipline only. This is certainly true — but it is also an oversimplification. Consequently, we should be careful before we think that obese people have no one to blame but themselves. A surprising experiment was performed with "gnotobiotic" mice, so-called "germ-free" laboratory mice without a microbiome. Feces of normal and obese humans were transferred to these mice. This resulted in lean or fat mice, according to the donor: feces from obese people led to fat mice and feces from lean people to thin mice. The feces transferred determined the weight of the otherwise identical recipient mice. However, when the lean and fat recipient mice were co-housed in a single cage, the lean mice "infected" the obese ones. "Thin is contagious!" The fat mice became thin — and not the other way round. The reason is probably that the mice feed on each others' feces inside the cage. The microbiome of the lean mice "wins" and dominates over the microbiome of the obese ones. The microbiomes of obese mice are more monotonous, less complex, less profusely replicating, lazier. The microbiota of the lean mice, in contrast, is more virulent, more heterogeneous, of higher diversity and dominate over the degenerate microbiota. Does that only work in mice?

There are names for the "good" and the "bad" bacteria, the *Bacteroides* inside the gut and the *Firmicutes* — we also had to learn that from our microbiome studies with the Zurich patient. The *Bacteroides* (B) are less efficient in exploiting food and less frequent in healthy people, about half as frequent. In contrast, the *Firmicutes* are more frequent in fat people (remember the F) and more efficient in exploiting the calories from food. The reader should also remember the ratio of the two, F smaller than B is better (F < B is better). It sounds like good and bad cholesterol, the ratio of which everybody tries to remember, "his" ratio, which "he" hopes will prevent a heart attack. Now we may soon have monitors or an app on our smart watches about the composition of our microbiome! "How are your *Bacteroides* doing?" may be the question in the future instead of "How are you?" What is the "correct" ratio, though? The good ones may not be good under all circumstances. So everything is more complicated.

Some day there will certainly be extracts or pills of the "good" bacteria. No matter where they come from — feces or laboratory cultures. That would be great, if it can be sustained — perhaps by a daily pill?

Recently it was investigated whether the microbial composition of the gut can contribute to psychiatric disorders. The results indicated that not only phenotypes such as body mass can be transmitted through feces, but also mental characteristics such as courage or "exploratory behavior" and the opposite, anxiety. Mice of one strain (NIH-Swiss type) are bold and those of another strain (BALB/c) are more timid. They can be influenced by their microbiome and their brain biochemistry, as the two communicate through the gut–brain axis. The mice were treated by feces exchange — and their characters changed! The courageous mice became anxious, and vice versa, the anxious mice increased their exploratory behavior. The test is based on a horizontal bar on a stick, like a gibbet in the Middle Ages. Only the courageous mice dared to move horizontally to the tip, then turning around to go back. The others stayed at the bottom. Simultaneously, some neurotransmitters were analyzed. Exploratory behavior correlated with higher expression of BDNF (brain-derived neurotrophic factor). Thus intestinal dybiosis (wrong microbes) can contribute to psychiatric disorders in patients with bowel disorders. The authors excluded inflammation as a cause. They speculate that patients with inflammatory bowel disease (IBD) and unstable bacterial composition and psychiatric diseases might profit from the probiotic *Bifidobacteria*, which could normalize both microbiome and behavior. Thus the "gut feeling" indeed comes from our gut and not from the brain! Such mental consequences of the microbiota composition are quite hard to believe. Recently, I had a norovirus infection, with severe diarrhea, vomiting and loss of fluid. I thought my electrolytes and mineral household were out of balance, and also my microbiome — but my brain? Yes, I had lost all my "exploratory" interests also! It took me quite some time to recover! So I went to the pharmacy of the Pasteur Institute in Paris to ask for *Bifidobacteria* — without the slightest success, no way, but this will come soon, I suppose! (Bioflorin is given in Zurich to patients to normalize the ecosystem of their guts.) Another new result correlated autism with the microbiota, but this was not consistently confirmed.

In general, to stay healthy or slim, or to be courageous or free of mental disorders — all this requires the right food — but what is the right food? It appears that this depends on local and family habits, environmental factors, personal metabolism, and age, so one cannot make predictions as to what the "right" food, the best microbiota might be — we simply do not know in general. What is "healthy food" — is Japanese food healthy for Japanese but not for Europeans? "Healthy" often seems to be the food we know from childhood, which indicates some co-evolution between microbiome and food. During a conference on "Nutrition" as part of a meeting series "The Future of Science" — held at one of the most beautiful locations on earth, the Palazzo Cini on the Venetian island of St. Giorgio in a monastery designed by the great architect Andrea Palladio — we learnt that an Italian or Mediterranean diet is one of the healthiest. This was the result of a nutrition study "Athena" described by Katja Petroni from Milan. This had also been described years before as being the least cancer-inducing diet: pasta, tomatoes, grapes, cheese, olives, red wine — but today young Italians prefer fast food, and gain weight. Do we need personalized nutrition now, or will we need it some day in the future? One experiment in Munich was publicly advertised: people could register and get their food composition analyzed. In no time, thousands of volunteers registered. People want to know what is good for them! And nobody really knows as yet.

"Stay slim" as a guarantee of a long life? That sounds like a safe goal — but, very surprisingly, the opposite also seems to be true: studies with *C. elegans* worms, which are often used as model organisms for aging, exhibited a longer lifespan if they were overfed, a result that one could summarize as "fat lives longer". A friend suggested: "survival of the fattest", which is a bit too drastic! Indeed, the experiment on the longevity of worms indicated that a greater body mass can extend their lifespan! This must be wrong — everybody thought. Perhaps the explanation is that not *any* kind of fat, not animal fat, but fats from maize, plants, olives (*the unsaturated fatty acids*) are allowed, even if they increase the body mass. No bacon for breakfast, as we have all learnt; olive oil on a salad is also not new — so the message is not entirely surprising. However, nonetheless, "fat against aging" — and not only consumption of vegetables with antioxidants — is worth remembering.

Habits and ideals may change. There are traditional food habits and beauty ideals. "Fat is beautiful" was the statement of a colleague about women from Senegal. There, body weight, especially of women, reflects a higher social status. One hundred years ago pale skin was considered beautiful in Germany, reflecting prosperity and also social status, distinct from working-class people who got suntans. The wives of local African tribe leaders are often recognizable by their body mass at meetings on African diseases; I noticed this during international congresses on HIV/AIDS. A conference in Dubai about hepatitis C viruses and cancer led to the conclusion, "we need help with diabetes". The traditional role does not allow women to work; I saw them in the world's biggest shopping mall, busily eating chocolates. So political changes liberating women could reduce obesity. After women start their own professional life their body mass index goes down.

One unexpected aspect of obesity is pointed up by antibiotics. They have been used to increase the body weight of soldiers (!), but also of animals, pigs and chickens. This is now forbidden by the EU because of the possibility that this can lead to antibiotic-resistant bacteria, which could be passed on to other animals or even humans, and also because of the possibility that they influence the consumer's body weight. Yet cows are allowed to be treated with antibiotics. Perhaps they do not get into their milk? In battery farms with 40,000 chickens the sick ones are normally not removed or selected for treatment; instead, the whole stall is fed with antibiotics, which thus may end up in our kitchens. Even worse is the consequence of antibiotic therapies in people, especially young people. Teenagers will often have received dozens of courses of antibiotics against infections — and this is today being seriously discussed as one important contributory factor to the increase in obesity of the younger generation.

Antibiotics were once exploited and given to people to help them gain weight and strength — body mass and muscles — and they were initially welcomed in the 1950s! Since then, over the intervening 60 years, the average American has increased in height by an inch and in weight by about 20 pounds. One-third of Americans are now classified as obese — depending to some degree upon the exact definition, as some people have recently expressed disagreement with the use of the Body Mass Index (BMI) to define lean and obese. In spite of the fact that a transfer of antibiotics from animal meat to its consumers is not likely, antibiotics in animal farming were forbidden in the 1990s by the EU, but still today they are added to the food of tens of thousands of chickens in battery cages even if only a few of the birds are sick. Currently we are discussing in Europe the import of "chlorine chickens" (in the TTIP US-Europe trade agreement). We receive our greatest exposure to antibiotics by the pills we take, rather than the food we eat! American kids are prescribed on average about one course of antibiotics every year. Martin Blaser, who wrote a book *Missing Microbes*, shows how high-calorie diets alone are insufficient to explain the ongoing obesity "epidemic", and that antibiotics could be contributing to it. Obesity is designated as an epidemic! Indeed, there are lots of microbes involved, including viruses. The use of antibiotics has been much more restricted recently, not only to reduce body mass but to fight multi-drug resistance. A Swiss colleague recently concluded that people, mainly recruits for the army are not growing much taller any more as they had done for decades, but from now on are becoming more overweight instead.

The Dutch Famine study

Who could have imagined that researchers would learn a completely new lesson about genetics from the Jewish refugees who went to Holland and suffered there from the Dutch Famine (November 1944 till April 1945) towards the end of World War II? Especially pregnant women were affected by the lack of food. For six months they only received 400 to 800 calories per day: two slices of bread and some potatoes. Women who were pregnant starved during critical periods of organ formation of their embryos, early gestation. Six decades later, their offspring are now being studied. They show a surprising increase in cardiovascular diseases (twice the usual incidence), metabolic diseases, diabetes, arthritis, obstructive airway diseases, and cancer (five times the usual incidence).

The lack of food did not influence the genetics of the embryos, but it did influence their epigenetics. Environmental factors influence the regulation of genes by chemical modifications (such as methylation), not the genes themselves. This was thought never to reach the next generation. Nature has a built-in security check; all epigenetic changes are eradicated from the genes at the very beginning of a new life, when it starts by the fusion of an egg and sperm. Then all temporary programs are set back to zero, just like a computer program when it is given the order to restart. Fertilization of eggs by sperm normally leads to reprogramming and eradication of epigenetic changes, reverting the cell to omnipotent stem cells. This is necessary because after fertilization an omnipotent cell must have the chance to develop into any of the 200 specialized cell types that are in a human body. Thus, epigenetic changes are normally limited to a single generation and are not passed down to the next one. This is a fantastic invention of Nature.

It came as a complete surprise that this precaution of Nature can be broken. The Dutch Famine Study has indicated that some transmission of epigenetic effects from a mother to the next generation is possible — not during fertilization, but later during pregnancy. This has consequences on the whole future life of the offspring. Not only one generation, but three generations are affected: mother, embryo (child) and the germ cells of the embryo (grandchild). All three of them are exposed at the same time to environmental factors such as malnutrition in a pregnant woman. This has been proven by the Dutch Famine Study.

Another event lends support to these results. Stress on pregnant women during the "Quebec Ice Storm" in Canada in 1998 had consequences similar to those of the Dutch Famine on the health of the next generation. About 3 million Quebecers were in darkness for 45 days owing to weather conditions, which constituted a stress situation. Offspring conceived at that time were surveyed later, as teenagers, and were found to have a higher risk of asthma, obesity and diabetes.

Today, headlines are attracting attention: Not only "Fat moms" influencing daughters and granddaughters by what they eat during pregnancy, but also "Dad's diet shapes children's chromatin" with lifelong consequences for their offspring. Fathers may have a different molecular mechanism than mothers: dietary extremes can alter the way genes are packaged in the sperm, shaping the embryo's and child's chromatin, affecting the metabolisms of future offspring for their entire life. Flies are now being used as animal models, to test what happens in male sperm and under conditions of extremely abundant food supply or lack of food, and to see how a high-calorie diet might influence the body mass of the next generation. Studies on flies now support the human paternal studies.

One of the genes influencing obesity in humans is the insulin-like growth factor (IGF-2), which plays a part in growth and development. Its gene was indeed shown to be under-methylated, which results in overexpressed IGF-2. Too much IGF-2 causes late obesity. The authors discuss the possibility that the body tries to store calories as efficiently as possible during starvation. However, once there is enough food this storage mechanism persists, with metabolic problems such as overweight as a result. This type of phenomenon has even been discussed in connection with the Neanderthals, our ancestors who lived about 100,000 years ago and underwent long periods of starvation, against which they were protected by obesity genes.

In mammals, about 1% of genes escape epigenetic reprogramming (*methyl groups survive by a process called "imprinting"*). These epigenetic changes can be passed on to the next generation. Then overweight women can transmit this predisposition to their granddaughters, and grandfathers to their grandsons, for two new generations. How frequently this happens we do not know. A British scientist such as Marcus E. Pembrey have found that a grandfather's starvation as a teenager changed the mortality rate of his grandsons, and a grandmother's that of her granddaughters. Epigenetic changes even before pregnancy or at the time of conception can lead to inheritable epigenetics. Even a father's smoking may influence the next generation. The effect of alcohol on the embryo, which today every pregnant mother is warned of, may also cause epigenetic changes.

The Dutch researchers are continuing with the analysis of microbiomes and nutrition. They started a new huge study, monitoring 140 people for 10 years with 40 million observations, protocols of food intake, health parameters, stool descriptions, etc. The data collected surmounted the capacities of their computers, so that the astrophysicists were called in to help. Those volunteers who drink buttermilk every day have already been correlated with a specific microbiome — not too surprisingly, except that nobody but Dutch people drink buttermilk or even know what that is!

Food, with its influence on epigenetic modification of our genes, has attracted much attention recently and has become a focus of research. One of the results seems to me to be very important: the effect of nutrition on epigenetic changes and on the jumping efficiency of transposable elements. Epigenetic changes in DNA target transposable elements and regulate their jumping frequencies. Thus, there is a link all the way from a dinner table to our genomic stability — to jumping genes. These changes can then result in genetic alterations, stable for generations. There is a lot to be learnt about what to do, whether or how we can influence our genes, and even possibly tame them. How would that work? We only learn about the negative or disease-causing effects. The other ones are not so noticeable, but they are there all the same.

There are even longer-lasting epigenetic effects possible, inheritance for more than three generations. It follows a different mode of action and has already a name: it is defined as "transgenerational" inheritance through many generations. Another name describing this phenomenon is "paramutation". It was already discovered 65 years ago by Alex Brink from Canada, who observed it in maize, at about at the same time as McClintock discovered epigenetics, around 1950 (see Chap. 7).

The phenomenon was demonstrated recently with *C. elegans* worms, which gathered at the site where a sex hormone was concentrated. This type of behavior was inherited for 50 generations. Unexpectedly, even without the sex hormone cue, when the smell was gone, they still gathered at the same site for many generations. No additional exogenous stimulus was required. This experiment was also performed with an ethanol cue alcohol addiction — the worms indeed remembered this cue also for more than 20 generations. How do the worms remember — and is this restricted to mating? Very surprising. Did the information get into the worms' genes? Not as far as we know. The mechanism involves a short piece of RNA, which is inherited through the sperm. (*The RNA is called PIWI-interacting RNA, abbreviated as piRNA, where PIWI is the molecular scissors for the process of gene silencing.*) The stabilization through many generations is attributed to some long-lasting changes in the chromatin.

The way from obesity to long-lasting inheritance, from regulatory RNA to piRNA in germ cells as a transmissible factor — this is going to be a new research area. What kind of future perspectives does all this open up? Do grandfathers influence their grandsons and also their great-grandchildren? Perhaps — but only in respect of body mass or mating behavior? (See Chap.12.)

I have a great-great-grandfather whose opera house became quite well-known (the architect Gottfried Semper), but I do not believe, unfortunately, that he transmitted any or at least noticeable amounts of his epigenetic or paragenetic experience — or his genius — to me! Perhaps only to male offspring? I am afraid so! It would have saved me a lot of trouble, if I had inherited his epigenetic or paragenetic experience to help me cope with some of my Swiss colleagues, which he had experienced also, building the ETH — that could have changed my life!

Elba worms outsource digestion

I never expected to run into anything as strange as the Elba worm. Its gut is turned inside out. Divers collected the Elba worm *Olavius algarvensis,* or *O. algarvensis,* in the shallow waters off the island of Elba. It is white, which normally matches a habitat in the deep sea. Yet it lives near the surface. It does not have a mouth or other openings; it has no digestive tract, no organs for secretion, no organs at all. Is it really a worm — just because it looks like one? It has genitals surrounded by a few strong hairs, and it can replicate. It lives in close symbiosis with bacteria. The gut-like structure is populated on the outside by microorganisms, which help with its digestion. These comprise at least five different bacterial types, a team of helpers, which guarantee food supply, the recycling of debris and waste disposal. Most surprising is the fact that the symbiosis between worm and bacteria metabolizes carbon monoxide and hydrogen sulphite, both toxic compounds. The whole complex moves up and down, activating oxygen-metabolizing bacteria where the water is shallow and sulfur-metabolizing bacteria where it is deep. One produces sugar and the other converts it back. The worm is a transport vehicle driven by the passenger bacteria. It just floats up and down. We too are colonized by bacteria, but the other way round: most of them are inside our gut, but they also help us with

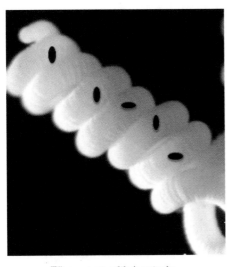

Elba worm with bacteria

Elba worm is populated from the outside with bacteria helping with digestion — just opposite to our guts, where inside and outside are reversed.

digestion of food, such as vitamins, feeding us, and by fulfilling essential functions, such as protecting us against foreign microorganisms. Our skin is also populated by bacteria, but they do not really feed us. The symbiosis of the Elba worm with the bacteria is so close that they cannot be separated physically. Therefore, the total system was sequenced. There were surprisingly many jumping genes, transposons, also the transposase with the "cut-and-paste" mechanism. Possibly this multiple symbiosis is rather recent, and there is still significant trial and error going on. The genome is adjusting to new living conditions. Jumping genes can be dangerous if too many mutations arise, but during adaptation this is useful. This is adaptation in real time — as suggested by Nicole Dubilier from the Marine Microbiology Institute in Bremen. Can this community be a model for other cooperative communities, even human society? This question is raised by the authors. What are its characteristics or analogies? Perhaps job sharing, win-win situations, mutualism, symbiosis, cooperation, interactions, dependences, subcontracts, outsourcing, interactions, no war, no stress, no fights, no animosities? All this is just as true outside of the Elba worm and inside our gut. Is this a model for interaction between people?

Jellyfish use similar interactions: a symbiosis of jellyfish as transport vehicles for algae, not bacteria. There is a jellyfish lake in Palau in the Philippines with millions of golden jellies in a symbiosis with algae, which constitute up to ten percent of the weight of the jellyfish. Algae plus sunshine leads to oxygen and sugar, food for the jellyfish. The jellyfish transport the algae for food production. The huge jellyfish circle around as swarms and attract researchers to study this unique biological system as an early stage of life with little oxygen. We have already mentioned that not only jellyfish, but also viruses and sponges, are light-sensitive and drive their hosts towards the sun. Are such multiple symbiotic systems frequent? Elba worms, jellyfish, sponges — ecosystems like us?

Ecosphere in a glass ball

I have some little housemates, four lively small crabs which inhabit a glass ball. (They remind me of my childhood and of the crabcutter "ERNA" which I once owned in Busum by the North Sea, where I grew up with lots of crab food). The ball fits into my hand — a whole, self-sustaining world, completely closed, with no opening for food or fresh water or air from the outside. There is water inside, also some little white stones, forming the ground. Some black plastic branches simulate plants at the bottom of the sea. In this closed glass ball there is life. This is an aquarium, but it is autonomous and does not need regular cleaning or feeding. However, it does need sunshine. That requires expert dosing: not too much light and not too little. If there is too much light, green algae start covering the inside of the glass wall. For that problem a little magnet is supplied, with a counterpart magnet on the inside, to remove the algal bloom. The glass ball has a lens effect and magnifies the crabs. My first act when I come home is to greet these cute little pets — and has been now for several years.

This is an "ecosphere" as was developed by the American Air and Space Agency, NASA twenty years ago. It was supposed to be taken into orbit, which may have happened, I do not know. On the Internet this is pilloried as cruelty. However, if the crabs swim in circles they may not realize that they are locked in. In this enigmatic ball other things must also be present beside algae and crabs; there must be bacteria, which one cannot see, and of course also viruses, which one has even less chance of seeing.

Light stimulates the metabolism of algae, the crabs feed on the algae and bacteria, and the bacteria digest the debris of the crabs, which again produces nutrients; the crabs and the bacteria produce carbon dioxide, which the algae need for production of oxygen. All clear? What a well-balanced equilibrium this is! Too many algae cannot be digested by the crabs, and if there are not enough algae, then the crabs will starve to death. One day one of the crabs had disappeared. Did it die of hunger, or lack of light? Whatever happened, it was certainly decomposed to molecules, which re-entered the metabolic cycle. This is death — a food chain.

Everywhere on earth this kind of well-balanced cycling between resources and waste is essential. A poorly dressed child on a mountain of garbage in one of the developing countries waves to a camera with a plastic Visa card carrying the name of Mrs. Müller from Wanne-Eickel, thousands of kilometers away! What kind of waste disposal is that? Or: the layer of oil spilled from Deepwater Horizon off the Florida coast in 2010 surprised scientists because of the speed with which bacteria lysed the spill. Again, the bacteria grow as long as there is food, and die out after having finished the cleaning, a self-limiting effect. However, oil clumps sank down — what will they do? Nature recovered with unexpected speed — at least on the surface of the ocean. This was also the case after the reactor incident at

Chernobyl and after the volcanic eruption of Mount St. Helens in the U.S. A healthy ecosystem can apparently survive disasters and start again.

The question concerning the viruses in the glass-ball ecosystem also needs an answer. For sure there are viruses. NASA did not mention them — because nobody knows about them. They also sit inside the oil-metabolizing bacteria, as phages, and also inside my crabs, in their guts and on their outside. That is known. However, they do not cause any diseases as long as everything is well-balanced. Therefore, we do not know about them. By the way, one can order the ecosphere on the Internet.

NASA is sending the tardigrade (*Acutuncus antarcticus*), a tiny slow-moving beetle-like animal with a wrinkled face to Mars. Many people may not have heard of it, as I had not until I visited my former Institute for Extraterrestrial Physics (formerly Pure and Applied Nuclear Physics), where 50 years ago I wrote my diploma thesis about cosmic rays. After World War II, air showers of elementary particles were analyzed there after they had passed through a seven-meter-thick concrete layer of a bunker equivalent to a second atmosphere, to slow down and catch primary particles of higher energy. Now tardigrade is going to take off there and fly into orbit. It had a spectacular resurrection. Some moss from the Arctics was frozen at -20°C in a freezer in Japan 30 years ago in 1983 and when it was recently thawed for research purposes, two of these beetles crawled out, one of them survived and laid eggs. These tiny creatures can survive freezing, thawing, heat and radioactivity, and they can become very old. They have a *Methuselah* gene. They serve as model for how to survive in the orbit. Are they crawling around on the earth also, in Chernobyl or on the volcano Mount St. Helens? Certainly they are a very different model for survival from the glass ball with the crabs. Do they also have LINEs and SINEs in their genomes, or real viruses? Certainly.

Some experiments were not even planned, but occurred by chance: spores that were fortuitously in a box made of expanded polystyrene were flown to the moon and back — and returned alive. I would send out seeds of the ginkgo tree into the orbit, because they survived the atomic bomb in Japan only a few kilometers away from the detonation. And how about viroids in a test-tube, pieces of RNA, probably the first biomolecules? They would be my favorite stowaways, and perhaps they will also come back intact — or start a new type of life elsewhere.

11 Viruses for gene therapy

Viruses against viruses

The principle of gene therapy is very simple: Nature provided us with viruses that can carry oncogenes, and virologists replace the oncogenes by therapeutic genes, whereby the oncogenic viruses cause diseases, and therapeutic viruses cure. Gene therapy follows the principle of all viruses: gene transfer. That is the primary role of viruses during evolution.

Viruses are so successful in causing diseases because they multiply to astronomical numbers within a body — and here comes the problem: therapeutic viruses are not allowed to multiply. So they are much less successful. This is a man-made restriction, because replicating viruses might be dangerous. They could enter germ cells and be transmitted to future generations. We dare not manipulate the genes of our offspring — nobody wants that to happen. It is quite difficult to design viruses that cannot replicate under any possible conditions, because viruses "learn" so quickly, and some may "escape". The viruses have to be made safe. Most cases of gene therapy have been performed with retroviruses, but many other viruses or synthetic viruses can be modified for this purpose. Gene-therapy viruses are complicated "designer viruses".

Infection of a normal cell with a tumor virus transforms a cell into a tumor cell, and the reverse is also possible: a tumor cell can be treated with a virus that carries therapeutic genes, in order to "cure" the cell, making it normal again. The therapeutic gene must be able to compensate for the mutation that caused the malignant phenotype of the tumor cell. Other genetic or metabolic diseases can also be repaired by gene therapy. The difficulty is to find the right gene and the best virus. Each virus has its pros and cons. The retroviruses are still the ones that have been most thoroughly studied as potential therapeutic viruses. Especially in laboratory settings, they are irreplaceable for transferring genes into cells. There is no molecular biology laboratory in the world in which retroviral vectors are not everyday tools for gene transfer. Actually, that is not quite true, because biosafety laboratory level BL2 is required. We had a precious room of that sort, and therefore helped with the synthesis of a great variety of different viruses for colleagues. Retroviruses integrate into the genome of the host. This was considered to be an important advantage, because long-lasting efficacy was expected. Yet it is not always long-lasting, and it can also be dangerous, because integration into the genome is often a genotoxic event, causing local damage to the DNA, which could then induce cancer as a late consequence of such a therapy. Thus treatment of cancer can cause new cancers. Even so, perhaps the patient may gain time or better health through the therapy.

Unexpectedly, animal studies with retroviruses indicated that expression of the therapeutic genes was not always long-lasting, but faded with time, in spite of integration. This is due to a counter-reaction of the organism to a foreign gene, leading to shut-off. The phenomenon is the silencing effect, whereby "foreign" genes are inactivated by an antiviral defense. The defense leads to epigenetic changes for inactivation of genes as described for the color genes of maize or mice. (*It involves methylation of DNA sequences or other procedures for gene inactivation.*) This silencing mechanism is apparently more prominent in adults, because a retroviral gene therapy in children against granulomatosis (a disease of the blood vessels) cured two children, whereas an adult died after two years. These studies are conducted with very few patients and the statistics are not satisfying. But it is possible that the therapeutic effect is longer-lasting in children than in adults.

There are alternatives: instead of replacing diseased genes, gene therapy can be performed by stimulation of the patient's immune system, to help him fight his tumor, to "help people to help themselves". Immune therapies can comprise genes for cytokines, which lead to a systemic — not only a local — effect. We used such a factor, the interleukin-12 gene (also called T-cell-stimulating factor), which activates killer cells and increases the levels of some types of interferon — a whole battery of components to fight against cancer.

The most important restriction on the use of viruses is that they are not allowed to replicate. No virus is allowed to get from one cell to another.

Each cell is a dead-end street for the therapeutic virus: to get in is allowed, but to get out is not. The virus should get into somatic cells — especially cancer or diseased cells — and kill or cure them, but it should never be released into the bloodstream, from where it might, by some accident, find a germ cell and be transmitted to the next generation. The chance of reaching a germ cell increases if the virus replicates to high numbers. This was strictly forbidden, originally by a self-restriction of researchers. Today there are strict rules laid down by legal authorities. Without replication, each tumor cell needs to be hit by one therapeutic virus. That is difficult. A tumor of the size of one cubic cm (one milliliter, or roughly one gram) contains about 10^9 cells. To treat such a tumor one would therefore need at least 10^9 viruses, if each virus finds one cell. However, only about one-hundredth or thousandth of this (20 million viruses in a milliliter) can be produced. So the volume is too great, and then the viruses do not reach all tumor cells. This already explains the limits of gene therapy using viral vectors: viruses have to replicate to be effective. That is exactly the reason why Nature developed them. Furthermore, it is never possible to reach every single cell, there always remains a "minimal residual disease". Even if only a few cells escape, they will often be sufficient in number to grow again, so that the tumor will come back.

To avoid producing replication-competent viruses, the viruses are often cut into pieces, their genes are fragmented into three to four sub-viral genetic pieces instead of one and then packaged into a virus particle. The pieces consist of DNA plasmids, and each of them may code only for a subset of genes. Lentiviruses were constructed to replace the simpler retroviruses. They are relatives of HIV. That seems frightening at first sight — not only to the reader, but also to the regulatory authorities. Lentiviruses have advantages because they are active in non-dividing (such as neuronal) cells. They can be made safe. One way to guarantee this is by deleting some viral genes, or by fragmentation and finding intelligent ways of preventing them from reforming to give an infectious, complete virus. They can be transiently put together and used to infect tumors, each particle hitting only a single cell. The system was developed by the Nobel laureate David Baltimore from the California Institute of Technology, Pasadena, California.

One day a summer student from Berlin surprised me in Zurich with the "Baltimore" virus. He simply wrote a letter to David in the U.S. asking for the therapeutic virus, and then carried it around in his backpack without any concern. I was shocked! We locked it away in a safety compartment as required. Most probably, the letterhead with my name on it had been his free ticket!

Another difficulty for gene therapy is how the therapeutic virus can find the tumor cell in the body. One can try to use a hypodermic needle and inject a suspension directly into, or around, the tumor. This only works with some tumors. We injected malignant melanomas, which form local skin tumors, with therapeutic DNA plasmids. The side effects for the patient are smaller, and a local dose is higher than a systemic dose, which would flood the whole body. This is, however, useless against generalized metastases, which migrate to distant locations. One can try to direct a virus even to a distant tumor or metastatic cell, by using surface markers, a kind of molecular address. This is, again, a principle identical to what Nature developed to aim a virus at its appropriate target cell. HIV finds its target cell by recognizing the receptor on the the surface of lymphocytes by means of the viral surface glycoprotein gp120, a ligand of the receptor, a very restricted entrance ticket for the virus.

Alternatives have been developed. One of them is to isolate tumor cells, treat them outside the body, produce more of the treated or cured cells in the laboratory and reinfuse them into the body of the patient. This "*ex vivo*" method is safer and more effective than the "*in vivo*" therapy.

Integration of a retrovirus is always a danger, because integration sites are random, and their modification could hurt or damage cellular genes. This phenomenon is called "insertional mutagenesis" or "promoter insertion", referring to the strong viral promoter, and it can cause cancer. Thus, cures can cause cancer. This was indeed the case with some immunodeficient children. Alain Fischer, in Paris, treated 18 children with a retrovirus vector, which expressed a growth factor for therapy against the inherited immunodeficiency "ADA"; this deficiency had made them so sensitive to infection that they had to live in sterile tents. The gene therapy cured 16 but led to the dangerous insertional mutagenesis in two of them, whereby the viral promoter induced leukemia in the children (fortunately this could be controlled). Thus, *ex vivo* gene therapy can also be risky.

Another step forward is the design of inducible retrovirus vectors. They can be turned on or off at will by the addition or depletion of regulatory compounds (such as the antibiotic tetracyclin). In that way one can

influence the activity of the virus and even control the dose of the therapeutic gene products. This may reduce toxicity or other side effects. Some
recombinant viruses can destroy themselves by a process called "SIN"
(Self-INactivation), once they have served their purpose.

There are also interferon-inducible genes, which a coworker of mine,
Jovan Pavlovic (a student of Jean Lindenmann, the discoverer of interferon)
developed in the Institute of Medical Virology in Zurich. That became
a focus of intense research of the Volkswagen Foundation for inducible
transgenic mice. In these mice one can turn genes on and off by interferon.
It takes about three years to breed such mice, and these projects were not
very attractive to the coworkers. Meanwhile the new and highly efficient
method CRISPR/Cas9 has been developed. Now the production of such
mice is only a matter of a few weeks.

Understanding the mechanisms of tumor formation is still a research
project, especially in transgenic mice. A question still open is that of how
the oncogene proteins Ras and Myc cooperate to induce a tumor — still
unanswered in 2015 even though these two oncogenes in tandem have
been known as strong causes of cancer for more than three decades. This is
very surprising. These two strong oncogenes do not easily form tumors in
mice — and nobody knows why not. What other factors are missing? This
puts a question mark beside the use of mice as tumor models.

In addition to retroviruses, also adenoviruses are being developed as
gene therapy. They have the advantage that they can infect many different
host-cell types and do not integrate, so that no genotoxic side effects occur.
They can also grow to higher numbers, such as 10^{11} viruses per millilitre,
for better therapeutic efficacy. In Philadelphia in 1999, at the beginning
of the gene-therapy era, a young patient who was not terminally ill was
treated by adenovirus gene therapy. The doctors gave him up to 10^{15} virus
particles, hoping that the more he received the better the result would be.
The opposite happened: the boy, Jesse Gelsinger, died of an overreaction of
his immune system called "cytokine storm" because he already possessed
antibodies against the therapeutic virus. A shock also went through the
whole field of gene therapy, and the principal investigator was relieved of
his duties. The lesson learnt was that patients should be naïve, free of antibodies, for the therapeutic viruses.

Most people have encountered harmless adenovirus infections in their
lifetime, and they harbor antibodies such as those that interfered with that

therapy. Therefore, today one administers adenoviruses that are not of human origin, but rather are from primates, because most people have never encountered these. Recently, Adeno-associated virus (AAV) was used. It was selected because of its ability to integrate very specifically at one site in chromosome 19, so that non-specific integration, with all the risks it entails, would be avoided. Unfortunately, the gene-therapy vector AAV has often lost exactly this advantage, but is further developed as one of the promising vectors.

The newest front-runners in cancer research are the oncolytic viruses, which infect, replicate and lyse only tumor cells (as their name indicates). Normal cells do not allow replication and stay healthy. Furthermore, the killed tumor cells release large amounts of viruses. So this approach overcomes the problem, mentioned above, that is inherent to replication-defective viruses. Oncolytic viruses can be injected, and will find their way to the tumor or metastatic cells throughout the body. Furthermore, the virus itself is a stimulus for the immune system as an additional immune therapy. The field of "virotherapy" has long been known from sporadic cures, in which unspecific viral infections or vaccines showed improvements in unrelated diseases. This therapy was not understood, and it was once abandoned, but it is now returning as a component of therapy with oncolytic viruses.

The use of oncolytic viruses against cancer cells is a very elegant approach, and it is being studied in many laboratories. Frank McCormick developed it about 20 years ago with his company Onyx, using genetically modified adenoviruses which replicate and lyse only tumor cells even though they infect also normal cells but there they cannot replicate, which they would normally do. The molecular trick is that the therapy virus is made sensitive to tumor suppressors in normal cells, which are absent in tumor cells anyway.

Meanwhile, about 20 clinical trials are in progress, even in relatively hopeless cases such as pancreatic and brain tumors. A drug against head-and-neck cancer was given approval in China several years ago. Drugs against malignant melanoma are also close to obtaining approval in 2016. At the German Cancer Research Centre in Heidelberg, too, this approach is under development with measles viruses. Other viruses can also be used, such as parvoviruses, polioviruses or modified herpes viruses — which are oncolytic viruses anyway. Let us hope, wait and see.

The biopharmaceutical company Amgen became rich from selling erythropoietin, which induces the replication of red blood cells and, alongside its legitimate uses, is often misused for doping in sport. Amgen bought Onyx for many billions of dollars and is now the first company to have obtained approval for oncolytic viruses.

Onyx also has Raf kinase inhibitors against colon carcinoma in its pipeline. Very recently, it received a substantial grant to develop inhibitors of the Ras oncogene — an effort that nobody has yet succeeded in, despite enormous expenditure on research and decades of effort by many drug companies. Mutated Ras is one of the most frequent oncogenes in cancer, but it has so many downstream targets that every treatment involving it has so far caused side effects. Combinations of different therapies will be required: double or triple therapies.

"Big Pharma buys Small Biotech". That has become the new strategy. Good for the small companies, which cannot develop drugs for clinical applications and meet all the regulatory demands, with costs amounting to a billion dollars. So Small Biotech prefers to be swallowed by Big Pharma as a lucrative fast-exit strategy. Small companies are more innovative, more flexible, more willing and able to take risks, to lose or win — whereas the big companies have somehow mutated into investment institutions that do not perform much research any more — perhaps just enough to follow new developments on the market, in order to take educated decisions on what to take over!

Frank McCormick was successful more than once. He was the main speaker at the first biotechnology meeting that I organized at the University of Zurich, in 1999, which I had learnt about as a part-time employee at a US biotechnology company during my former research position at a Max Planck Institute in Berlin. Some colleagues in Zurich resented that they were not first! When I set up this meeting, there was still a significant fear of biotech, fear of a commercial influence on science, involving money and patents. Science used to be public, free, with no strings attached, and non-disclosure of results was considered unacceptable. I remember the new last paragraph on the form for grant applications to the Swiss National Fund: How do you plan to put your results into practice? My coworkers did not understand the question, so far away was their thinking on application for future drugs! That had been totally unheard of before in biomedical research, but was normal in chemistry or engineering. Today, every year one can participate in competitions for support of spin-off companies at almost every university. McCormick's biotech company was one of the most successful ones. We learnt from him, but probably not enough to have similar success.

A leaky door, Lipizzan horses, sheikhs — and publish or perish

In Philadelphia, USA, as Chief Scientific Officer (CSO) of a biotech company named Apollon, I guided the development of a vaccine against HIV. It was based on a new principle, on naked DNA plasmids ("naked" means DNA without a coat or packaging into particles). This was tested on HIV-infected volunteers in Zurich and Philadelphia in the 1990s. The vaccine is a DNA plasmid, a circular double-stranded DNA, which transfers genes into muscles. It imitates a viral infection in the patient and leads to the production of HIV antigens and then, as we hoped, to protective antibodies. This was the first of its kind in Europe. It was without any side effects for three years. Safety is extremely important for a vaccine, because it is normally applied to healthy or young people. We injected a patient with a "wrong" HIV vaccine, which did not match the strain of his infection. At first we were shocked. However, a Phase I Clinical Trial is only for testing safety, and not efficacy, so it did not matter — but was it embarrassing enough. Now this vaccine, in combination with another vaccine against HIV, is under investigation in the U.S. Army, because the DNA vaccine is not very effective by itself.

We also applied this approach to cancer patients with malignant melanoma, injecting DNA at the site of the tumor, which expressed an immune stimulator (interleukin-12) as support for the patient's immune system to fight against cancer, malignant melanomas. We determined good tumor reduction in mice — and even in white horses. Two animal models are required by the regulatory authorities. We chose white horses because they get malignant melanomas, even though they do not die from them. The tumors shrank under our treatment, and the animals escaped slaughter. This brought us invitations to the White Horse Breeding Farm Piber in Austria, where the Lipizzan horses of Vienna Castle are bred. But we can only reduce individual tumors and cannot offer systemic cures, which was not considered good enough, sadly so.

To treat patients one needs material made according to Good Manufacturing Practice (GMP), which makes the material very expensive to produce (about one million dollars). This is the bottleneck for new therapies. The company Berna, in Bern, Switzerland, kindly supplied the material but produced it twice — because a door did not close properly the first time. This was a waste of time and money — but one has to follow the rules. Some tumors in the patients were reduced in size, and even a foot was saved from amputation. We went a bit proud of our results to Hoffmann-La Roche, who said they owned the patent rights. They were not interested in any follow-ups, to our disappointment. They had used the compound before, at a thousandfold higher concentration, and found that it killed patients. We knew about the high concentrations and had avoided them. But even so, nobody there wanted to touch it any more.

Later on, some more trouble followed, as a surprise: The clinical investigator had ignored the age limit for the selection of one of the twelve patients — only by a few years, but enough to cause severe trouble with the Swiss authorities. He had decided this in the patient's interest, but against the clinical trial protocol, and was severely reprimanded for doing so. Do not take a drop of tears, saliva, semen or urine from a patient for research without the permission of an ethics committee! The rules are strict. (We recently used semen given *anonymously* from the scientific co-workers to test the effect on a microbicide, which we are trying to develop — it took months to settle this point with the legal authorities and to proceed with the reviewing process of the results for publication.)

Then, a colleague liked our clinical results on skin cancer and offered the therapy to some sheikhs in the Emirates — without informing us, even though we were the inventors. This was brought up publicly and terminated our contact with the colleague — which was very unfortunate, because we could have cured people. Instead, fights were initiated about authorships. A "mediator" (I had not previously known what that is!) was appointed by the university to solve the problem. I buried a beautiful summer under a pile of patients' protocols and data, to combine them all by myself for a publication. This worked out — however, with no consequences for further studies, because meanwhile the regulations by the European authority EMEA had changed and had become too strict and expensive for a research-driven clinical trial such as ours. In the future, all patient analyses will have to be paid for separately — as will the clinical investigators, the study nurses — about 100,000 dollars per case. Hopeless. Also at the Charité in Berlin, we failed to negotiate a cheaper deal. Four years were lost. Since then the precious compound made under expensive GMP conditions, has been stored in a safety refrigerator, because I cannot decide to throw it out.

It may be worth mentioning that the novel vaccines against Ebola and Zika viruses that have to be developed very quickly in 2015 and 2016 are based on this type of DNA vaccines, because all that is needed is knowledge of the sequence of the virus, the ability to synthesize patient-grade DNA and some biotechnological expertise. Except that the vaccines are often rather weakly immunogenic and may need additional immune stimulating components. However, they are safe, stable, convenient to ship, and easy to store in remote places.

And the papers — "Why are they so important?", some readers may ask. They are the basis on which scientific success is judged and applications for more funds are decided upon. Assessment of a paper is often based on highly controversial criteria. The quality of a paper is often guaranteed by the prestige of the journal that accepts it after critical reviewing by anonymous referees, whose names will never be disclosed. (Very few journals are trying to change this type of policy by "open peer review".) This reviewing process often takes a year, by which time coworkers may have left and the focus of the group may have changed — which can also raise problems. Journals are ranked by so-called "impact factor" ranging from 1 to 60, which is the average number of citations per year by other articles, whereby *Nature, Cell, Science* and some medical journals rank the highest. The quality of a paper can be quantified by their "science citation index" of the journal. The journals publish this number to attract scientists to choose the journal for submitting a paper. (The "h" factor reflects the number of quotations an author gets and is less frequently used). These parameters have to be indicated on applications for jobs from postdocs to professors and are the basis for research funds. One is not always so lucky as to receive a huge "core grant" for cancer research from Dr. Mildred Scheel personally. She was the wife of a former German President and founder of the German Cancer Aid, and died of cancer herself, sadly so. How lucky I was to receive a cheque for HIV research from a Swiss banker after a television interview, without writing any grant applications, as once happened to me.

Here are some numbers: Over many years, I wrote up to eight proposals a year, assuming that each one would be funded by 50% to about $200,000 or $300,000. A paper in my field of research costs more than $600,000 and takes two years of intense research on average. Taxpayers are financing a large part of our research — my hobby! I am grateful.

To become a professor one needs about 50 to 100 papers, with several very prestigious ones among them, and a research area with promise for the future. Fifteen papers, with the applicant as first or last author, are needed for a *Habilitation* (qualification in the German-speaking countries to teach at a university). In the past, women needed more papers than men; now they get away with fewer for a professorship! The list of authors is rarely alphabetical, as it often is in physics. The scientist who made the greatest experimental contribution is first and the research-group leader last. Having provided money is no longer a criterion for coauthorship. A boss is no longer automatically the important senior last author. Every author should understand *all* data, not only his contribution, and their correctness — a recent new obligation to reduce scientific fraud! Recently it has become customary to specify everybody's contribution and to declare any commercial interests in a paragraph at the end of the paper. I recommend one more declaration: the authors should confirm that they have dealt with the literature correctly. It is extremely "useful" (but constitutes scientific misconduct) to omit the contribution of a competitor, hoping that the reviewers will not find out — and thus gain an unjustifiably high-ranking publication, as if the results were novel! I have mentioned a few cases that I have experienced this. The acquisition of grant money is important for university appointments. Everybody has to help with financing his research and often also some of the infrastructure of the university. Support from commerical companies always raises some concern. Teaching is also evaluated, often by a vote of the students, and then: good luck. After a successful selection I wish nobody the revenge of the losers — they often have better knowledge of the university structure and local networks and take more time for fights. At least that happened to me: discrimination, false allegations published anonymously, and direct threats, even by top "colleagues" with the biggest prizes. I know what I am talking about, and I wish everybody better luck.

"Mosquito vaccination" against viruses

Gene therapy recently also targets mosquitos. I do not mean the mosquitos that bother us on mild summer evenings at barbecues close to lakes or still water. Those are normally not dangerous — but others are! One of them is the "yellow fever mosquito" (*Aedes aegypti*), which transmits the dengue fever virus (DFV), also other viruses causing West Nile fever virus (WNV), yellow fever and Chikungunya (which is mainly restricted to the French island La Réunion). Very recently, Zika viruses were discovered also to belong to this family of flaviviruses. These mosquitos like it hot and humid, so we are safe in Northern Europe — though, even then, a related striped "tiger mosquito" (*Aedes albopictus*) was recently observed as vector in Frankfurt.

By chance it was noticed that mosquitos infected with bacteria could produce antiviral substances, which can kill viruses. (*The bacteria probably carry phages to produce the toxin.*) This combination has to be introduced into the dangerous mosquito. How? By microinjection — that sounds totally unrealistic. This is indeed not trivial to do for mosquitos, but done! The hope is that some manipulated mosquitos are sufficient and will overgrow the others. However, often such "designer insects" grow more slowly. Nevertheless this project using designer mosquitos containing Wolbachia bacteria against dengue fever virus, is supported by the Bill & Melinda Gates Foundation in Australia, but many other countries also declared interest: Vietnam, Thailand, Indonesia, and Brazil, and the US Army to protect their soldiers who suffered during the Vietnam War. Fifty million people fall ill with the virus every year, 600 in Germany. There is a new trend: the tiger mosquitos are coming from Southern Europe via Switzerland to Germany — further north than ever before; this is attributed to global warming. Four types of dengue fever virus are known, whereby a second infection increases the risk of severe secondary responses such as hemorrhagic fever, which can be lethal. The term "mosquito vaccination" is used, but the name is wrong. This is no vaccination, but an infection with bacteria with phages producing toxins (remember EHEC!). The method is cheap and simple, and no pesticides are needed that could lead to resistant viruses. How successful it will be remains to be seen. Recent results in Australia look promising.

Today people are using gene manipulation to produce sterile male mosquitos, which mate without producing progeny. The population should die out. This is called SIT — sterile insect technique. The sterile males are unfortunately lazier than the wild type! Alternatively, they may even transfer a gene, leading to the name "gene-mosquito". Genes in the males should influence the females so that they only produce males, which would stop progeny production as well. There is also a fungus under investigation for killing the mosquito larvae. Which of these fantastic designs will be applied to the new Zika virus threat besides spraying insecticides?

It is important to note that insecticides are frequently used, and these tend to fail in the long run. The bacterium *Bacillus thuringiensis* (Bt), which carries a DNA plasmid for a toxin, is authorized in the EU for use as a pesticide, because the poison is fully biodegradable — but resistances arise.

The mosquitos house in rainwater reservoirs, water cans, flower vases in cemeteries — and are a test of the new approach for field-wide research in Europe by a "citizens' science project": whereby citizens are encouraged to collect mosquitos and send them in for characterization. Such a project will give us important epidemiological answers about the migration of the mosquitos. Do they get north of Frankfurt?

A similar approach as that in use against the tiger mosquitos is being pursued against the *Anopheles* mosquitos, which transmit malaria.

Recently, vaccination with one of the DFV proteins was tested in mice. The vaccine consisted of a recombinant non-structural protein (NS1) of DFV and protected mice against the virus. The protein was newly identified as causing leakage of blood vessels, and it may have been overlooked for a long time.

Vaccines against tropical diseases may now be produced more rapidly since the latest Ebola outbreak in 2015 and subsequent decisions by the G7 summit in Elmau, Germany, also probably against Zika. DNA vaccines can quickly be synthesized and produced but need supplements.

Viruses for therapy of plants

How can all the people on Earth be fed in the future? One hundred years ago, nitrogen fixation and fertilizers were developed by Fritz Haber in Berlin and today they are essential for food production. We need another discovery for our survival. Are gene-modified organisms (GMOs) the future? We have genetically modified maize and tomatoes, which many European consumers refuse in theory but buy anyway because the products are cheaper. How long can we afford such a rejection? The EU has decided that maize treated with *Bacillus thuringiensis (Bt)* carrying a toxin gene against insects is permitted, and has published that decision. Then there is the herbicide-resistant maize, which is not affected by weed-killers. Also special starch-rich potatoes, not for food but for industrial purposes, are permitted as gene-modified organisms, or recombinant products, the GMOs. More recently, it is no longer up to the EU but to the individual countries to decide about GMOs. Remember, phages sprayed onto potatoes increase the crop fivefold. A future option in food production without GMO? I prefer it.

Gene therapy of plants can be performed with special bacteria, which cause tumors in plants. It is surprising that plants can have tumors — indeed, the bacteria have the meaningful name *Agrobacterium tumefaciens* and produce "crown galls" in plants. The bacteria are infected with a phage-type DNA, the tumor-inducing Ti plasmids. These circular DNA plasmids do not integrate into the bacterial genome, but are transferred by them into the plant cell, where they integrate into the genome and induce tumors. Gene therapists are known to turn things around and to replace tumor genes by therapeutic genes. This is how retroviruses with oncogenes become viruses with therapeutic genes. Analogously, tumor-inducing genes are replaced by therapeutic genes such as toxin-producing genes in Ti plasmids, directed against parasites in plants. Such gene-modified organisms (GMOs) scare many farmers and consumers and keep legal authorities busy to come up with acceptable regulations. About 80% of soya, cotton and maize are GMO plants in the US, with a significant increase in yield. The new trade project TTIP would strongly influence such applications.

Now comes a shock: GMO maize was offered as food to rats — which went on to develop tumors within two years. That was the opposite of what had been expected, even though the Ti plasmids were true to their name and induced tumors in the rats. This result put not only that particular GMO plant on the line, but also the whole system used for gene therapy, the Ti plasmids as carriers of therapeutic genes. Whether, however, humans will have a problem similar to that of the rats cannot be concluded immediately. Perhaps the rats were overfed, overdosed with a single kind of food. This is possible, but it is a common test for drug development: definition of the "maximum tolerated dose". Nobody will eat that much. So, would a bit of the GMO maize not be dangerous then? The maximum dose is tested to include the possible risk for children, elderly, sick or weak people. Could one reduce the dose of therapeutic genes in maize as an alternative? That would be a good question. Maybe the breeders overexpressed the toxin.

The publication about the rats developing maize-induced tumors caused a scandal. Protesters pointed out that the statistics were not good enough and argued that the number of animals tested had been too small and that the data did not sufficiently support the conclusions. The authors were told to withdraw the paper — but they did not. They argued that their manuscript had been given a scientific evaluation and the referees had accepted it for publication. Why then should they withdraw it in retrospect? The discussion continued: objections were raised that obvious mistakes had been overlooked, and finally the publisher declared the already published paper to be withdrawn. What was wrong? The data? The statistics? Were the reviewers incompetent? Indeed, some controls were missing! This was a unique case. What was going on behind the scenes? Was some coercion involved? Was this science or commerce? Consumers become sceptical.

The newest approach to modify genes and to revert diseases is based on the CRISPR/Cas9 method, an antiviral defense system in bacteria against phages. Genes can be modified in a very short time, albeit — interestingly — not by inserting or recombining — a piece of DNA, but by supplying a guide sequence and letting the cell do the correction by reading the "guide" sequence, a procedure defined as "editing". No recombinant technology is involved, as defined by the GMO law. Is a new law required for this novel technology? Until such time as a law comes into effect, a lot of modified organisms can be generated! I think that the authorities have not yet noticed this deficit!

Can viruses save the chestnut trees and bananas?

Chestnut trees are dying of fungi. Can they be saved? Can viruses help in preventing their death? In Neustadt in Holstein, in Northern Germany, the city of my childhood, the hundred-year-old elm trees were all cut down, leaving behind a huge, deserted market-place and a naked red gothic brick church, the size of which nobody had noticed before. Moreover, the alley along the Baltic Sea was bare and had lost all its charm. The Place des Vosges, the Palais Royale, the Avenues Parisienne, parks everywhere — nothing but clear-cutting. The "Napoleon Elm" died in 1977, although at least a slice of it got into a museum in Bonn. Oak trees have disappeared. Will the beeches be next to vanish? How did the end of the elm trees come about? A fungus is introduced by a beetle, the elm-splint beetle. It breaks open the bark and drills its way inside, carrying spores of fungi. The larvae of the beetles then bore mines, channels, into the interior of the trees, where the toxins of the fungus start to work. They plug the sap channels, so that no water or nutrients can ascend, and the trees are strangled. The fungus came at the beginning of the last century, from Asia to the US and then to Europe, and it caused severe damage. Did the fungi hide in patches of water in old car tires during their transport to Europe? There are some anecdotes that suggest this. What can be done, except to fell the trees? The dying of chestnut trees is a similar story. The disease is "chestnut blight". America was hit by it 100 years ago. Chestnut blight has destroyed four billion trees in the US, whole forests. Chestnut trees are the icons of the American east coast. Fan clubs and tree enthusiasts went into action and tried to save the chestnut trees. The fungus, *Cryphonetria parasitica* or for short *C. parasitica*, was introduced from China in 1904. With and without the beetles, even without the "mining", it succeeds in destroying meter-thick trees. Birds, insects, raindrops or the wind can transport the spores to the next tree victim. Everything was tried to stop the disaster: fungicide, sulfur gas, radiation; even some ceremonies were performed. Perhaps the Mayas had exactly this kind of problem in agriculture and sacrificed humans to please their gods.

New superbugs popped up all at a sudden in Austria, Asian ticks, and as a precaution against their spreading, all the maple trees within reach were cut down — I saw this disaster in 2015. The ticks migrated in palettes, packaging material from China that should have been disinfected beforehand but was not. Now dogs are being trained to smell the bugs.

In China, the chestnut trees are resistant. Therefore, their genome was sequenced and the resistance gene was identified. This gene is now transferred by the Ti plasmid to bacteria, *A. tumefaciens*. Also the poison of the toad was cloned into the plasmid and is expected to destroy the fungi; similarly, genes for bitter spinach are on the list for horizontal gene transfer by Ti plasmids.

Swiss scientists have developed a gene therapy for chestnut trees using a virus, called *hypovirus*, which paralyzes the fungi of the trees. This strange virus is relatively unknown. (*It consists of naked double-stranded RNA and has no real coat, only one or two coding genes, and it is specific against fungi.*) Unfortunately, the virus only grows inside the fungi and moves from fungus to fungus. Therefore, the virus-infected fungi are used to treat the trees. There the virus hopefully moves to the naïve fungi to kill them — if possible all of them! Every tree needs individual treatment, a personalized medicine for trees! This is too cumbersome even for Swiss environmentalists! Chinese researchers are trying to save their maize and rape seeds from fungal infection in a similar way, by spreading virus-infected fungi. Roots, too, are being treated.

Yet the safest way is the old-fashioned breeding approach, by crossing the resistant Chinese trees with the endangered species in the US. Six percent of Chinese genes suffice to confer resistance upon the US trees. Even then, they do not look too much like the Chinese trees, and they do not have hairy leaves, which Americans do not like. But this takes time. In Germany, we also see brown chestnut trees and worry about them, but "mining" here is not quite as bad, because only the leaves and not the stems are affected, so only the leaves fall off and the whole tree does not die. The red chestnut trees, which are so typical of Berlin in May around Breitenbach Square, are less endangered. Lucky Berlin: no waiting for gene therapy by Ti plasmids with poison or spinach genes — the only protection available against the fungi is the old-fashioned collection of the leaves in the Fall for immediate burning.

Bananas are very important as food. They are also threatened by the fungus *Sigatoka*. The genome of bananas, which has been sequenced, turns out to be incredibly large, with 32,000 genes — bigger than ours (we have about 20,000 genes) because of many gene duplications. Only a few resistance genes against fungi were detected. The plants in Africa are treated up to 50 times each year with insecticides, but this fails against the disease (*Fusarium oxysporum*). Crossing and breeding is not possible, because of asexual reproduction. Therefore, the bacterium *A. tumefaciens* with Ti plasmids is the only hope, and it is already being tested in Africa. Yet, worldwide, the bacteria are not well accepted in food. A new variation is to use a pistol to inject the Ti plasmid directly into the plants, without using the bacteria for transfer. "Gene guns" are an option — which we once also used to shoot DNA plasmids against cancer into muscles of human patients, with some success. Syringes with needles were more efficient in our hands — and so the expensive gun ended back in its expensive-looking box, lined with red velvet, as delivered by the manufacturer.

Gene food also includes gene-modified fish. The insulin-like growth factor (IGF-1) was inserted into the genome of fish, and now they grow much faster with less food. However, nobody wanted to eat such giant fish — almost just one big tumor, ten times the size of a normal fish. The pharmaceutical industry is now using it to test new therapeutics — if these giant fish are relevant as an animal model at all. "Gene food" is a subject of lively discussions. The German Nobel Prize laureate Christiane Nüsslein-Volhard discussed in public with a professional gourmet chef about gene-modified food. Again in 2016 she argued in favor of gene-modified food as the only way to feed mankind in the future. More than one hundred Nobel laureates signed a petition in favor of GMO food recently. More research is needed though. And transparency. Both of them have published very good cookery books. Gene-modified tomatoes have entered cheap tomato ketchup — and the consumers do not care about scientific controversies, but go for the lower price.

We should remember the phages.

Fungi have sex instead of viruses

What about fungi — do they have viruses? Sure enough, fungi too have viruses! There is no life on our planet without viruses. In Zurich, my little office is located next door to the University's "yeast lab" for hospital diagnostics. There they can diagnose "only" about 1000 yeast types, when in fact there are millions of species. A dermatologist commented "Most of them do not make us ill." Thus, it is the same story all over: fungi, viruses, archaea and bacteria do not primarily cause diseases. Of course, there are some pathogenic fungal infections. These are sometimes unpleasant, and one wants to get rid of them when they occur in the vagina or underneath the toenails. They can also cause severe lung infections, with 1.5 million deaths per year worldwide. Fungi can cause unhealthy mold in bathrooms, cellars, and refrigerators — and the cabins in Yosemite National Park in the US. A measure against them is a sudden draught or movement of fresh air, which is therefore included as a necessary daily duty in flats. Didn't archaeologists die when they opened Egyptian pyramids for the first time after thousands of years — by inhaling spores from fungi? There was never any ventilation there. The spores that fungi use for their propagation can be bad for crops, and caused "St. Anthony's fire" in the Middle Ages. Some medieval paintings on the Isenheim Altarpiece in Colmar show suffering patients praying for help. Fungal contaminations are found on nuts, especially peanuts, on dried fruit and in spices. A combination of viruses and fungi, especially the poisonous aflatoxins from *Aspergillus fumigatus*, can be a dreadful combination for the development of cancer in people.

Fungi also affected politics when Ireland was hit by the "potato blight", leading to the Great Famine in 1845, when two million Irish people emigrated to the US and many others starved to death.

The herpes virus "Epstein–Barr virus" (EBV) leads, together with fungi, to nasopharyngeal cancer in the Far East and, together with hepatitis B virus (HBV), to hepatocellular carcinoma (HCC), a form of liver cancer. Aflatoxins on the top of jam or in bread are unhealthy and these foodstuffs, if affected, should be discarded. Yet we need a subgroup of fungi such as the single-celled baker's yeast, *Saccharomyces (S.) cerevisiae,* for bread and also for beer — and wine-making — and do not forget: fungi supplied us with the antibiotic penicillin.

Fungi are difficult to classify. They are eukaryotic (nucleus-containing) microorganisms, close to the higher organisms, closer to animals than to plants. The great systematic organizer Carl von Linné did not know where to put them in his classification scheme. They are separate from the other life kingdoms of plants, animals, protists and bacteria.

Fungi can be small and unicellular, like yeast, but they can also reach the largest size ever observed for a living organism. The roots of fungi can spread to areas as large as football fields, weighing tons. Fungi are very cooperative, undergoing a symbiosis with almost all plants and helping them to extract nutrients from the soil. Together with viruses they can make plants heat-resistant, so that they can survive longer in droughts. The virus is called *Curvolaria thermal tolerance virus*, and it lives in a triple symbiosis with the fungus and the plant, allowing the plant to survive 65°C. Viruses against global warming should attract attention! This should be advertised; perhaps it can help save fossil water. Farmers in California are using precious fossil water by drilling thousands of meters down into the ground to irrigate their almond trees, even though the water is irreplaceable. Four billion people will experience water shortage soon. I do not buy almonds anymore, because each almond needed 4 liters of water.

Fungi can become very old — up to 2400 years (perhaps inside Egyptian pyramids?). They can form spores, which make them the longest-living species on our planet — and perhaps even elsewhere. They are resistant against radiation, high pressure, vacuum, heat or cold. They may have been on Earth for the last 1.5 billion years. Spores have been enclosed in amber for 30 million years, and recently some of these were brought back to life — if that claim is really true. Some sceptics do not quite believe it, arguing that this could have been a laboratory contamination. Apollo 12, on its journey to the moon, carried by accident some spores in a box made of styrofoam. They returned alive, in spite of exposure to vacuum and cosmic rays, which are known to mutagenize genetic material.

Baker's yeast was already selected for sequencing in 1996, because of its small genome comprising only 13 million nucleotides. The title of the publication was: "Living with 6000 genes" (we have 20,000). They are distributed on 16 chromosomes. Baker's yeast and humans diverged from a common ancestor about one billion years ago, and we share several thousands of similar genes. About 25% of baker's yeast genes are related to human genes. Therefore, yeast was used as a model system for eukaryotes. The oncogene Ras, known from retroviruses, has a relative in yeast and was characterized there, showing some differences. Now yeast genes have been replaced, one by one, by almost 500 human genes within the new field of research that is called synthetic biology. Despite extensive sequence divergence, they were found to fulfil similar (orthologous) functions, fitting into metabolic pathways. That was very surprising. Yeast is going to be a "microlab" to produce human nutrients, drugs such as taxol against breast cancer or artemisinin against malaria, or aromas such as vanillin. In the future one will be able to "humanize" yeast by replacing entire pathways for compound production, analysis and drug screening — using yeast instead of mice or humans! That would be simple and fast. Yeast has numerous advantages over bacteria: it is safer, its components are easier to isolate, and its production can be scaled up.

Viruses of fungi, the *Mycoviruses*, are rather inactive; they never leave their host, they do not carry a protein coat, and they barely replicate, but rather rely on their host's cell division. Free virus particles outside of cells are rare or do not exist at all. They have almost no extracellular life, but persist inside the cells. Fungi can transmit the viruses without replicating. The unicellular baker's yeast is infected by *Narnaviruses*, which — as their name indicates — contain RNA. They are tiny, only half the size of polioviruses or picornaviruses, comprising 2900 nucleotides. (*A narnavirus consists of single-stranded RNA with a stabilizing hairpin-loop structure protected at each end by a cloverleaf structure; it has no coat, and codes for only one multifunctional protein that is responsible for replication, for packaging and for RNA protection, a surprisingly minimalistic multifunctional variant of a virus. Fungi produce thousands of viral RNA copies, that persist inside cells.*)

Yeast cells have it all: sequences from former infections by retroviruses, transposons, Ty elements abbreviated with y for yeast. Ty1 to Ty5 elements are retrotransposons, with Ty3 being very similar to retroviruses (*with LTRs, gag, reverse transcriptase, nuclease and integrase but no coat*), similar to the retrotransposons in insects such as "*gypsy*" in flies, or in plants. The yeast Ty3 elements may have moved into ancestors of their current hosts, insects and plants, by horizontal gene transfer. Ty5 is a model to study the architecture of telomeres with their telomeres at the chromosomal ends related to reverse transcriptases and the mechanism of integration (Chap. 3). By one interpretation the transposable elements in yeast are the precursors of retroviruses — this was proposed in the 1980s by one of the discoverers of reverse transcriptase, Howard Temin. Or the other way round, degenerated retroviruses?

Recently, mobile elements of budding baker's yeast were used for drug discovery against HIV and yielded 60 new lead compounds for potential drugs, as described by Suzanne Sandmeyer from Irvine, California. Interacting proteins, 120 of them, also influence the mobile elements and potentially HIV. They are human host-cell factors, interacting with HIV as potential drug targets. What a surprisingly close relationship between baker's yeast and human cells — close enough for drug screening against HIV! This is certainly something to remember when we are eating bakery products or drinking beer!

A yeast specialist, Joan Curcio from Elmhurst, New York, explained to me why viruses are so rare in yeast cells. Yeast cells (they can be male or female) mate extremely frequently, leading to new progeny and new genetic material. In many other organisms, the task of refreshing gene pools is often fulfilled by viruses. They are the drivers of evolution. Not so in yeast. One can conclude for yeast: their cells have sex more frequently, and they therefore have much fewer viruses.

(Or could one turn this around: fewer viruses allow more sex?) Sexually transmitted diseases are not known for yeast. Another reason to enjoy a beer.

Almost every laboratory carries out the "yeast two-hybrid assay", an elegant approach to search for unknown proteins as interaction partners of target proteins: take a protein as "bait" and select for unknown binding proteins as "prey". If the binding is strong enough, a color signal is activated. This is a safe way to find new interaction partners — if one does not have enough imagination for new research topics. However, if one indeed has fished out a new protein, then that is when the hard work begins: one has to characterize it. Then the editors and reviewers of a manuscript normally ask all conceivable questions and demand all the answers at once. The amount of data grows and, in parallel with it, the numbers of co-workers involved. Ultimately, nobody wants to work on the project any more, when the co-workers who supplied previous results are lining up to be included in the list of authors in a publication. That is when you need to be a diplomatic boss.

Stem cells — almost tumor cells?

A dream of mankind is eternal youth, or to be young again — to get a second chance, to start all over again, to try again. There is a medieval painting in which a bath in a pool is enough to rejuvenate very old people. And a Greek myth even describes the regeneration of organs: Prometheus was punished by the god Zeus, because he wanted to create a human being from clay, and for his hubris he was chained to a rock. Every day an eagle came to feed on his liver, which grew back every night. How did the Greeks know, thousands of years ago, that the liver is indeed the organ most capable of regeneration in humans? In a museum exhibition on stem cells in Zurich, visitors were asked: have you ever thought about how old you are? Your only cells that have the age as on your birth certificate are the eye lenses and a few tendons — cells without blood supply. All the rest is younger and is constantly being renewed: skin cells in a few days (the dead ones are shed by the gram); hair grows one cm per week, fingernails 4 cm in a year, white blood cells live for one day and red blood cells for 100 days; cells in other organs, even the brain, have a turnover. That leads to the question: Am I constantly changing? Who am I? In the *Miracle of the*

Delphic Sibyl, known as the "Theseus' paradox", exactly that question was raised: if all parts of a ship are completely replaced by new ones, is that then still the same ship? How old is it? Likewise, wooden Japanese temples are ancient in structure, but the wood they are built of has been renewed repeatedly for centuries.

Growing hair or fingernails — is that self-renewal? Some stem cells must be involved. Pull out a hair from the skin, and a clump of cells sticks to the bottom of it. Some of them are stem cells which can develop into adult cells. The stem cells are hidden in a growth cone at the bottom of the follicle. During migration to the skin surface, stem cells differentiate by stimuli secreted from their surroundings. Alternatively, human embryos are a source of human embryonic stem (ES) cells. They are the most versatile ones, *totipotent* and can develop into any type of adult cell depending on the stimuli. They could lead to a new person! However, they have to be derived from human pre-implantation embryos, leftovers from *in vitro* fertilization. Their isolation is controversial and in some countries restricted or forbidden. Scientists wanting to work on these cells have to move to other countries where the experiments were allowed. In Germany, the ES cells must have been isolated before the year 2007. *Pluripotent* stem cells develop into several different organs or cell types, such as liver. Then there are the *multipotent* adult stem cells which replace cells in organs for their regeneration, and can grow only into certain cells, they originate from skin as mentioned above, bone marrow, or umbilical cord blood. Heart and brain show not much renewal and have almost no stem cells. Haematopoietic stem cells, which are isolated from bone marrow or blood can be transplanted to patients, which is successful in treating certain leukemias. Stem cells are also being discussed in connection with cancer development: whether they are involved at all, and what part they may play for therapies is open. Instead of talking about "tumor stem cells", names such as "tumor initiator" or "precursor cells" are preferred. A surprising fact is that some parents store cord blood cells from newborn babies for decades as a potential "health insurance" to treat diseases later.

The breakthrough for the study of human stem cells came with a "cocktail". Its ingredients allow one to reprogram a differentiated adult cell to become a stem cell. The dream of mankind thus becomes fulfilled, because this cocktail can turn adult mouse cells back into embryonic cells. The cocktail can "reprogram" the adult cell to "start again". All that is needed is a set of four proteins as components of the cocktail. They are named c-Myc, Oct4, Sox2 and Kif4. They are all transcription factors, and each of them can regulate numerous genes, hundreds in fact. That adult cells can become embryonic again is then proven by implanting the cocktail-treated cells into a mother mouse, whereupon a new mouse develops. This experiment is really fantastic: a fully specialized (differentiated) cell is transformed back into an embryonic, universally pluripotent cell. This is possible, because in every cell, also in an adult cell all the genes are present, but only some of them are actively expressed, thanks to their cellular programs. There are two hundred different programs for the two hundred different cell types in an adult body. The rejuvenating factors must be able to undermine all these programs and to reactivate all genes again, to make the cell pluripotent. Since the four factors "induce" this state, the cells are designated as "induced pluripotent stem cells" or in short "iPS" cells. These cells can then start all over again and can develop into any of the 200 different cell types in the body by specific differentiation factors. An easy method is to place an iPS in the neighborhood of the tissue one wants to generate: for example, existing muscle cells help to make new muscle cells from iPS cells by secreted muscle-specific stimulating factors. This is the basis of regenerative medicine.

The iPS system was developed by the Japanese Nobel laureate Shinya Yamanaka from the University of Kyoto. He systematically tested which factors could perform this reversion, and tested 24 factors in various combinations. The result was the four transcription factors. Retroviral vectors transfer the genes for the factors into human adult skin fibroblasts and activate hundreds of genes. The transcription factor c-Myc stimulates cellular growth, driving the cell through the cell cycle and cell division. However, c-Myc also occurs in many tumor cells, in brain or lung cancers, and was originally discovered as an oncogene of retroviruses as mentioned above. Thus, if it is deregulated, and if too much of it is produced, the cells can become tumor cells. This demonstrates the problem with stem cells. They are in dangerous proximity to cancer cells. A tumor cell cannot be stopped from growing. How can one stop a stem cell? First one wants cells to grow then one wants to terminate it. Tricks are being developed to turn the four factors on or off in a defined way, as required. They also need to be controlled and expressed in the right dose. This will reduce the risk of tumor development.

Yamanaka received the Nobel Prize in 2012 "for the discovery that mature cells can be reprogrammed to become pluripotent", which means the development of iPS cells from adult, differentiated cells. He shared the prize with John Gurdon, who is a generation older, and who in 1962 transplanted a cellular nucleus from the intestines of a frog into a frog egg cell. Out came a real frog. This experiment evoked scepticism, and even some disgust, as I can well remember. Later it became the basis for cloning sheep, cats and dogs. We all know about the cloned sheep Dolly, the most famous sheep in the world, a sheep without a father. She was produced by Ian Wilmut in Scotland in 1996 by isolating a cell from the udder of a six-year-old sheep. He pulled out the cell's nucleus and implanted it into a denucleated egg cell. This became the sheep Dolly. However, her genetic material was already six years old. Therefore, the telomeres at the ends of her chromosomes were already shortened, as in old sheep; the rejuvenation was incomplete. Dolly soon developed age-related diseases such as arthritis, and died in 2003. Her death was due to infection by a sheep virus — a retrovirus closely related to the one that helped to generate the placenta of humans, about 40 million years ago (the *Jaagsiekte Sheep Retrovirus*, JSRV). (I was surprised that this virus is still today around!). After all, Dolly had offspring. One could well have imagined that Ian Wilmut would have joined Yamanaka and Gurdon to Stockholm. Rudolf Jaenisch from the Whitehead Institute in Boston, a friend from Hamburg many years ago, has helped us to understand the reprogramming of embryonic stem cells, work for which he has received many prizes.

The iPS cells make the isolation of human ES cells obsolete. Today, adult somatic cells can be reprogrammed to iPS cells. Many improvements are under investigation.

Stem cells divide by a special mechanism, by asymmetric division, whereby one part of the cell remains attached to a "niche", while the other part is exposed enough to be reached by growth or differentiation factors to produce new cells that are released. One may think of a piece of ice breaking off an iceberg in the Arctic. The cells left behind are characterized by self-renewal and can undergo unlimited numbers of divisions. The stem cells released can be induced to differentiate into one of the body's 200 specialized cells.

Only recently people became aware of the fact, that the germ cells are transmitted for ever and seem to have eternal life. In *C. elegans* worms this is now under investigation. The germ cells must have rejuvenation mechanisms but become mortal, if the animals become infertile by mutations. The goal is now, to find out, which chemical pathways might cause rejuvenation and immortality.

Dreams of future applications stimulate the researchers. One hopes for spare parts, new heart valves, new discs to treat back problems, better hips, younger brains or new skins after burn accidents — all this may be generated from iPS cells. Artificial skin can now be developed already with pigments — but it still takes too long for a victim to wait for his newly grown skin. The immune system would not protest against the "own" cells, and no rejection is to be expected, because the patient itself will be the donor. Moreover, skin- or hair-cell donation does not frighten anybody and nobody would have to wait for traffic accidents and deaths for organ transplantations.

However, before these methods will have matured enough for application to people, one could imagine simpler applications. One could use iPS cells from patients and grow them up under laboratory conditions into specialized cells such as brain, liver, heart or tumor cells and use them to test their response to drugs. One could determine whether a particular patient can tolerate a certain drug and whether it might help him. This may take a few days for laboratory experiments, but it could reduce potential risks and the cost for expensive drugs, some of which against tumors cost hundreds of thousands of dollars. This is one aspect of future personalized medicine. In the US, about 50,000 people die every year as a consequence of inappropriate treatments, instead of being cured. In England, health insurers already request pre-tests before underwriting extremely expensive treatments.

It would only be one step further to inject iPS cells directly into the heart of a patient, where they could become transformed into heart-muscle cells by specific differentiation factors possibly supplied by neighbor cells. Direct injection into the heart, without the long passage through tissue cultures, has already proved successful in mice. The scar tissue after a heart attack is a focus for rejuvenation. Perhaps a syringe with transcription factors such as c-Myc, applied directly to the scar tissue and supported by local differentiation factors might suffice? That would be fantastic. Regenerative medicine is full of promise. Cardiology is a focus — a main cause of death in the Western world. But it will take time.

Together with an American company, Neuronics, we analyzed pluripotent stem cells in a Petri dish and studied the effect of differentiation factors in activating cell-specific programs. We were able to recognize neurons, and we saw cells beat like a heart! But a cure for a tumor was way out of reach. The founder of the company had a brain tumor and was trying to support research for a new therapy. He attributed his tumor to the excessive use of one of the first mobile phones, when they had just entered the market. I remember that the ear got warm during use — which stimulates the growth of cells — possibly promoting cancer.

I remember that fliers were distributed in German trains in the 1960s, advertising "fresh-cell therapies", and the rumor went the rounds that our — then very old — German Chancellor Konrad Adenauer, as well as the Pope, was being treated with rejuvenating cell extracts, obtained from embryonic calvesor sheep. This corresponded perhaps more to a "hormone replacement" treatment than to a stem cell replacement, if bovine and human hormones are sufficiently closely related — which some of them may be. This treatment is no longer allowed, because diseases can be transmitted.

According to recent newspapers treatment with human stem cells was reported to be successful in San Francisco. However, it turned out to be a fraud! The same kind of misconduct happened in Japan. Why were two frauds in a row perpetrated in stem-cell application and research — when it was clear that they would immediately be uncovered, because everybody would try to repeat the observation? Are the falsifiers fanatics, or are they sick? It has happened before. Whole mice were once fraudulently painted with black ink, and in another case spots were painted on autoradiograms with a felt-tip pan. The researchers may end up selling gasoline — as in one case I know of. This was a young postdoc sitting next to me at dinner during a scientific conference in the US. He was the hero of the symposium; his results decorated the front cover of the abstract booklet. He seemed abnormal to me, more ill than a genius. And, indeed, he got his supervisor, a well-known biochemist, into real trouble and quit science. I once spent most of my Christmas vacation stealthily in the laboratory, going through the notebook and protocols of a co-worker, to analyze some data and make sure that they had not been manipulated. No, luckily nothing was wrong. But it happens surprisingly frequently, even at such elite places as the Federal Technical University, the ETH, in Zurich. Computer manipulations

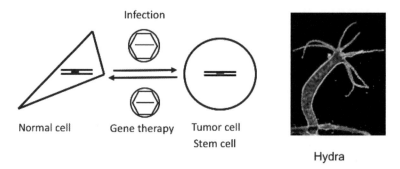

Normal cells are transformed to tumor cells by virus infection which is reverted by viral gene therapy. Tumor cells and stem cells are related, the (proto)-oncogene Myc is one factor, which can cause brain tumors and leads to hydra's new head.

are especially difficult to detect. I would not be able to investigate possible fraud during the Christmas holidays any more, because I do not know how to run some of the fancy computer programs on my own! I decided that I should no longer be the corresponding senior author on papers where I cannot judge the biostatistics myself and cannot guarantee the correctness of every detail. I wonder whether I am the only researcher with this rather novel type of problem.

Hydra's new head

What is so exciting about hydra? Hydra occurs in the Greek legend, where Hercules had to cut off the heads of Hydra as one of ten tasks — but, each time he did so, several new heads regrew.

This ancient legend has some knowledge behind it. One can cut the animal hydra, a small sweet-water polyp, into halves and each half will regrow and replace the lost other half. Even if is passed through a sieve, each of the pieces will grow into a complete, living hydra — a self-organizing system. Hydra can also regenerate some of its nerve cells, even though they are more primitive than ours. How can tentacles regrow, and form a new head or a foot? We are today very interested in replacing or regenerating parts of our body. This is the topic of iPS cells and tissue or organ regeneration. We need new spare parts. Only the liver seems to do well in regeneration and homing of cells. "Homing" means that if one injects a liver cell into a vein in the arm, it will find its way to the liver.

At the MPI at Tübingen, Alfred Gierer celebrated his 80th birthday. His favorite research object is hydra, only some millimeters long, and often sticking to the walls of the aquarium. Hydra dates back 500 million years. A "head activator", a neuropeptide, has been reported to play a part in generating a new head. Activators have some history in Tübingen, where the Spemann–Mangold activator was first discovered in frogs. Gradients of growth factors have been predicted by Gierer and may have inspired the later Nobel Prize laureate, Christiane Nüsslein-Volhard, a former neighbor of Gierer's office in Tübingen, to investigate how a head and a tail form in the larvae of flies. Gradients of factors, morphogens, are the keys. Positive and negative cues regulate differentiation. This probably happens in hydra too, but is this enough? Perhaps hydra never becomes an "adult", but stays in an embryo-like state and therefore can regrow missing parts for ever (my speculation!). Then human adults may just have been unlucky, and have no chance to regrow missing parts.

I could have worked for my PhD with Alfred Gierer, but I did not have enough courage or vision, and ended up next door in the virology department. Gierer later examined my thesis on retroviruses. Retrovirus-like elements play a role in hydra too. Hydra's genome was completely sequenced by Jarrod A. Chapman *et al.* in 2010, together with the Craig Venter Institute in California, with 70 co-authors and 20 research groups. The result justifies including hydra in this book. First of all, the title of the article is worth mentioning: "The Dynamic Genome of Hydra". Hydra's genome is also continually changing, there is intense gene-jumping going on. About 57% of hydra's genome consists of retroelements, which follow the rule "copy-and-paste" requiring a reverse transcriptase helping to enlarge the genome by duplications. The retroelements were not only active in hydra many million years ago, but they are also still active today. Then there are the simpler DNA transposons, which perform "cut-and-paste" and make up about 20% of hydra, also active till today. The reader knows all that by now! New to me was, that one can define three periods when jumping genes were particularly frequent in hydra. These may have corresponded to particular events during the history of the earth. What was going on — eruptions of volcanos, ice ages, meteorites? Then the population development of hydra went through a bottleneck, many hydras died off, and the few left behind passed special traits down to their progeny. Jumping genes replenished the genomes. Their size increased threefold to about one billion base pairs now (one-third of the size of the human genome). Also, hydra was found to be in close contact with a previously unknown bacterium. The researchers could not get rid of it for sequencing. It is in a stable symbiotic relationship. Thus, hydra may be similar in this respect to the Elba worm mentioned earlier. The symbiotic bacteria help digest food for the host. Can one learn from hydra how to grow a new head? Hydra does not have the four transcription factors required for iPS cells to grow, but one of them only, Myc (more precise c-Myc). Yet there are four different Myc proteins — which is fun for a Myc specialist. Can several Myc proteins help hydra to grow new heads — is that hydra's secret? Myc, on the other hand, is an indicator of human brain tumors, which overproduce neuro- or N-Myc by a factor of up to 10,000. Myc stimulates growth and contributes to the most aggressively growing human brain tumors. Thus, there is only a narrow line separating a new head and a mortally ill head — both regulated by Myc. There are other signaling pathways activated in hydra, also known from cancer research, such as the "Wnt" pathway, which plays a role in breast cancer in mice — strangely enough! Such genes are everywhere, in hydra, mice and man — we are all related! (*Wnt is also an oncogene, active as a proto-oncogene in human stem cells; "wingless" and "integrated" are fused to "Wnt".*)

We cannot yet read in the DNA how hydra can grow a new head, but possibly the transcriptomes, the transcripts from all active genes, will tell us soon. There is another unique property in hydra: it does not age, it lives forever. Are hydra cells mainly — or all — stem cells? A "Methuselah gene", a gene for eternal life, has recently been found in hydra by researchers in Kiel. It is a variant of the transcription factor Foxo3, which also exists in the worm *C. elegans* and, much more importantly, in long-lived people. The oldest living animal on our planet is reported to be 10,000 years old. It thus lives ten times longer than the turtles, which many people think are the oldest among animals — Darwin's turtle just died! Yet it is a sponge in the deep sea, it does not metabolize much, it lives a very restricted and monotonous life and consists of around three cell types. These cells must have something in common with stem cells. Such a reduced, boring lifestyle appears to be the trade-off for longevity — is that true in general and a prediction for us?

Just like hydra, planarians recently attracted attention because they can also regrow from pieces. Better known are lizards, which can regrow their tails. The genes involved are not known yet — does Myc help? I bet it does!

And re-growing teeth, as crocodiles and sharks can do, would also be an interesting research project. Sometimes in humans a third tooth can grow by accident. Does it grow out of a stem cell at the tip of the root of the tooth, which may be on the way to becoming a tumor and has, just in time, turned a corner to became a new tooth instead?

A small baby was shown on TV not long ago. It had lost the tip of a finger and was treated with powder derived from the extracellular matrix, a substance that builds up cell-cell contacts. The doctors were very surprised: the fingertip grew up again and even a fingernail reformed. It probably arose in the same way as the first one, by gradients of growth factors or differentiation factors. A critical journalist asked whether the powder was essential. He did not get an answer. However, the regrowing of the finger may have been due to the fact that the baby was still young enough to have embryonic stem cells left. Maybe the cocktail of four transcription factors would have done an even better job than the powder — or was one factor present in the powder? In gene-therapy studies, too, children have done better than adults, in whom the therapy genes were switched off. We need to find out about the differentiation programs and reactivate them, as a possible new start for treating severed fingers in adults. And we need more than just fingers as spare parts. And we may also learn about longevity.

12 Viruses and the future

Synthetic biology — dog or cat out of the test-tube?

Much to his own surprise, in 2002 Eckard Wimmer in Stony Brook, Long Island, made headlines. Why? He had succeeded in the complete synthesis of the poliovirus genome in the laboratory, and had published his work. The synthetic virus was infectious and was able to replicate. However, the public did not regard this as scientific progress but, laymen and scientists alike, they screamed in protest against it. "Such a paper should never have been published!" "This is a cookbook for bioterrorism!" All the reagents could indeed be obtained commercially, without license or restrictions. Even the sequence of the poliovirus was freely available in the Internet. Anybody could now reproduce this synthetic virus — and distribute it as bioweapon — so there was considerable public disquiet. No, this synthesis is certainly not a trivial task. But sometimes there are even expert scientists who turn into bioterrorists, if they are dissatisfied, if they run out of grant money for research, if they get frustrated or go crazy. This was the case once, when a bioterrorist sent out anthrax spores in letters, through the post — he was an anthrax specialist, who was angry.

Poliovirus is small, only 7500 nucleotides long. To make the synthesis simple, Wimmer synthesized the viral genome as DNA, because DNA is more stable and the desired RNA will then be copied from it automatically inside the cell. Meanwhile even the extremely dangerous SARS coronavirus isolated from bats, comprising 29,700 nucleotides, has been completely synthesized. It was subsequently "humanized" by introducing mutations. This isolate can then be transmitted from humans to humans — which is always a fear. Poliovirus is comparatively harmless. What about the reconstructed synthetic retrovirus Phoenix, dating from 35 million years ago? Influenza viruses, too, can be produced synthetically, even faster than in cell culture or by breeding the virus inside eggs, and they can be mutated into more dangerous isolates, even though a synthetic influenza virus as a future vaccine is the goal. For a vaccine one tries to "codon-de-optimize" the virus sequence and insert mutations, 27 of them, to slow down virus replication and reduce risks. The very successful, life-saving polio vaccine harbors only a single mutation, which can back-mutate to the dangerous wild type. Even though mankind has almost got rid of poliovirus infections, this vaccine is regarded as too dangerous according to modern standards. Today it would never be approved by the authorities; the less effective but safer dead-vaccine is still preferred. Bill and Melinda Gates are supporting the novel poliovirus eradication program. Synthetic, artificially mutated virus vaccines do not only apply against polio, but also against influenza or other viruses. The approach is called SAVE, Synthetic Attenuated Virus Engineering.

Synthetic biology is still in its infancy. Synthesis of a real living cell, the smallest minicell, is the highest priority of the visionary Craig Venter, who is always a little ahead of his contemporaries. He started with the smallest bacterium *Mycoplasma (M.) genitalium* and reduced stepwise the total number of genes to identify a minimal set of essential ones. Out of 482 genes he could get away with one hundred fewer, only 382. A third of the reduced genome has no known function; about 206 genes were related between species. These genes regulate replication and survival. Venter and his colleagues synthesized the DNA *in vitro*; however, to everybody's surprise no living bacterium arose. The synthetic DNA had to be inserted into a "degutted" (emptied) bacterial host. That was cheating! What was swimming around inside the bacterial belly? Press reports went so far as to make artificial life or synthetic life out of it, even calling it "playing God". Wrongly so. Venter himself was more realistic and called his effort "transformation": to make a cat out of a dog. It was also termed genome transplantation. He had brought back to life an emptied bacterium, that was all. Nobody, however, had ever had any doubt that this would be possible. Yet a bacterial DNA is about 100 times larger than polio DNA. Therefore, this was not a trivial technical challenge. Venter also had to find tricks to inactivate defense mechanisms in the recipient cell, the restriction endonucleases, which degrade foreign DNA. Venter describes his creature as "first species to have its parents be a computer." Now here comes indeed the first really living synthetic creature: in 2016 Venter succeeded in generating the first minimal bacterial cell from a synthetic genome of 531,000 base pairs with 473 genes (438 protein-coding and 35 RNA genes) which grows in laboratory media. It looks ugly, a clump of balls of different sizes, no symmetry, no self-organization? The authors call its appearance "polymorphic". Again, 149 of the genes have unknown function. It replicates in 3 hours, a bit slower than the original organism, so something important is still missing. The team used a new methodology, "transposon mutagenesis", and that will lead to future results. The authors stress that there is no real "minimal cell", because this depends on the environment. The richer the growth medium the simpler the cell!

Once a functional synthetic cell is made one can do research and study pathways for the production of useful products such as drugs or industrial chemicals with numerous applications. What is life? The chemistry will be solved soon — still many open questions remain.

By the way, the first author of this earthshaking newest paper by Craig Venter, Clyde Hutchison, is 77 years old — which makes me wonder about my obligatory retirement at the age of 65.

In 2014 a groups of 60 undergraduate students enrolled in the the project "Build a Genome" under Jef Boeke at New York University. In a seven-year effort they pieced together small DNA snippets of synthetic DNA to 1,000 base pair stretches. Instead of a bacterial cell with synthetic DNA they used a yeast cell, and inserted novel synthetic genes. Yeast is a eukaryote, closer to human cells and much more complicated than bacteria: it harbors a nucleus with 16 chromosomes, only one of which he produced artificially to replace the natural one with. The cell could grow, which proved that so far no mistake had been made. Only 3% of the yeast DNA was synthetic. Furthermore, about 500 mutations were included, and 50,000 nucleotides out of 317,000 were removed as a safety measure in case some yeast cells are released unintentionally into the environment. First, everything was designed on the computer and then it was generated by synthesis — 10 years of hard work. Thus yeast was "humanized"; human genes are not identical to yeast genes, but they are similar enough to fit into metabolic and signaling pathways. This I find surprising. It is a breakthrough for future studies replacing human cells. This will be useful in the future for screening drugs against human diseases. There are numerous visions.

Also Venter wants to use his tailor-made bacteria for metabolic studies, drug screening, and production of biogas, a cornerstone for bacteria-based energy production. He also envisages their application in producing vaccines or therapeutics, and degrading garbage. For food production also.

Some futuristic visions embrace designer-cells not only for detection, but also to produce new compounds, such as artemisinin against malaria. A new hope is "editing" the genetic modification and cure of defective genes by means of the phage-derived CRISPR/Cas9 system. This could help to fight multiresistant bacteria — but it could also lead to first steps towards "designer babies". Genetic modifications of human embryonic cells have already begun. But to stay away from germ cells and inheritance is still a very strict obligation.

Fancy biological switches, which can turn reactions on or off, depending on stimuli such as light, already exist; that topic is not futuristic enough. In the future we may swallow not preformed pills, but sensors that will determine needs as well as "producing" drugs, supplying therapies on demand depending on the patient's requirements. Insulin could be supplied automatically when, and in the quantities, necessary. Engineers, medical doctors and evolutionary researchers will have to collaborate, because the new concepts are multidisciplinary. Most important is that they have to be compatible with real Nature. Nature has developed certain equilibria over billions of years and will counteract unknown or unfamiliar artificial influences. New patent laws are on the way in which biological building blocks have the "freedom to operate", but not the methods for constructing them. The trend is that natural genes cannot be patented, whereas artificial ones can. Indeed, Timothy Lu, a synthetic biologist at the MIT is constructing viruses or phages with genes, biological machines, designed to destroy bacterial layers, biofilms, which are 500 times more resistant to therapy than free-floating microbes are. They populate hospital equipment, from catheters to waterpipes. He also wants to transfer phages into bacteria or cells to produce substances. This will perhaps one day make insulin injections through a hypodermic needle obsolete, supplanted by automatic supply. Tim Lu also started modulating *Bacteroides* in the gut to sense and respond to stimuli. He uses bacteria as "hard discs" (as journalists put it), memories for information storage. Stimuli lead to mutations, which can be stored for up to 12 days and then be read and quantified. Would they one day signal "stop eating"? That may even be an easy task! Tim Lu designs "living materials that build themselves" — almost a kind of biological 3D printer! And he targets multidrug resistant bacteria — a most urgent need.

We will swallow sensors, nanotools, robots, micromanipulators, and hope that they can do it all — what a vision! Remember, viruses are excellent sensors for stress, such as herpes viruses, which creep through nerve fibers and activate chemical responses such as lip blisters. Thus viruses can do most of this already in special cases. They are good inventors and teachers for us to copy them.

More than 50 years ago the highly original thinker and Nobel Prize laureate Richard Feynman foresaw all this. He stated: "What I cannot create I do not understand"- which will keep Craig Venter busy for some more time. Feynman`s great vision was: "Swallow the doctor"! Are we almost there?

Which came first — the virus or the cell?

This question I raised in 2013 in Davos, Switzerland, during the annual RNA meeting. I talked to colleagues, to many virologists, and to pioneers for RNA, for chemical evolution and so forth. They all gave the same reply, as do almost all textbooks, the reviewers of my articles in international journals, and perhaps also the reader. They all answered immediately: cells came first, as viruses need cells. I was disappointed that the top RNA and ribozyme researchers argued this way, even though ribozymes are the first active biomolecules, which can replicate, evolve, cleave, and join. Ribozymes are the core of ribosomes and thereby key molecules for protein synthesis, and circRNA and recently piRNA are the chief regulators of genes.

Ribozymes are viroids and viroids are viruses. Then viruses came first — or with a bit more modesty: they were around from the very beginning of life until today. We were surrounded by other scientists and young students, who listened curiously, quite surprised. I got carried away, stressing the giant viruses, many of them bigger than bacteria, which can do almost as much as living cells — not quite as much, but they are close! They can harbor other viruses. They can even "almost" see! Viruses bigger than many bacteria and viruses as hosts of viruses — that was unheard of until recently. There is no sharp separation between viruses and cells. What are we discussing, viruses or cells? The students were surprised; they had never heard this. Then there are 10^{33} viruses on our planet, the most successful biological entities on earth, the biggest biological population there is. No living being is free of viruses. Do all organisms contain viruses, even in their genomes? I bet they do! So the viruses must have been there from the beginning; no infections could have achieved that afterwards. Did my "speed-talk" convince the listeners, or Tom Cech? Viruses first? I do not know whether they finally agreed.

There are more arguments for "viruses first". The most recent sequence analyses show that even though some viral genes have similarities with cell genes, there are many more viral genes than cell genes. A front-runner in respect of the role of viruses for the evolution of life was Luis P. Villarreal, Center for Virus Research, Irvine, California. He is describing the dominance of viruses on our planet as "virosphere" and uses the word "replicator" for the proto-RNA, the first replicating RNA biomolecule. Then sequence comparisons, performed by the bioinformatics specialist Eugene V. Koonin, support this concept. The Virus World — more precisely, the RNA Virus World — was first, followed by retroviral elements, and then the DNA viruses arose. The discoverers of the giant viruses put their giant viruses at the bottom of the tree of life, but these viruses are far too big for an origin and this discussion was not generally accepted! Even the discoverer of the phages, Félix d'Herelle, placed the phages — intuitively — at the root of the tree of life. This evolutionary development one could designate as "bottom-up", from simple to complex, from small to big or the "virus-first hypothesis", stating as it does that viruses were first to inhabit the prebiotic world. The main difficulty with this idea is to accept that the first viruses could do without cells, living in niches, Darwin's "warm little pond", or some compartments with inorganic components and perhaps clay as a catalyst. Energy was required, yes, but no protein synthesis yet and no cell because chemical energy could do!

New chemical analyses confirm that the three major building blocks of life, nucleotides, amino acids and lipids could be synthesized in Darwin's "primordial soup" with an energy supply from the surroundings. The chemist John Sutherland, in the U.K., can produce in a "one-pot" synthesis in a single test-tube all three building blocks for life starting only from simple precursors such as HCN, P, H_2S, H_2O, and UV light. This could have been the origin of life. Only when typical cells entered the scene could the viruses change their lifestyle and become intracellular parasites — a well-known evolutionary trend depending on environmental conditions. They interfered with their competitors for cellular resources and established early primitive antiviral defense mechanisms, cellular immunity. Viruses may even have supplied the nucleus for the first eukaryotic cells.

The definition of a virus I have modified and broadened compared with the one given in classical virology textbooks where viruses are parasites and need cells. Instead, I include viroids, poly-DNA-viruses, plasmid DNA, and even prions as being "viruses" or at least virus-like. Then viruses comprise everything from RNA-only, through RNA/DNA, to DNA, with or without proteins, to protein-only — that is my proposal. So, what is a virus? An entity, which can replicate, evolve, and interact. Its information would most often be "genetic", but even structural information could do, therefore I do not include the word genetic in the definition. Today viruses are mostly parasites but primarily not. I find it logical and simple, but not everybody agrees and I do not want to be dogmatic about it. This scenario reflects an evolutionary timeline. Then viruses, bacteria and archaea, symbiotic combinations resulted ultimately in complex mammalian cells — and, highly surprisingly, the bulk of our genomes turned out to be derived from the ancient Virus World.

Another concept describes the other direction: cells first, viruses later: viruses came from cells, big structures became small, "top-down". This is the most frequently accepted concept, viruses as "satellites" which arose by "cell degeneration" described as "escape" theory. Viruses are parts of cells; they split off and "steal" cellular genes. A piece of DNA with some genes got wrapped up in proteins or cell membrane as protection and became a virus. Indeed, some of today's viruses, especially the big DNA viruses of mammals, such as herpes viruses, seem to support this theory. Tumor viruses, too, recruit their oncogenes from the cell. (They also bring them back in, so the genetic information goes back and forth.) The name "reduction hypothesis" describes this phenomenon, by which viruses arise from free living cells.

One problem is that there are not enough genes: the sequence space of cells is small, whereas that of viruses is gigantic. Viruses "know" more than they can have ever learnt from cells. There are too many viral genes without cellular counterparts. Viruses are variable and can increase their sequence complexity during error-prone replication. Yet this does not explain the astronomical abundance of genes. There is much more information in Nature than even the viruses carry around and much less so the cells.

An important question is: where did the first cell come from? The first and simplest minicell, which was produced by C. Venter, is extremely complicated, having several hundreds of genes and more than 200,000 nucleotides. That is huge! Where does all this information come from? The defenders of the "cells-first scenario" do not offer any explanation. The viruses supply offers.

Freeman Dyson suggests two origins of life in his book *The Origins of Life*, combining metabolism and genetic information, machines and programs, hardware and software — closer to the second scenario. Increase of information encoded in genes he attributes to "*Schlamperei*", a German word for messiness. He speaks German, especially words from children's vocabulary, because his wife came from Berlin and he had to cope with a disorderly household that included five daughters and one son. What he means is: Nature is not precise. What is "almost good" can be improved to become better and better. Nature tries everything. This increases genetic diversity.

Max Delbrück also used the word "sloppiness": mistakes, imprecision, trial and error, as a basis for innovation. We sometimes call our sloppiness "creative chaos". Manfred Eigen calls it "quasispecies" in which many species are present at the same time, from fit ones to unfit ones. Perhaps "cloud" would also be a good word. Progress evolves from mistakes. Even Charles Darwin pointed out the importance of mistakes.

And here comes my credo: the error-prone reverse transcriptase is one of the most successful "inventors". It is "sloppy"! It makes "mistakes". With mutation rates of 10 per 10,000 nucleotides per round of replication, because of the poor fidelity of the reverse transcriptase, it leads to new information from the outside — just like newspapers. And it is abundant, present in almost all biological systems, including bacteria and archaea, and yeast (not — or no more? — in some DNA viruses or phages) and it runs the show in plankton. Indeed, this was a recent result from the "Tara Oceans" sampling experiment: 13.5% of plankton proteins are reverse transcriptases (RTs), compared with only about 5% in human genomes, but there too the RT is the most abundant of all proteins. Why is there so much of this specific enzyme in the unicellular eukaryotes in the oceans? The abundant reverse transcriptases originate from numerous retrotransposons. Are so many retrotransposons required in the oceans for innovation and survival? Are there so drastic environmental changes selecting for it?

In a third scenario viruses are responsible for horizontal gene transfer (HGT). Then genes and viruses move horizontally and serve to shuttle genes. We know that there is horizontal gene transfer from viruses to cells and cells to viruses, in both directions. That would correspond to many parallel horizontal lines. That cannot have been the beginning but must have come later.

Formally, another variant is in theory possible: that viruses and cells arose in many parallel vertical lines, evolving in parallel. Then there would have been many origins. Patrick Forterre from Paris suggests three origins, ribosome-containing cells he calls "ribovirocells" which produce proteins and viruses. Viruses invented the DNA in the three domains of life independently. How did the RNA cells emerge?

Did there ever exist a "primary" or "primordial" virus, a "proto-virus", one single ancestral virus as origin of all viruses — similar to Johann W. Goethe's "primary plant"? This question is often raised. It could have been the most primitive RNA in the form of a viroid, but viroids have no specific sequence and RNA of only a 50-nucleotide long RNA allows 4^{50} (or 10^{30}) possibilities, the sum of all possible sequences as a quasispecies. That was not a real primary virus. The sequence space, the sum of all possible sequences, described as the "gene pool", has to this very day not been exploited by the sum of all biological systems on our planet. We have just learnt that RNA can easily be produced in a test-tube, as shown by Sutherland. But this is not a single-sequence-containing initial "pre-pro-proto-type" RNA virus.

An extreme position was assumed in an article in *Nature Reviews Microbiology* in 2009: "Ten reasons to exclude viruses from the tree of life". This article, with its aggressive title, provoked ten vehement counter-reactions. Not a single one of the ten authors agreed with the original article.

Here comes my — personal — final apotheosis on the viruses. Organ players are familiar with the fugue following the prelude in Johann Sebastian Bach's music, which ends in a superposition of all voices and volume with the greatest possible number of pipes, the *tutti*. Here comes my *tutti* — a summary — on the viruses:

Viruses were around as replicating entities such as viroids from the very beginning as first biomolecules. They tried everything and evolved by combinatorial replication and self-interactions, generated diversity as quasispecies. We can align all today's viruses and find most developmental stages. That is most astonishing, because almost all life and species in between have become extinct. Virus genomes cover more information than all cells together; genome structures of viruses are more complex than in any other species, covering everything — RNA and/or DNA, single or double-stranded or both, linear, circular, or fragmented, structured, coding, non-coding. Compared with such a variety of genome structures our double-stranded DNA has a surprisingly simple configuration. DNA is a stable conservation of successful products perpetuated by the viruses as the most imaginative experimentalists. DNA is not a stable "end product", because transposable virus-like elements guarantee that evolution never ends. Also, the modes of replication of viruses are numerous, as well as their regulation. Many mechanisms or molecules were discovered first in viruses and later in cells, such as splicing, reverse transcriptase, scissors, frameshift, etc. The viroids can fulfil as many functions as hundreds of proteins in human cells, such as replication, cleavage, joining, evolution, defense. Viral genomes span more than five orders of magnitude (about 10 to 2.5 million nucleotides), with particle sizes ranging from molecule-sized, uncoated pieces of RNA to giant viruses of bacterial size, with numbers of genes ranging from zero to 2,500. "Zero genes" refer to the viroids, which cannot "yet" code for proteins and can fulfil many functions without the genetic code, suggesting that they existed before the genetic code evolved. They were necessary to develop proteins and the code. The other extreme is the *Pandoravirus dulcis,* with five times more genes than are present in many bacteria. A few curious chimeric viruses have been mentioned, some as strange relics or transition forms and as witnesses of intermediate or developmental stages; these include the very rare retrophages or retroviroids, as well as strange proteins with "forgotten" RNA tails. Viruses are the motors of evolution, the designers of our genome — our body-builders.

Viruses and cells learn from each other, exchange genes, recombine them — in both directions, they co-evolve. In recent years we have more than once been surprised by the observation that there is no war in diverse ecosystems, from the human gut all the way to the microbiome composition of the oceans. I heard a report on the ocean expedition in 2016 by Eric Karsenti. He was surprised by the peaceful co-existence within the microbiome in the oceans — and I was surprised that 73% of the microbiome of oceans and human guts, which includes viruses and phages, are functionally related and that the most abundant protein was the reverse transcriptase.

The error-prone reverse transcriptase of retroviruses or retroelements is probably the most generous supplier of novel information. Recent evidence supports the notion that the retroelements are retroviral precursors. Retroelements can fill 75% of the genome in maize and 50% in rice. With a coat they become mobile and can leave the cell. Reverse transcriptase and RNase H are the most abundant proteins in the world, and there is almost no difference between their respective structures in retroviruses, yeast, bacteria, plankton, plants, mammals, or humans.

A personal surprise was that I witnessed the discovery of the reverse transcriptase and its importance till today during 45 years, and that my research on the cleavage enzyme RNase H has led me from retroviruses through the entire evolution and immunological defense systems. Viruses build the antiviral defense. Integrated retroviruses, DNA proviruses, are immediately available as cellular genes for antiviral defense.

Both use the same toolboxes. Viral components lead to antiviral effects such as RNA interference, silencing, interferon, and immunoglobulins, all with highly related mechanisms of cleavage by the RNase H type. Retroelements are the drivers of evolution. Perhaps even the eukaryotic nuclei originated from viruses. They may have generated the eukaryotes. Finally, the speed of viral replication is a million times faster than any other replication process. There are so many viruses. They are very successful.

So, are viruses alive? Almost. More yes than no! There is no sharp boundary. Here my organ recital ends.

Fast-runners and slow-progressors

The distribution of viruses in various species is strange: RNA and DNA viruses are unevenly distributed, in plants the RNA viruses (mainly single-stranded) dominate, and in the bacterial world of phages almost all contain double-stranded DNA, similar to the giant viruses in algae or the DNA viruses of archaea. Some exceptions to these "rules" need to be admitted.

Humans have many different virus types. Why do plants have RNA viruses and phages have DNA viruses? How do such preferences arise, in the hosts or in the viruses? Nobody could explain this to me, and textbooks of virology do not even discuss this issue. There is a colorful poster showing the taxonomy of viruses and published by the International Committee for the Taxonomy of Viruses (ICTV). It shows an ellipse with segments for the hosts and their viruses, and makes a very nice decoration on the doors in almost every virology institute around the world. I called the producer to find out what the basis was for the organization of the poster — "taste and space" was the answer, the artist's personal taste, beauty, esthetics, not science.

Perhaps this is an explanation: if RNA came first on our planet and then DNA emerged, then the DNA viruses must have overcome the RNA world. The DNA viruses may initially have been RNA viruses and then became DNA viruses, leaving behind the earlier RNA world. The RNA viruses are the "slow progressors" and the DNA viruses the "fast runners". Is there any evidence for this? Perhaps yes: the speed of replication, the doubling time of the DNA viruses and their hosts. Indeed, bacteria and their DNA phages replicate in about 20 minutes to hours, depending on the nutrients and growth conditions. Plants, on the contrary, can take up to 3,500 years for doubling (like the mammoth trees in the US), or even longer (like gingko or Norway spruce, which attain ages of 9,500 years). Plant viruses are predominantly RNA viruses. They can persist in trees, slowly replicating without ever leaving their host. The doubling time of the hosts, numbers of generations of the bacteria in comparison with plants may be up to a million times higher. The growth of bacteria and their viruses is reflected by some numbers: there are 10^{30} bacteria and 10^{33} phages on our planet, all mainly with DNA genomes. Nothing is more successful in our biological world than the DNA viruses and their bacteria in the oceans, in the soil, in our guts, in the ecosystem of the human body, in all ecosystems in general. Even in our genes! The algae-infecting giant viruses also fit into this proposed model based on replication rounds. Algae are plants, so they should have mainly RNA viruses, but many algae are unicellular organisms growing fast and reaching very high numbers, so that after all DNA viruses can be expected to win because of their speed of reproduction. Unicellular plankton (nanoplankton) can divide rapidly and, accordingly, harbor double-stranded DNA giant viruses. Thus, the common denominator for explaining the dominance of RNA and DNA viruses in various species appears to be the speed of replication and the number of divisions.

RNA was first and the more generations passed by, the higher was the chance to become DNA — progress in evolution!

We, as mammals, range somewhere in the middle between the two extremes. We have both RNA and DNA viruses. This is my hypothesis.

There is another proposal to explain why plants did not develop DNA viruses, but mainly RNA viruses. They cannot easily migrate from cell to cell, because DNA is too big and stiff. The transport system in plants would not allow passage (through small connections the plasmodesmata). This was suggested by Eugene Koonin — and I have not checked the diameter. But, couldn't some kind of adaptation have taken place, smaller viruses or broader vessels in plants? Then it might have worked out.

One could find more evidence in support of the assertion that not only the plant viruses, but also the plants themselves, appear to be slow developers. They still have the active "cut-and-paste" DNA transposons, which exist mainly in plants, while they no longer exist in higher organisms. In plants such as maize or rice, 85–90% of the genes are actively jumping genes. This is also true for tulips. These are, however a bit exceptional because of their intense breeding, which may have brought about their extremely large genomes, ten times bigger than the human genome; it is a paradox that genome size does not correlate with complexity. Yet plants are not especially efficient, or fast, in adjusting to new environmental conditions by their lifestyle. Plants do not move around to pick up new information; moreover, their viruses are not very active. They cannot form particles in most cases, and they are not very mobile. In most cases they are chronic, persistent viruses and do not actively replicate. And not even stress can activate the hidden viruses. Therefore, DNA transposons generating two cuts may be advantageous in plants to create new information. Retrotransposons lead to only one cut in the DNA, they are predominant in our genome but also present to a large extent in plants. But the frequency of retro- and DNA transposons is not yet really clear, it is always described within broad ranges and it can vary.

When does jumping of DNA transposons or retrotransposons happen?

In humans, jumping of DNA transposons stopped about 35 million years ago. In our genome only 3% are derived from DNA transposons, and that was a long time ago. We have active retrotransposons, and some are still active — predominantly during human embryogenesis — potentially resulting in cancer, or in engendering geniuses. 20 to 200 people encounter one retrotransposition in their germ cells or embryonic cells in a lifetime. More frequently this occurs in cancer or brain cells.

The DNA-containing viruses from bacteria or algae are, according to the concept presented here, the fast-runners during evolution, and the RNA viruses together with the plants are the slow-comers. Human viruses rank somewhere between these, with an average human doubling time of about 30 years between the two extremes of the bacterial viruses and the plant viruses. Since the beginning of the Christian calendar about fifty human generations have passed. One round of replication in humans corresponds to about a million doublings of bacteria. Compared with the "average" human virus, ranging from the RNA plant viroid-like hepatitis delta virus to large double-stranded DNA viruses such as poxviruses, the phages progressed to the DNA world and probably left the world behind. With many RNA molecules in key positions in our metabolism and replication, we lie somewhere between the RNA and DNA worlds. RNA is still surprisingly important in elementary processes in the DNA world, for example as starters for DNA replication, as sites of protein synthesis, or as safety back-up in gene regulation. Yet the more DNA and proteins arose, the faster our development progressed. RNA splicing can be performed autonomously, solely by ncRNA — but in humans 100 proteins are required for this. However, human spliceosomes are faster and can perform many and more diverse special functions, for example during embryonic development or gene regulation.

The regulatory RNA comprises ribozymes/viroids/circRNA /piwiRNA, which are all functionally and structurally related, and the properties of which have remained almost unchanged from the beginning of the RNA world up to this very day in all our cells — that is something to be really surprised about. The structure is so robust. Perhaps the primordial soup in the early world was the most dangerous milieu and therefore generated this type of molecule — the fittest forever!

The Nobel prize laureate Sidney Altman, who together with Tom Cech discovered catalytic RNA (the ribozyme), protests against the dominance of proteins in today's concept of life and strongly emphasizes the importance of RNA. He defines our world as an "RNA–protein" world, which is less defined by DNA than by RNA and proteins. The DNA is the memory, the preserver — but do not forget, DNA can jump, causing innovations; this needs to be borne in mind.

Monsters in the test-tube

Two extreme developmental processes are surprising: for bacteria and phages the motto is: "small, many, fast, simple" and for mammals and humans: "big, few, slow, complicated". Why did such differences arise — some stay small and the others become big? Are the microorganisms, phages, viruses, or bacteria a kind of optimum, so successful in replication that no increase of complexity takes place? How and why did we complicated humans arise? Why are we not still as small as bacteria and phages? Here is my speculative answer: During evolution there are two forces, the tendency to increase in complexity and another, opposite, tendency toward a decrease in complexity. What decides which gains the upper hand? I searched for an answer to this and finally concluded: the environment.

There are even experiments supporting this: if one allows RNA and replicating enzyme of the phage "Qbeta" to multiply in a test-tube, then what happens will depend on the growth conditions. Under luxury conditions the phages will become smaller, will throw away some of their genes, and replicate faster and faster. How? They incorporate incorrect nucleotides in a sloppy way, picked out of the rich supply pool, and accumulate mistakes. Some genes die out because of too high an error rate. This is the error catastrophe described by Manfred Eigen. Qbeta becomes "minimalized". The number of genes is of course not reduced to zero, but to a minimum, the smallest number that is enough for replication. What is growing faster and faster, and continually becomes smaller, is the so-called "Spiegelman's monster". After a few hundred rounds of replication in the test-tube, the monster had shrunk from about 4,500 nucleotides to about 200 — it had become a dwarf! Most incredible is the fact that during this process the RNA became non-coding RNA. There were no ribosomes to read a code, so the triplet-coding capacity got lost. Evolution, all the way back to the beginning, had gone into reverse, in only a few days. With the non-coding RNA, everything started. Surprising. Replication and evolution of the phage Qbeta in the test-tube was first performed by Sol Spiegelman in New York City about 50 years ago. Manfred Eigen, in Gottingen, pushed the experiment further and simplified it by omitting the RNA and the replication enzyme, the replicase, and supplemented the test-tube only with nucleotides. After some time, RNA formed by itself, replicated and evolved; the nucleotides assembled and formed an RNA strand. From chaos, the first RNA molecule arose. Spiegelman's monster created itself. These are two ingenious and spectacular experiments: Decrease and increase of complexity.

For these experiments the conditions are of central importance. Only under paradisiac growth conditions is gene reduction possible. Growth accelerated further, the more genes were thrown out. "Small is fast". If one extrapolates this concept to the other extreme, "complexity is slow", then one will end up with the most complex system — humans — as the other extreme. "No more Paradise", we are told in the Bible! Humans responded to difficult growth and environmental conditions with increased complexity. "Big is slow". Starvation forces individuals to learn new tricks, to pick up new genes, to specialize, to guarantee survival: "Necessity is the mother of invention". We know that archaea are highly inventive survivors, adjusting to new environmental conditions; however, they need to develop new pathways, which takes time. So they are small *and* slow!

Here I take the liberty of inserting a little anecdote: Spiegelman, who sadly died at an early age, made spectacular presentations at meetings in Cold Spring Harbor about 40 years ago. He was introduced by the chairman Charles Weissmann, his friend from Zurich, both of them specialists in Qbeta phage. Weissmann started with one of his famous jokes: a Rabbi from Warsaw introduced a Rabbi colleague from Lodz by praising his wisdom, education and generosity, when the visitor touched his sleeve and whispered into his ear "Don't forget the modesty". By this allusion, Weissmann was telling his friend Spiegelman not to be immodest — and, quick-witted as he was, Spiegelman replied "The same to you." They could compete for a whole evening to see who knew more jokes — I witnessed that as a young student, and was surprised how competitive scientists are even during a party! Weissmann was sure he would win — which he did. He remembered this and related the Rabbi joke again to me in a recent e-mail, 50 years later!

Spiegelman's "modesty" manifested itself in a paper in the journal *PNAS*, where he tried to prove that retroviruses can transmit cancer through breastfeeding milk — he "cheated" a little, showing graphs with units of "10 counts per minute" instead of the universally accepted convention "cpm" (one count per minute) — hoping that nobody would notice that this made his result look ten times better than it was. His result was wrong anyway. So much for the characterization of two very distinguished and creative scientists.

Writing about the decrease of complexity of systems under conditions of abundant supply reminds me of the newest data on the microbiome in our guts, where similar observations were made in connection with obesity. Rich food leads to exactly the same phenomenon, reduction of microbial complexity: the gene counts go down. Conversely, a low food supply, a restricted diet, leads to high microbial complexity. This complexity, the number of gene counts, is almost a measure of healthy microbiomes now. An isolated tribe of native Africans revealed an unexpectedly high microbial intestinal diversity — but they also had a shortage of food and needed to eat a highly varied diet. How? The loss of complexity is regulated by viruses! They exit from integration during obesity by losing repressors and lyse the cells. In the oceans, too, the local population density of the virome correlates with high diversity, perhaps due to sparse food supply?

Let us return to replication rates: humans seem to be replicating rather successfully, having reached a world population of about 7.4 billion in 2016. However, this is very few compared with the astronomical numbers of viruses and bacteria. The microorganisms will win the race for survival, and they will still exist when humans have disappeared. Nothing is more effective in the biological world than the viruses and the bacteria — a success story of viruses.

Gene reduction or gain is often achieved by symbioses. Giant viruses growing in reaction vessels show a surprisingly fast loss of genes — perhaps an interesting model system for gene reduction in a rich growth medium? Simplifications often result in specializations and peak performance. A familiar and prominent example is provided by our mitochondria, which specialized in the energy production of the cell, the powerhouse, after delegating 90% of their genes to the cellular nucleus. Or think of geniuses, who are highly specialized yet often fail in everyday life.

How big can we grow? What is the opposite extreme to the smallest living cell — the biggest cell or organism? There are only speculations about limiting factors for size. Amoebae are among the biggest cells. There are limits such as the distance that has to be traversed by signal-transducing molecules, the stability of bones and their scaffolding function, design, geometry, form versus function, and natural enemies. Galileo Galilei speculated in his famous book *Dialogues*, which he wrote under house arrest in 1638, about what places limits upon form and size — "gravity" was his answer. Bigger animals need relatively thicker bones. Trees would grow much taller if gravity was lower. On Mars gravity is one-third of that on Earth, so would this permit larger trees or animals? Mass increases by a cubic power law, height by a square law. Therefore, ultimately the weight of the biggest animals would break their legs and they would collapse. Has anybody calculated that for humans? Why could stronger bones not have co-developed by evolution? Were the bones of dinosaurs stronger? This is a serious question, but only one out of many.

Formation of compartments is an important element for growth and could overcome some of those limits. Size was replaced during evolution by multi-cellularity. My new big suitcase has many compartments, and I use bags as well to organize the chaos. Are amoebae the biggest cells? It seems so — and at the same time they harbor the biggest viruses we know of. Is that obvious? And these giant viruses in turn host smaller viruses! What are the smallest and biggest living beings — bacteria, nano-plankton and dinosaurs? Elephants became smaller on the island of Sicily when food was becoming scarce because of the rising Mediterranean Sea level. Can one reverse that? More food makes bigger animals, as can be concluded from the biggest dinosaur ever found, now being exhibited in Berlin, the size of 14 big elephants! Yes, indeed there is this theory that the dinosaurs became so big because they were such good food exploiters. Then their size may also have been determined by genetics. Their heads were not so heavy, allowing them to reach high up. This is not my speculation but a serious theory. True? Giraffes are big as well!

Is there an unlimited size with sufficient food? Humans grew within a few decades by about 10 cm. What are the determining factors for our size? Recent data from Zurich University suggest that our growth in height slows down and turns to overweight as tested in recruits. Absence of jumping genes limits growth, as has been suggested by a recent theory. This means lack of innovation and slow adjustment. Two forces counteract one another for optimization: too much precision is not innovative enough, but sloppiness or too many mistakes leads to malfunctioning. There are two extremes, with a golden mean in between: Too many errors will make reproduction too imprecise, with fatal outcome, and too much fidelity will be too rigid, will fail to allow enough innovation and will cause death because of lack of adjustment. Transposon activity with jumping genes is influenced by epigenetics, by environmental factors, even by nutrition. Thus under limiting growth conditions we may activate transposons and invent survival strategies. This is a good and promising prognosis and should help to allay fears about the future. Other limits of growth are rather more precisely predictable: the maximum age of humans is 120 years, and the maximum world population is 12 billion. Who knows?

The end of dinosaurs came through the Yucatán meteorite, another type of catastrophe. Only small animals survived. Aha — big is risky. We are big!

Lucky so far — but what about the end of the world?

Who makes predictions about the future? The church? Perhaps politicians? Then U.S. space organization NASA certainly does! NASA took the initiative and arranged — as early as in 1982 — a congress about dangerous events threatening our world through planets and asteroids, and how we might be able to defend ourselves. Almost nobody noticed it or took it seriously. The calculation showed that we are surrounded by half a million asteroids, 8,000 of them in dangerous proximity to the Earth and 70 threatening us each month — the Near Earth Objects, NEOs. Most of the asteroids have a diameter of only 10–100 m. In 1908 a meteorite 40 m in diameter exploded just outside the earth and destroyed through the pressure wave an area 40 km in diameter, including 80 million trees, in Siberia. In 1913 a meteorite injured about 120 people in Timbuktu. An asteroid hit Russia on February 15, 2013; it had a diameter of 18 m and a speed of 70,000 km per hour. It exploded 20 km above the Earth. Only a few windows broke. Astronauts recently detected a crater 30 km in diameter in South Africa, which had not been noticed before. NASA is worried about a planet approaching the earth, because we shall only know about it at short notice, ten years before the collision. Then we have to hurry to save our planet.

There are some scenarios about our future. We could try to park the Earth somewhere else to avoid planets threatening us. With 10^{15} tons of explosives the earth could be moved like a spaceship; one could shift the orbit of the earth by 500,000 km, and a meteorite would then miss us. We may already have enough atomic bombs to do that. The Swiss astrophysicist Fritz Zwicky had the vision 50 years ago of moving our whole solar system by using bombs. Freeman Dyson, too, developed a plan for a spaceship he called *Orion*, as big as an ocean steamer and housing hundreds of people.

How seriously is this to be taken if such geniuses make pronouncements about it? Dyson, after all, did the calculations for quantum electrodynamics (QED), which helped others to the Nobel Prize. Now President Obama wants to build an *Orion*, although Dyson insists that it is not like the one he designed. And Zwicky discovered hundreds of unknown supernovae and calculated that there was a deficit of matter — which is still missing, the dark matter. An astrophysicist in Zurich, Ben Moore, suggests that we should migrate to another planet. I do not believe a word of it, but perhaps he is already practicing for the move by organizing the 'love parade" in the streets of Zurich.

The former NASA astronaut Ed Lu created a foundation called B16 (reminiscent of Antoine de Saint-Exupéry's *The Little Prince*) as a system for early detection of asteroids. "Astropyhs", 270 m in diameter, will hit the earth in 2020. Recently a meteorite 15 km in diameter just missed the earth — it would have been worse than the meteorite 10 km in diameter that hit Yucatán 65 million years ago and destroyed about 75% or more of all living species. That did not extinguish all life, as small animals survived — and ultimately we developed from them. There were five mass extinctions, called the "Big Five" starting 540 million years ago caused by supereruption of volcanoes. The most "recent" one was 71,000 years ago, the Toba catastrophe in Indonesia. By some theory mankind went through a bottleneck and the few thousand survivors led to a rather homogeneous Indian population. More than once humans barely survived.

There is a research project in progress to find out how life could go extinct on the earth. The Green Lake near New York is one of the models — the name is sheer sarcasm. No life is left there, no oxygen, a toxic stinking broth, which people can only approach to analyze if they are wearing safety overalls. Lack of oxygen killed the fish in the lakes of the former East Germany, with frightening-looking white fish floating belly-up. However, surprisingly enough, the regeneration of these lakes took only five years. Extremely slow aerators were enough to regenerate the water by mixing and aeration. Lake Zurich was regenerated in the same way. The movements of our oceans have been demonstrated by a most unusual experiment: A container ship burst in 1992 in the Pacific Ocean and released 26,000 yellow plastic ducks, children's toys from China. They showed up seven months later 3,500 km away in Hawaii, in the Arctic in drifting ice, and near Eastern Europe. Such an amusing experiment nobody would have thought of. It was not only very informative but also cheap. Movements of the oceans are important. Water currents are under intense investigation, to record and predict climate changes — what will Europe be like without the Gulf stream? Maybe toys can help to follow the changes caused by El Niño or the Humboldt current? Indeed, such an experiment was recently initiated, with 1500 black balls distributed over all the oceans, termed "Drift Bodies" and equipped with lots of detectors in a "Global Drifter Program".

Can we extrapolate from our past, and from the causes of the ends of various civilizations, to predict our future? There have been about 20, some people say even 50, ice ages, caused by physical effects, perihelion precession, or catastrophes such as meteorites, cosmic events or volcanic eruptions. Humans have a poor chance of survival, because they cannot adjust very fast. Archaea are unfortunately also only slowly developing new ways of survival. Again, some life could be protected in niches — as it was 65 million years ago — and start again. Could there be a re-run of the beginning 3.9 billion years ago, or even another Big Bang like the one 14 billion years ago? Da capo!

"Social" viruses

The following thoughts may be perplexing: Viruses display social behavior. I refused to use the vocabulary of war. Humanizing descriptions of viral behavior are also inadequate. Nature cannot be characterized as good or bad. Life is simply geared to survival and replication, accompanied by adaptation or mutation and evolution. Viruses are the experts on this aspect, they react direct and simple, so that their behavior is easy to understand. Yet there are fancy descriptions of viruses, such as opportunists, extremists, minimalists which I find most important, they are designated as selfish or egoistic based on the famous book by Richard Dawkins, *The Selfish Gene*. Viruses are mutualistic with benefits for both sides, and others. The mutualism is often based on contributions of the virus to survival of the host — and thereby to increased progeny for the virus. The most frequently used type of viruses in my research were the "helper" viruses, all cancer viruses are defective and need helpers to supply coats and become transmissible from cell to cell or among hosts. As mentioned before, viruses can feed "mother-like" foreign offspring of insects, promote the survival of their hosts by repairing genetic defects, protect some plants against desiccation — for the benefit of others as well as their own, often as mutualism or symbiosis. Viruses can even repair defective genes of a dead host, again for their own benefit, can adjust to their environment and have offspring — as all life does. Typical viral behaviors can be described with molecular definitions such as complementation, interaction, interference, transdominance, cooperation, with the capacity to become part of the host. Viruses have two modes of interaction with a host — integration or lysis. Both extremes are linked by "stress", lack of food or space, which can induce lysis of a chronically or persistently infected cell, and terminate viral integration.

They cooperate and compete — both of which are basic mechanisms of life. Two viruses do better than one alone, see Hepatitis Delta virus with hepatitis B virus, or tulip breaking virus together with lily mottle virus — "better" means they cause death of their hosts. This may have evolved by accident. Everything can happen. Yet, before viruses destroy all their hosts, they find new ones, just for statistical reasons, and they will not eliminate all hosts, because they simply do not find the last one if there are only a few left. Those can regrow. Most importantly, several animal models were mentioned, where viruses turned to be harmless — remember, SIV does not kill the monkeys any more. What can be extrapolated to other populations or societies?

It may be worth considering how virus-host systems interact. Phages with their bacteria or viruses with algae as hosts have related lifecycles. Invaders entering some populations have two potential outcomes, either integration occurs, which is a persistent or chronic co-existence or alternatively lysis, disruption of the host and release of progeny. Both mechanisms are possible depending on whether the host is "permissive" or not for infection, replication or integration or lysis. This also depends on environmental factors and a chronic infection can lead to a lytic response by stress induction, such as lack of nutrients or space. How does this translate to human societies?

"Social behavior" can even be observed in dead matter, all the way down to molecules, which can bind, repel, stabilize, or interact. If life started with RNA, then RNA must be capable of transmitting some kind of useful information. Indeed, RNA can destroy other RNA recognized by base pairing (this is for example what ribozymes do), if competing for the same resources. Attraction and rejection we know all the way down to molecular or atomic interactions. The microbiomes of individuals are perhaps a contributing factor for social interactions. Kissing may be an equilibration of microbiomes between two partners — how disillusioning! From information transfer to social behavior is a big jump, but may be a logical consequence. Similarities between the micro- and macro-worlds have already been mentioned. Structures such as clover leaves resemble transfer tRNAs, or the tumor suppressor p53 with "hands" or "fingers", which identify the mutations and give time to the cell for repair!

What about the phenomenon "camouflage", outfits mimicking landscapes used as a typical human trick by soldiers in combat to deceive their enemies? "Do viruses show this type of behavior?" I was once asked. Yes, very much so, it is almost a hallmark of viruses to pretend to be something that they are not. One can call this cheating. The entry of viruses into a cell is often based on mimicking a normal cellular step or component. HIV binds to a receptor CCR5, as if the virus were its regular ligand. The viruses interact with host cells by using normal cellular mechanisms and imitating cellular gene products. Camouflage is the basis of almost all virus–host interactions.

RNA can distinguish between related and foreign RNA by sequence homology. Cells occupied by RNA do not admit other related RNA to enter, as mentioned for the ribozymes, viroids and viruses as antiviral defense. Do we project our thinking into the nano world, talking about structural similarities or camouflage from the macro to the micro-world? The principles look the same, we are related.

A fantastic new "genetics" with sex hormones

Toward the end I would like or raise some hope and describe some miracles, fantastic new developments — which reflect our present state and also lack of knowledge.

Fat moms have fat granddaughters — and smoking grandfathers affect their grandsons.

We have learnt about epigenetics, which is not due to genetic changes, not to mutations but due to environmental factors influencing activity or expression of our genes. Toxin or heat or food can lead to some RNA which activates chemical modifications (*methylation or acetylation of DNA or histones*) and modify gene expression. This is beyond genetics and therefore designated as epigenetics. This is normally erased during mating. The next generation can start all over again. So this procedure is at best a life-long learning effect — and not transmitted to the next generation.

Yet this goes on: all of a sudden we need to learn that something is transmitted from grandmothers to granddaughters or from grandfathers to grandsons. What grandparents eat can leave behind changes in the genes of their grandchildren. This information transfer is most likely restricted to time points during pregnancy. For three generations, epigenetic modifications of the genes are passed on from pregnant mothers to embryos and the embryo's germ cells, as evidenced by the Dutch Famine study.

This is now a two-generational effect, inheritance without mutated genes, a more sustained epigenetics. The colors of the maize kernels Barbara McClintock described as epigenetic changes. Already when she described the epigenetic effect, again the maize kernels indicated a further phenomenon, with effects not for a lifetime but several generations. This was observed by the Canadian plant geneticist Royal Alexander Brink already in 1956, which he designated as "paramutation", it was no real mutation and no epigenetics for one life span. We now describe this phenomenon also as transgenerational effect, because several subsequent generations are affected.

Here is an example: mutated mice have white stripes on their tail. If one mates mice, the white stripes again do not obey the laws of inheritance. Instead of crossing the mice, one can even go further and take out RNA from the mice with white tails and inject it into the veins of black-tailed ones. Their offspring can have white tails — a completely unexpected result. Through a piece of RNA! The name of the RNA is "piRNA" and the name of the gene is "kit". This gene is known by oncogene researchers, its proto-oncogene is a stem cell receptor tyrosine kinase, a signaling molecule. This signaling molecule is shown to go through the sperms to the next generation. It can also be isolated from the sperms. This was demonstrated in 2006 by the scientist Minoo Rassoulzadegan from Nice, a friend of mine and colleague — who struggled and still does, to convince her sceptics about her surprising results. She estimates that about 200 out of the 10,000 mouse genes could behave this way. Do they have mainly effects on reproduction? I wonder!

White stripes are a characteristic feature of zebras and also of zebrafish. They have been studied by the German Nobel laureate Christiane Nüsslein-Volhard for developmental properties. The stripes are not caused by a virus or environmental effects but by mainly two or three cell types and are based on genetics. Mutant zebra-fish can exhibit perpendicular stripes, turned by 90 degrees! Even spots can result from mutations.

Darwin suggested that colors have consequences on sex appeal and reproduction. Also colors play a social role, natives in African tribes painted their skin to indicate their membership to a community. We humans have to come up with paints, uniforms, dress codes and the like, because we do not have patterns, but only melanocytes for light or dark skin, perhaps one more cell type for red hair. All stripes caused by genetics or epigenetics, involve periodic pattern formation with activator and inhibitor. The famous British mathematician Alan Turing, known for solving the ENIGMA code of the German submarines during World War II, as well as James D. Murray from Princeton in his article "How the leopard gets its spots" describe how patterns arise. Here I got excited, because there is a Tulip Breaking virus causing striped tulips, *Potato* or *Papaya Ring Spot viruses* for stripes, spots, rings, mottles, even the tiny viroids participate in pattern games, the *Potato Spindle Tuber Viroid*. Perhaps mutations and paramutations and viral infections all share some common features. The molecular mechanisms including silencing are not well understood.

Transgenerational effects also exist in plants. One can grow tomatoes or potatoes and place caterpillars on their leaves, which they fight against by production of toxins, and the next generation of both plants then does that also without ever having encountered a dangerous caterpillar. Even the rather poor wheat plant Arabidopsis remembers for generations how to cope with too much salt in the soil or heat. These are fast response and memory for a few generations. Could such a fast transient correction precede a long-term memory? What about more than three generations?

At the end one could speculate about some "miracles" — things we do not yet understand but nonetheless wonder about. *C. elegans* worms are attracted by a drop of pheromones, sex and mating signals on a culture dish in the laboratory. Now comes the miracle: some generations later the offspring meet at exactly the same spot, even though the smell of the pheromone has long disappeared; no cues are required. This has meanwhile been shown for up to 50 generations of worms. That sounds incredible. A second experiment supports this. *C. elegans* worms like to live in orchards, such as those in Orsay close to Paris where they were studied. Apples tend to fall down in the orchards and the worms feed with delight on the rotten apples — containing some alcohol. The alcohol "addicted" worms remembered the spots where ethanol was spilled on a Petri dish without any alcohol cue left after 30 generations.

The suspect is a small RNA, which presumably causes changes in the chromatin and stores this information. This behavior raises the question of whether grandsons will find and visit the bordellos that their grandfathers frequented a long time ago? No! Humans are too complicated. Are the worms more than an exception — are they perhaps a model for other organisms — alcohol addicts?

Bees are well known for their strong social structure, and now are being analyzed for transmitted RNA. Only the queen bee gets royal jelly. The royal jelly has been reported to contain an ingredient for social behavior in the next generations: the components of royal jelly are currently being dissected and analyzed, to search for piRNA. We can buy royal jelly for 5 Euros in German supermarkets — for better social behavior? That would be too simple. But food leading to transgenerational or paramutation effects has been reliably proven to exist in plants and worms, not so long-lasting in humans as yet.

The sperm RNA has already got a name, it is "PIWI-interacting RNA" or piRNA, indicating that RNA can interact with the protein PIWI, a scissors, an RNase H-like molecule. What do the scissors cleave? Transposons, jumping genes, the RNA of retrotransposon is "silencd", cut. The piRNA is a bit longer than siRNA, is specific for germ cells, is responsible for "taming" jumping genes, and is the "guardian" of our genomes in germ cells. It is necessary for fertility. Perhaps it prevents an "error catastrophe", too many mutations. It also has an effect on learning and memory — on brain cells, on neurons — and on inheritance of learnt and social behavior and on longevity? And finally: is there a link to sex, mating, production of progeny? I have this suspicion. (Here young researchers may want to get involved. A place to go is Cambridge, UK to Eric Miska.)

piRNA may be everywhere. piRNA is perhaps specialized to mating and survival of the species.

This rings a bell: it reminds us of inheritance of acquired properties propagated by Jean-Baptiste Lamarck, the French botanist and zoologist (1744–1829). He published a book *Philosphie Zoologique* (1809), disputing with Charles Darwin and his book *The Origin of Species* (1859), based on adaptation by selection. This is now being actively discussed again. Perhaps the contradiction between the two may be less dramatic than it seems: Could it be, that acquired properties, the paramutations, may become genetically fixed after some generations as real mutations? RNA can become DNA, that we have known for some time now. The magic and ubiquitous reverse transcriptase could help.

Perhaps this can also happen with the piRNA — but this is my wild speculation. This closes the circle from the primary pre- or pro-RNA, viroid-type RNA, to ribozymes, to circRNA, and now also to sperm piRNA, all with hairpin-looped RNA structures — so robust that it has resisted all dangers from the very beginning till today for about 3.8 billion years. This closes also the circle of this book.

But many questions remain open. What about other strange phenomena? What are the Global Positioning Systems, "GPS" for information storage in various animals? The turtles swimming across the oceans for 12,000 kilometers, all alone, for 20 years, to lay their eggs on one specific sandy beach? The offspring will follow the same transit route and deposit their eggs at the same beach without learning it from their mothers, who have died before they hatched. In the Berlin Zoo, the animal keepers had to learn to never move a single stone in the cage of a turtle, if they did, the turtle would not reproduce. Every spring storks find their nests in Linum, a tiny village close to Berlin, coming all the way from Africa, first the male and a bit later the female — finding each other. The monarch butterflies died when the trees on which they were *used* to convene for mating were cut down. We all know about the trout or salmon that remember where to lay their eggs and put their sperm. Magnetism, smells, chemotaxis, magnetism, light or polarized light, color, food, or enemies, all kinds of environmental factors may contribute — perhaps mainly for reproduction and progeny — since the beginning of life, 3.8 billion years ago. Could piRNA contribute to the first step, fertilization of eggs by sperm in addition to genes in the form of DNA? RNA has been there since time zero; piRNA is a variation of the theme, information is inherited as RNA in the RNA world and later as DNA, but the RNA has stayed with us.

Viruses for predicting the future?

Can we learn anything from the viruses and phages and their hosts about the continuation of life and the future or the end of mankind? Infectious diseases present the main cause of death on our planet. About 100 million people died, under devastating conditions, of influenza during World War I. About 37 million, predominantly in the developing world, have died of HIV/AIDS. 15 million people die annually of infectious diseases worldwide, about 10 million of cancer — of which 1.5 million cases involve viruses. A new "epidemic" is obesity, affecting about 300 million people, and some researchers predict that up to one-third of mankind will soon be obese — with no known "infectious agent" involved. There is no virus known *as cause* — but what about the bacteria? Antibiotics, after all, are given to animals to increase their body mass. It is estimated that teenagers today have had a dozen cycles of antibiotic in their lifetime, which modifies the microbiome of their guts. Could this explain the so far totally unclear rapid increase in obesity, which seems like an epidemic. This possibility is being seriously discussed. Obesity was shown to increase cancer and change of stem cell differentiation in the intestine. What are the potential consequences on the bacterial composition in a person's ecosystem also in the long run? If the bacteria are affected, the viruses/phages are naturally affected as well, since after all they lyse the bacteria under stress conditions, which include obesity and the microbiome in the guts. What about the archaea and yeast? Nobody knows.

I have stressed that microorganisms do not make us ill if our body is in a well-balanced ecosystem. Such stable systems do not permanently exist, however — otherwise we would never get sick. We even need to get sick once in a while to train our immune system, if we do not want worms for training it.

Balances become destabilized. Lack of space, oxygen, and nutrients, by the proximity of humans and animals, may be difficult to avoid, and these cause stress. Stress can activate viruses. Infectious diseases are among the primary causes of death in developing countries. Yet microorganisms or infectious diseases will not eliminate mankind. They were not even discussed as one of the dangers for mankind in the NASA study. They may regulate population densities, though.

Microorganisms and viruses will certainly survive us mammals and eukaryotes. They were here long before us, since about 3.8 billion years. We came much much later; Lucy came out of Africa about 3.2 million years ago. Only those who could cope with the microorganisms in the existing environment survived. We even need the microorganisms and depend upon them. We are invaders in the world of bacteria and phages — not the other way around. They help us to digest our food, populate our body, and to protect us against unfamiliar germs. Life itself will survive without us in the form of the microorganisms. Bacteria and phages are autonomous, microbes do not need us, but we need them. The most versatile elements are the viruses, the bacteria, and perhaps the archaea; the latter at least give us an idea of what extreme conditions life can cope with. They all can start again after some life-threatening catastrophe. They may have done so several times. However, the waiting time for a human being to become an extremophile that, like the archaea, can survive heat, high salt, poison, or other extremes, would be too long. We are too complex, too slow, too big, unable to adapt quickly.

Humans may be sufficiently intelligent and innovative to protect themselves against doom. We may be intelligent enough, but perhaps too egoistic, not sufficiently mutualistic. Behavioral studies have indicated that humans do not behave mutualistic in groups bigger than 150 people — that corresponds to an extended family with friends and neighbors. When humans live in megacities, as many will do according to a prediction for the world population, could then microorganisms and viruses destroy us and eradicate mankind? As an answer, here is an example: When the hantaviruses in New Mexico did not find any more mice as their hosts, because of environmental changes, they jumped to humans instead. This can continue as long as there are alternative hosts. Our increased mobility accelerates the finding of new hosts, as we have seen with SARS coronavirus; for it to spread from Hong Kong to Vancouver took just a few hours. But if there are no more hosts available, what would happen then? Where would the microorganisms go? They would change their strategy: They would no longer exploit or even kill their hosts, if there are no new ones, but would find arrangements to adapt and to develop co-existence. This is constantly happening. Examples are SIV-resistant monkeys and gibbon ape leukemia virus GaLV-resistant koalas, which endogenized infecting viruses into their genes, where they not only were tolerated, but also protected the host against outsiders. Virus-like sequences entered the eukaryotic genome as protection against outside viruses. They used to be exogenous viruses, as was demonstrated with the reconstructed virus Phoenix, which became endogenized and protected us once against exogenous ones. Then microorganisms would not be deadly dangers for mankind, but they protect us from the day of birth. Can one extrapolate this behavior to humans, with enemies becoming friends — as the best or only way of survival? This is good news if we get scared of pandemics. The death tolls may be high, but they are never unlimited.

There are analyses of our future chances, which predict that our modern civilization cannot survive. One such study is called HANDY (Human and Nature Dynamics) from 2014. It includes several topics for the worst-case scenario, such as an increase in population density, a lack of resources (water, food, energy), natural catastrophes such as climate change, and finally social and political instabilities, with economic inequality, the gap between rich and poor. These factors were considered to be the main causes of the fall of empires: the Roman, the Mayan, the Cambodian. Collapse is inevitable, that was the prognosis. The study used differential equations for predictions of future scenarios. This was too pessimistic even for NASA, and it stepped back from this prognosis.

It is interesting to note that this study did not take microorganisms into account. They are not on the list of threats to life on our planet. Diseases, microorganisms and infectious agents were excluded. Population density was first on the above list — it will have a strong impact on infectious diseases, but perhaps more on a local than on a global scale.

What about food? Here viruses become involved. Nutrient production has doubled owing to technical advances in the last 100 years — first and foremost by the German chemist Fritz Haber, who in 1915 developed a method to fix nitrogen as fertilizer, which still today allows mankind to be fed. We need novel technologies and luckily enough we succeeded already in an important innovation: Genetically modified food will certainly be a primary possibility to help feed mankind. We are lucky that it exists. But we need to learn how to use it properly. So we may not need to eat "stones" as was discussed by Dyson, or grow fur again and need less heating. Eating insects or algae would not be so bad. Some calculations predict that limits of growth are reached with 12 billion people. Such prediction do not include innovative steps and are unreliable. Remember, one hundred years ago it was foreseen that horse droppings would block the streets of Berlin — but then came cars and no horse apple catastrophe! Cars in turn have given rise to new frightening extrapolations.

On a cosmic scale, the sun will finally terminate everything. It will get too hot first for us and later for the microorganisms. I believe, as a nuclear physicist, that ultimately the sun will blow up, following the stellar cycle (the Hertzsprung–Russell diagram) and first become a "red giant" and then a "white dwarf". The red giant will burn up our planet. The time until then is half over, so we have a second half of another 4 billion years ahead of us — but only if other cosmic events do not eliminate us earlier.

However, can we turn the argument around? The world of microorganisms could raise some hopes, too, if we think about it. Every fall algae bloom and bacterial lawns on the oceans are terminated by viruses, which contribute to population dynamics and recycling of nutrients. There are always survivors, starting with small numbers initially, and increasing exponentially. Next summer the algae will bloom again if there are sufficient nutrients. Could that be some hope? Even if many species will die, subpopulations of diverse species could escape through a bottleneck of population density — transposons my get activated — and help to develop new properties. Such phenomena may have happened five times during the past of our planet. More recently it has often been local. "Deepwater Horizon", the biggest oil spill ever, demonstrated to the surprised scientists that microorganisms grew up, digested the oil and disappeared again when the oil was used up and they ran out of food. Similarly, after the accident at Chernobyl some plants grew better on the radioactively contaminated soil. There are probably more surprises ahead of us. Mutations constantly happen. We learnt to drink milk, we may eat different food, legs may degenerate or more fingers grow to use computers. We have already lost 6,000 receptors for smelling and tasting, and are left with 60, since we can get away with having only that few.

The world is very heterogeneous; it offers niches where life can hide and be preserved. Thus what happens with the world of algae and the Red Tide could occur with us humans. Overgrowth is followed by virus-induced destruction and a restart in the next season. There may always be a niche somewhere, a warm little pond at the beginning as described by Darwin. That also happened in Yucatán. Life cannot easily be eradicated completely.

Our aids to survival are the viruses, supplying us with new genes. Viruses including mobile elements are innovative, they have contributed and will in the future continue to contribute to all life, for adaptation, survival and biodiversity. We should remember, we can influence the viruses, the jumping of transposable elements, by environmental factors promoting or dampening down the jumping activity. We are also responsible. Retrotransposons vary greatly in their jumping activities.

In the oceans alone, 10^{27} genes are exchanged daily between viruses and their hosts. There are many new genes for us, and the high viral mutation rate and gene exchange will help us to survive. The viruses, mainly the retroviruses and their related retroelements have contributed to the evolution of life, to the survival of all species on this planet. Without the viruses we would not have this progress, this diversity; in fact, we would not exist. Yet, thanks to the viruses, we do.

And where is God? This question was addressed to me at the end of an interview about viruses on Swiss television one Sunday morning. The question came as a surprise. My spontaneous answer was and still is: wherever we encounter the "unknown" we feel inclined to believe in a God. Miracles may be explained by science. But each solution opens new questions and points to new miracles. There will always be miracles, no matter how much knowledge we accumulate. We can always attribute the unknown to the activity of a Higher Power, a God Father if you wish. I cannot believe that, but those who can are better off than me.

Glossary

AA/amino acid: Component of proteins that are polymers of AAs produced by ribosomes, they read the messenger RNA (mRNA) in the cell, and synthesize the AA polymer, the protein; each triplet (codon) of the mRNA sequence corresponds to one AA, transfer or tRNAs transport the individual AA to the ribosome for assembly, 20 canonical AA exist, only about 2% of human genome code for proteins.

Ab: Antibody, also known as immunoglobulin, secreted by specialized B cells, have hypervariable region as antigen binding site, need time to form, diversity arises by rearrangements involving RNase — like nucleases, the RAG enzymes, recombination-activated enzymes 1 and 2. Vaccines work by inducing antibodies, especially neutralizing antibodies which prevent infection by neutralizing surface proteins of viruses, antibodies are part of passive vaccines (in contrast to active vaccines, where the body needs to produce the antibody after immunization with an antigen), best way to fight viral infectious diseases.

ADA: Adenosine deaminase deficiency, a metabolic disorder that causes immune deficiency, children cannot cope with pathogens and have to stay in sterile tents, model disease for gene therapy.

Agouti mouse: Mouse whose fur color is indicative of environmental effects including carcinogens, yellow fur and obesity indicate carcinogenic food which induces epigenetic changes, they are genetically identical to black and lean mice, which are healthy.

ALS: Amyotrophic lateral sclerosis, inheritable disease that causes progressive paralysis.

Alu: Repetitive, short regulatory DNA element without genetic information, can exert regulatory effects on neighboring genes, belong to the SINE family of retroelements, 1.1 Mio copies in the human genome (10.7% of the sequence), mainly in primate genomes.

antibody: See Ab.

apoptosis: Programmed cell death different from dying (necrotic) cells, a regulated program that is activated when cells are no longer needed or are dangerous such as cancer cells.

archaea: One of the three kingdoms of life, besides bacteria and eukaryotes, former name was archaebacteria which is not used any more, archaea (plural) may be the oldest cells in the tree of life.

Argonaute: Enzyme for silencing of gene expression consisting of PAZ and PIWI domains (like reverse transcriptase and RNase H of retroviruses) in the RNA-induced silencing complex (RISC), Argonaute proteins bind non-coding small regulatory RNA, which then guide the complex to a target RNA for cleavage.

bacteriophage: Same as phage, virus of bacteria.

base or base pair: Abbreviated as bp for double-stranded nucleic acids, or only base as abbreviation of a nucleotide consisting of base and sugar, the building blocks of single-stranded DNA or RNA. Pair rules in double-strands are symbolized by A-T and G-C for DNA or A-U and G-C for RNA.

BDNF: Brain-derived neurotrophic factor, a human signaling molecule that plays a role in communication between guts and brain cells.

bisulfite conversion: Detection method for methylated nucleotides in the DNA that is an important marker of epigenetic regulation, methyl-modified and non-modified cytosines (C) can be distinguished, treatment converts C to thymine (T), while methyl-modified C is unaffected by bisulfite treatment.

bnAB: Broadly neutralizing antibody against broad spectrum of HIV strains for virus neutralization, blocking cellular uptake of the virus and infection.

BL: Burkitt's lymphoma, see EBV.

cancer: Cells growing uncontrolled or indefinitely due to genotoxic events caused by combination of many environmental effectors, radioactivity, toxins, oncogenes, some viruses, genetic defects can be inherited, multi-stage or multifactorial effects, activation of cancer genes or loss of tumor suppressor genes, can have 18 introns (normal average is 5 to 7), leading to high complexity, can develop over 30 years.

Cas9: CRISPR-associated protein 9, molecular scissors cleaving the DNA moiety in RNA-DNA hybrids, defense against DNA phages in bacteria and archaea, containing RNase H-like DDE amino acids (a DNase H that cleaves DNA in RNA-DNA hybrids).

catalytic activity: Proteins with catalytic activity are enzymes, RNA with catalytic activity are ribozymes or often viroids, they both speed up chemical reactions and can be reused.

CCR5: Receptor for signaling molecules (chemokines) on the surface of human cells, co-receptor besides the CD4 receptor for HIV to enter a host cell; a CD4 lymphocyte or macrophage with CCR5 is first target cell, later cells with CXCR4, another chemokine receptor, are infected.

cell cycle: Life of a cell, comprising duplication of the genome or DNA synthesis (S-phase), growth, and division (mitosis or M-phase).

Central Dogma: Information flow inside a cell, according to F. Crick: "From DNA to RNA to protein", information transfer from DNA to RNA is termed transcription, important role of RNA as first biomolecule suggests that the opposite direction emerged first during evolution: from RNA to proteins and from RNA to DNA, then the so-called "reverse" transcription is the real (original) one.

chromatin: Complex of cellular DNA with proteins such as histones, around which the DNA winds, can be changed by epigenetic modification (such as DNA methylation and histone acetylation).

chromosome: Genetic material (genome) of eukaryotes is distributed on linear DNA molecules with associated proteins, together called the chromosomes, humans have 22 common chromosomes each present in two copies, plus two sex-specific chromosomes (females: XX, males: XY), a total of 46 chromosomes.

clade: Groups of HIV belonging to certain surface protein characteristics, M (major group) pandemic, C prevalent in South Africa.

codon: Triplet of three nucleotides coding for one amino acid, more triplets exist than aminoacids (redundancy), "coding DNA or RNA" means it can be translated into a protein, non-coding RNA/DNA cannot.

colors: Can help to "see "effects of mutations, epi- and paramutations in maize, in Agouti mice, in white-tailed mice, in tulip stripes, in plants with mottles, stripes, ring spots, spindle tubers, but also in zebras, leopards, and as I assume also in kois. Color patterns arise by an interplay of essentially two cell types, chemicals or "morphogens", an activator and inhibitor, with influence by geometry and frequencies. This is described by two differential equations by the famous British mathematician Alan Turing in 1952. The "Turing mechanism" refers to an activation-diffusion system or activator-inhibitor system. The activator is perhaps sometimes a virus and the inhibitor may be the silencer siRNA which causes white patterns in petunias.

CONRAD and MTN-030/IPM 041: Acronym (Contraception Research and Development), supports clinical trials against HIV infection, such as the MTN-030/IPM 041 trial testing vaginal microbicides.

CpG: Abbreviation of the two nucleotides C (cytosine) and G (guanine) linked by phosphodiester bridges in DNA molecules, an important motif for epigenetic gene regulation as targets of methylation.

CRISPR/Cas9: Clustered regularly interspaced short palindromic repeats/ CRISPR-associated protein 9, an antiviral defense system of bacteria and archaea against invading DNA phages, exploited for gene deletions and gene modifications, named "editing", a novel method with enormous potential for gene-changes in all species, simple to use, also fast and cheap, no recombinant mechanisms involved, since a "guide" RNA prescribes the new sequence which is then copied (edited). No genetically modified DNA, not under GMO law, raises ethical discussions.

CSH: Cold Spring Harbor, Laboratory and Press and conference venue near New York City that holds the Symposia on Quantitative Biology.

CXCR4 or CCR5: Receptors on lymphocytes allowing HIV surface glycoprotein (gp120) to bind and enter the cell, at early and late stages of the disease.

DNA: Primary genetic material in most organisms, composed of deoxyribonucleotides, in most cases double-stranded, known as double-helix, stable, some viruses contain single-stranded DNA or RNA as primary genetic material.

Drosha: Enyzme with RNase H fold, cleaves double-stranded RNA for processing of siRNA (small interfering RNA).

dysbiosis: Microbial imbalance in the guts due to illness.

EBV: Epstein-Barr virus, a herpesvirus causing three diseases, mononucleosis, a non-malignant "kissing" disease in the Western world, BL: Burkitt`s lymphoma in Africa and nasopharyngeal carcinoma in China, depending on sequence and co-factors.

EC: Elite controller, people with natural resistance or control of HIV load.

EHEC: *Escherichia coli* bacterium causing the enterohemorrhagic disease, can arise by phage-mediated acquisition of a toxin (Shigatoxin) originating from animal feces and fertilizers, harzardous for human health, EHEC-contaminated sprouts caused the biggest food scandal in Germany.

Eliava Institute of Bacteriophages: Microbiology and virology institute in Tbilisi, Georgia, founded by Félix d'Herelle and Georgi Eliava around 1930 with a focus on phage therapy, exists until today, increasing importance for phage therapy, spectacular cures of diabetic toes.

EMBO: European Molecular Biology Organization, based in Heidelberg, Germany, offering top quality scientific meetings, scholarships, stipends and scientific training, expanding from Europe to worldwide programs.

EMEA: Recently just EMA, European Medicines Agency, evaluation of medicinal products, protection of animal and human health, approval of clinical trials.

Emiliania huxleyi: A carbon alga (coccolithophore), abbreviated as EH, forms huge lawns of blooming algae, relevant for climate, carry algae viruses, named EhV or EHV, which terminate the bloom by lysis as stress response in the fall at lower temperatures, leading to nutrient recycling.

endogenous (retro)virus: (Retro)viral sequence in the genome originating from viral infections of germ cells, normally do not form particles, inherited vertically to subsequent generations, endogenization process can be observed in real-time in Australian koala populations.

epigenetic changes: Non-genetic DNA/chromatin modification, not by Mendelian-type mutations, triggered by environmental factors, attachment of methyl groups to nucleotides of the DNA or modification of the DNA-associated proteins by acetylation or others, and packaging of the DNA normally for one generation caused by environmental influences, erased during fertilization, but sometimes also transmitted to the next two generations through germ cells mainly through stress during pregnancy such as famine.

epigenetics: Transient changes of DNA/chromatin without mutations, non-Mendelian genetics, chemical modification by attachment of methyl groups to DNA (preferably at CpG sites) or acetylation of histone proteins, caused by regulatory RNA and activation of RNA-dependent DNA methyltransferases, the methylators, often causing inactivation of promoters, normally not inheritable, colors as visible indicators of epigenetic changes, detected first in maize, epigenetic changes can be involved in cancer, require new diagnostics such as "bisulfite conversion", whereby non-methylated C in the genome is changed to T that can be detected afterwards by sequencing.

Erb-B2 or HER2/neu: Oncogene (cancer-promoting gene) of avian erythroblastosis virus, epidermal growth factor receptor, stimulates cell proliferation via the Ras-MAP-kinase pathway, overexpressed and indicator of human breast cancer, 25% of breast cancer patients have mutations in Erb-B2.

error catastrophe: In a population of replicating molecules, the error frequency (e.g., mutation rate of RNA molecules) can get too high and detrimental, also a possible therapeutic approach proposed by Manfred Eigen, driving replicating viruses into too many errors for termination.

ERV: Endogenous retrovirus, integrated DNA in the genome of eukaryotes, inherited as DNA provirus through germ line cells, not infectious anymore, often defective, originates from infection of the germ cells by formerly infectious viruses, which often went extinct, can lead to an antiviral defense against similar viruses, endogenization is presently ongoing in koalas in real-time, identified by conserved integration sites, while *de novo* infecting viruses integrate at random sites, see Phoenix.

ES cells: Embryonic stem cells, can develop into one of the 200 special (differentiated) cells of the human body by biomolecular stimulators, often forbidden for research use, differentiated cells can be reverted from differentiated adult cells back to the embryonic state by a combination of four transcription factors (see iPS).

ETH: "Eidgenössische Technische Hochschule", Swiss Federal Institute of Technology in Zurich, also named Poly for Polytechnicum, one of the leading universities in Europe.

EU: European Union with framework programs to support multicentered research programs.

eubiosis: Describes the healthy state of the gut microbiome.

eukaryotes: All organisms whose cells contain a nucleus, such as plants, algae, fungi/yeast, animals and mammals including humans, in contrast to prokaryotes such as bacteria which do not have nuclei.

exon: In eukaryotic genes, exons constitute the protein-coding sequence, interrupted by introns which are not protein coding, viral and most pro-karyotic genes consist of only exons.

foamy viruses: Pararetroviruses with a slightly different life cycle than ret-roviruses. They perform the reverse transcription step from the viral RNA to the proviral DNA just before leaving the cell, not just after entering it. Thus the packaged viral genome is DNA, not RNA as in retroviruses. Other pararetroviruses are hepatitis B virus, HBV or plant viruses like the cauli-flower mosaic virus. They package DNA into virus particles and therefore do not look like retroviruses at first sight.

FT: Fecal transfer, transplantation of feces from one person to another as therapy against *C. difficile* infections.

FTO: "Fat mass and obesity"-associated protein, one of several genes asso-ciated with obesity, a long stretch of 47,000 nucleotides, not one protein, many splice variants, acts together with obesity-associated transcription factor IXR3.

gag: One of the retroviral genes, group-specific antigen, name is based on immunology of viruses.

GaLV: Gibbon ape leukemia virus, a monkey retrovirus, which killed many koalas before they got resistant by endogenization of the virus.

gene: Unit of the genome coding for proteins that build our body and metabolism, one gene can code for more than one protein, in eukaryotes, genes are composed of exons and introns, the exons are the protein-coding regions, introns can be eliminated by "splicing", cleaving and the joining of RNA transcripts allows for many combinations, basis of the human complexity.

genome: Entire genetic information of a species, normally double-stranded DNA, inherited and present in each cell, composed of genes but also non-coding genetic information, which means not coding for gene products, such as regulatory information.

genome compositions: Humans (3%DNA transposons, 45% total transposable elements), yeast (mostly retrotransposons, very few DNA transposons), *C. elegans* (mainly DNA transposons, few retrotransposons), plants (variable amounts of DNA transposons and retrotransposons).

genome sizes: *Amoeba dubia* (700 Bio bp), the plants *Paris japonica* (150 Bio bp), tulip or maize (30 Bio bp), humans (3.2 Bio bp), fungi, algae, worms (50 Mio bp), protists (10–100 Mio bp), bacteria (1 to 10 Mio bp), archaea (around 6 Mio bp), giant viruses (0.5 to 2.5 Mio bp), *Mycoplasma mycoides* (1 Mio bp), *Nanoarchaeum equitans* (0.5 Mio bp), *Mycoplasma genitalium* (one of the smallest living cells, 580 kbp), poxviruses (350 kbp), herpesviruses (100 kbp), influenza or retroviruses (10 kb), poliovirus (7.5 kb), viroids (0.3 kb).

giant or giga viruses: Unusually big viruses, their hosts include amoebae and algae, "almost-bacteria", with names like: mimivirus (for mimicking bacteria with their unusually large size), names: mama-, marseille-, pandora-, pitho-, samba-, molliviruses, large genomes with up to 2.5 Mio bp, algal giant viruses include EhV-86 infecting *Emiliania huxleyi* and *Chlorella* viruses, giant viruses host smaller viruses (virophages), contain components for protein synthesis, hypothesized to be either degenerated bacteria or incomplete, unfinished evolutionary precursors of bacterial cells.

GMO: Genetically modified organism, to which laws on recombinant DNA technologies apply, tight regulations before application for human use and crops need to be fulfilled.

gnotobiotic mice: Mice with non-natural gut microbiota, models in research, grown and raised in sterile conditions and instilled with foreign fecal microbiota or defined bacterial species, laboratory mice to study functions of the microbiome.

group II introns: Large catalytic self-splicing RNAs and mobile retro-elements that encode a reverse transcriptase, an intron-encoded protein, colonize the genomes of bacteria and organelles (mitochondria and chloroplasts), are precursors of the reverse transcriptase.

gypsy: name of a retrotransposon closely resembling retroviruses, present in *Drosophila*, equivalent to Ty3 in yeast, contain LTRs, *gag, pol, nuclease, integrase*, but no *env* gene; they can transfer genes.

HBV: Hepatitis B virus, pararetrovirus, replicates via a reverse transcriptase, but packages DNA, not RNA into virus particles, causing in some cases hepatocellular carcinoma, vaccine available.

HCV: Hepatitis C virus, can cause at late stage hepatocellular carcinoma, new therapy available, very successfully eliminating the virus but costly ($50,000).

HDV: Hepatitis delta virus, infecting human liver, a ribozyme, perhaps of plant origin, together with HBV severe diseases, rare.

Herceptin: Breast cancer therapeutic, only applicable to 25% of patients with defects in hormone (EGF receptor) signaling.

HERV: Human endogenous retrovirus, about 450,000 copies in the human genome, rarely forming infectious particles, no disease directly correlated, but with their promoters, the LTRs, they can influence neighboring genes, role in cancer and genetic diseases with unknown frequencies, relic of former retroviral infections of germ cells.

histone: Protein that packages the DNA in eukaryotic cells, main proteins of chromatin which condensate DNA, forming chromatin, their acetylation plays a role in epigenetic gene regulation.

horizontal gene transfer (HGT) or lateral gene transfer (LGT): A prominent role of viruses or phages to shuttle genes from one cell or organism to another, each genome consists of genes from many other species that have been acquired by HGT.

HPV: Human papillomavirus, mainly two of many types can cause cervix carcinomas within decades with co-factors, vaccine recently available for young girls and soon for boys.

immune system: Defense of cells and organism against invading pathogens, in the RNA world: silencer or siRNA, in the DNA world: restriction endonucleases and CRISPR/Cas9, in the protein world: interferon and IgG (immunoglobulins), all immune systems use RNase H-like molecular scissors, transposable elements with their enzymes play a major role in the development and diversity of the immunoglobulins (antibodies).

INRA: French Institute for Research in Agriculture, located near Paris, among the front-runners of microbiome analysis.

intron: Word fused from "intragenic regions", intermediate regions between exons, genes consist of exons and introns, introns are non-protein-coding, missing in viruses, only exons lead to proteins, RNA transcripts of genes are processed by splicing which removes the introns, allowing numerous combinations of exons, most abundant in humans permitting greatest combinatorial complexity, special groups are self-splicing introns removed by self-catalytic ribozymes, RNA catalysis, many hairpin-loops, in fungi, plants, protists, or bacteria, perhaps evolutionary related to spliceosomes.

iPS cells: Induced pluripotent stem cells, rejuvenated adult fibroblast or other differentiated cells by a set of four transcription factors (Oct3/4, Sox2, Kif4, and c-Myc) inducing proliferation and de-differentiation, one of the four is the proto-oncogene of the viral "oncogene" myc, iPS cells are similar to embryonic stem (ES) cells, both grow indefinitely and maintain pluripotency, iPS cells replace ES cells in research, which are forbidden to obtain. iPS cells can be driven into one of the 200 specialized human cells such as brain cells, nerve cells, muscle cells, the hope is to regenerate dead or sick tissue for replacement.

IRX3: Transcription factor involved in obesity together with FTO.

jumping gene, transposable element (TE): Transposon ("cut-and-paste") or retrotransposon ("copy-and-paste"), move inside the cellular nuclei, influencing genomes, defective viral elements without coats, locked inside cells, 45% in human genome, jumping causes changes in the genetic material, jumping activity is regulated by epigenetic effects, can be innovative or dangerous for the genomes and organism, controlled by silencing mechanisms (siRNA), which restrict movements.

Kit: Receptor tyrosine kinase in stem cells and also an oncogene in a cat virus. Kit plays a role in white-tailed mice inherited by paramutation and a role in stripe formation of zebra fish.

Lamarckism: Inheritance of acquired properties proclaimed by botanist and philosopher Jean-Baptiste Lamarck.

LFI: Loeffler-Frosch-Institute, biggest institute for veterinarian viral diseases, north of Greifswald, northern Germany, two early German virologists, in their honor a medal of virology is awarded.

LINE or LINE/L1: Long interspersed nuclear element, DNA regions in the genome, transposable element or jumping gene, coding for a two "open reading frames" (ORF1 and ORF2) corresponding to two proteins, an RNA-binding protein and a reverse transcriptase/endonuclease (RNase H-like), retrotransposon which acts by "copy-and-paste" including an RNA intermediate, a copy by reverse transcriptase and nuclease and reinsertion into host DNA, leading to gene duplication, no LTR, 850,000 copies in human genome (21%), help the SINE elements for proliferation, in humans only 100 L1 elements are mobile with potential for mutagenesis and diseases (some neuronal), small interfering RNA (siRNA) control and suppress retrotransposition, LINEs are drivers of evolution, are actively supporting innovation, also active in germ cells, there controlled by piRNA, the abundance of retrotransposons makes the reverse transcriptase to one of the most abundant proteins.

LTR: Long terminal repeat of retroviruses and some retroelements, one of the strongest promoters known, bind transcription factors, direct the cell machinery towards viral replication, two LTRs are generated during replication of retroviral RNA with duplication at termini of DNA provirus, important for integration, can also promote neighboring genes and can cause cancer (promoter insertion mechanism), can be left as "solo LTRs" in genome as footprint of deleted retroviruses.

LUCA: Last universal common or cellular ancestor, first cell during evolution.

lytic phage: Phage which lyses bacterial host due to stress or other environmental factors, causing derepression, with production of large number of progeny virus particles, destruction of bacterial host, in contrast to "lysogenic" phage, where the phage DNA is integrated, chronic or persistent.

metagenome: The sum of all genomes of an organism or any sample or ecosystem, sewage, feces, lung, etc.

MetaHit: Metagenomics of the Human Intestinal Tract Consortium, 13 research centers in 8 countries, 2008 till 2012 by European Commission, 7th Framework Programme, sequencing of bacterial strains from human intestinal tract.

metastasis: Tumor cell, which leaves the cellular context and settles at other sites of the body, displaced tumor cells growing to tumors at other locations, late stage of cancer.

microbiome: Describing the collective genomes of all microbes in a given sample, in humans amounting to 10 times more bacterial cells than body cells, including about 5 Mio bacterial genes in addition to the 20,000 original human genes, our second genome.

microbiota: Refers to the community of microorganisms, formerly called (micro)flora, populating a certain habitat such as gut, vagina or skin, largely homologous with microbiome.

mitochondrion, mitochondria (pl): "Powerhouse" of eukaryotic cells, supply cellular chemical energy, derived from a bacterium, taken up by symbiosis, with gene loss up to 90% and high degree of specialization instead.

miR: Micro RNA derived from a double-stranded RNA precursor processed by a protein complex RISC with molecular scissors (named Argonaute comprising Paz and PIWI domains), which separate the strands, one strand is discarded, the other one binds sequence-specifically to a target RNA to be destroyed. RISC is the RNA-induced silencing complex.

MPI: Max Planck Institute, research institute of the Max Planck Society, about 100 different institutes mainly in Germany.

MRSA: *Methicillin-resistant Staphylococcus aureus*, also multi-drug resistant hospital bacterium, increasing danger as hospital contaminations and threat for patients, increasing death toll.

Mu: Transposable phage, a model for DNA transposon and retroviral integration mechanism, highly mutagenic (name Mu), related to retroviruses, served for identification of HIV integrase inhibitor such as Raltegravir.

mutant: Genetically changed gene, spontaneous or acquired, also experimentally in test-tubes possible, can lead to resistance against therapies against viruses or cancer.

Myc: Myelocystomatosis viral oncogene in chickens, proto-oncogene as transcription factor in normal cells, overexpressed in cancer cells (neuroblastoma, lung), driving cell cycle progression, is a downstream effector of kinase cascade, DNA-binding protein, nuclear localization, role in cell cycle regulation, heterodimerizes with a shorter Myc-related protein Max, modulating the function.

***Mycoplasma genitalium* and *Mycoplasma mycoides*:** The bacterium *M. genitalium* has one of the smallest genomes in nature of 580,000 base-pairs and 482 genes, first bacterial genome to be synthesized *in vitro* using chemicals only, but to date the synthetic genome has not been shown to be functional in a living bacterium. By inactivating genes of the natural genome one by one and testing if *M. genitalium* can still grow, a trial and error process just as nature does it, 382 essential genes were identified and 100 were dispensable. *M. mycoides* is the second bacterium whose genome has been synthesized. Transplantation of its natural genome into a related emptied bacterial cell created a bacterium that behaved like *M. mycoides* and was able to multiply. New in 2016 a fully replicating viable synthetic cell, even though slowly growing, was generated by Craig Venter, comprising 473 genes (with 149 of then with unknown functions) and 531,000 base pairs.

NASA: National Aeronautics and Space Administration.

NIH: National Institutes of Health, American health authority, in Bethesda, Maryland.

non-coding: Non-coding ncDNA or ncRNA, without triplets, therefore no information for proteins, not coding for amino acids, often regulatory functions, structural information and sequence interactions.

nucleic acid: Polymer composed of nucleotides, deoxyribonucleotides or ribonucleotides, forming DNA or RNA, respectively, chimeras possible, single or double-stranded, linear, circular, fragmented genomes.

nucleotide: (nl) also designated as base or base pair (in the case of double-strand), building block of nucleic acids, ribonucleotides in the case of RNA, deoxyribonucleotide or deoxynucleotide for DNA, consists of a sugar moeity (ribose or deoxyribose) and one of four bases, adenine (A), thymine (T), guanine (G), cytosine (C) for DNA and A, uracil (U, instead of T), G, C for RNA.

oncogene: Cancer or tumor gene of retroviruses, such as src, raf, ras, myc, about 100 known, modified cellular proto-oncogenes, often growth promoting genes or gene regulators (transcription factors), RNA oncogenes are micro RNAs (miR), also T antigen of SV40, a DNA virus.

paramutation: Inheritance of genetic information without mutations, non-Mendelian inheritance, therefore designated as "paramutation", first described by Royal Alexander Brink, a focus of new research, transgenerational or paragenetic effects are distinct from epigenetics, because longer-lasting and heritable, by changes in chromatin packaging, traits in *C. elegans* maintained for 30 generations, exists in plants, described in mice, not known to exist in humans, theory of Lamarck (inheritance of acquired properties).

pararetrovirus: DNA virus, uses the reverse transcriptase for replication such as hepatitis B virus (HBV), the plant virus cauliflower mosaic virus, or foamy viruses, form a large pregenomic RNA inside cells.

PAZ: Protein domain (about 110 AA long), related to retroviral reverse transcriptase, name based on three proteins containing this domain (PIWI, Argonaute and Zwille), component of RNA-induced silencing complex (RISC).

PCR: Polymerase chain reaction, highly sensitive amplification method for DNA molecules, can be used for diagnostics to detect nucleic acids including mutations, allows for quantitation of DNA, performed in a thermocycler, a machine in which a DNA-containing sample is first melted to make it single-stranded and then copied by means of a DNA polymerase enzyme to generate double-strands from each strand (with the help of a starter/primer). Melting and copying is repeated for 20 to 60 times and will amplify the number of copies exponentially (1, 2, 4, 8, 16, etc.) A quantitative reaction including reverse transcription of RNA to DNA beforehand, as e.g. for detection of HIV RNA, is the "qRT-PCR", a method of analysis in very frequent use. A heat-resistant DNA polymerase is the Taq-polymerase, with Taq standing for the bacterium *Thermus aquaticus* improved the efficacy of the PCR.

PDZ domain: Protein module found in some tumor suppressors that binds to hydrophobic terminal amino acid motifs, Src has such a PDZ-binding motif, if it is mutated Src loses contacts to PDZ domain-containing proteins and cells become metastatic. PDZ: P for PSD95 (*post synaptic density*), D for *discs large* in *Drosophila*, and Z for *zonula occludens* — all three proteins are tumor suppressors.

PEP: Post exposure prophylaxis, requires antiretroviral therapy within 4 to 24 or 72 hours depending on the drug to prevent HIV infection in spite of sexual contacts ("morning-after-pill").

peptide: short protein.

phage or bacteriophage: Mostly double-stranded DNA virus of bacteria, can have spikes, can be integrated as DNA prophage into the host genome (temperate or lysogenic cycle) or become episomal leaving the host DNA and lyse bacteria (lytic cycle) for virus/phage replication, most abundant viral type, in oceans, soil, air, etc.

Phagoburn: EU program, Framework 7: The main objective of Phagoburn is to assess the safety, effectiveness and pharmacodynamics of two therapeutic phage cocktails to treat either *Escherichia coli or Pseudomonas aeruginosa* burn wound infections. From 2013 until 2016.

Phoenix: A retrovirus derived from cellular endogenous retroviral sequences, proving that the endogenous sequences are indeed derived from retroviruses which went extinct. Endogenous viruses can protect against exogenous viruses. Human endogenous retrovirus HERV-K(HML-2), related to mouse retroviruses, can form particles and infect animals and replicate, Phoenix was generated from HER-K(HML-2) sequences by Thierry Heidman, 3-million-year-old virus, 8% of human genome are remnants of HERVs but became inactive long ago by mutations.

phycodnavirus: Giant virus, large icosahedral double-stranded DNA virus of algae (phyco), immense genetic diversity, can be lytic and lysogenic, *Chlorella* viruses can destroy algae bloom such as "Red Tide" perhaps mentioned in the bible, affecting geochemical cycling processes and weather patterns, no known pathogenicity for humans.

piRNA: PIWI-interacting RNA, PIWI is an RNase H scissor-type enzyme, which cleaves retrotransposons in germ cells preventing transposon activity (jumping), forms RNA-protein complex, piRNA has 26 to 31 nucleotides, larger than the typical siRNA or microRNA (21 nucleotides), transmitted through semen, present in germ cells, where piRNA counteracts retrotransposon formation, therefore called "guardian of genome", if piRNA is absent the sperm is infertile, leading to para-mutation or transgenerational non-Mendelian inheritance.

PIWI: "P-element-induced wimpy testis" originally in *Drosophila*, protein, RNase H-like enzyme binding to double-stranded nucleic acids or

RNA-DNA hybrids, cleaves the RNA moiety in hybrids or one RNA strand of double-strands for RNA silencing, RNA can come from infections (viruses) or endogenous non-coding RNA regions **as regulatory RNA,** RNase H is an Argonaute protein in the RNA-induced-silencing complex RISC for silencing, expressed in spermatogenesis in testes of mammals but also ovarian cells, perhaps resulted from retrotransposons, other name is rasiRNA (repeat-associated small interfering RNA), PIWI-interacting RNA or piRNA is thought to silence retrotransposons, inhibit jumping to prevent interference of retrotransposons with germ cell formation, are candidates for transgenerational or paramutational changes via chromatin alterations and present in germ cells and sperm, thereby associated with an inheritable form of RNA, piwiRNA or piRNA. Thereby described as the "guardians of the genome".

placenta: See syncytin, cell layer separating immune system of the mother from the embryo.

plasmid: Naked (without associated proteins, see chromatin) circular double-stranded DNA, inside of cells (mainly bacteria), can transfer genes from one cell to another, one example is the Ti plasmid of *Agrobacterium tumefaciens* that induces tumors in plants and is basis for plant gene therapy, plasmid DNA vaccines encode proteins/antigens that are produced by the recipient after injection.

platypus: The only mammal that lays eggs instead of giving birth, it has a duck-like bill and a beaver-like tail, endemic at the east coast of Australia, close to going extinct.

poly-DNA-virus: Polydnavirus, insect virus carrying 30 DNA plasmids of host DNA coding for toxins and other proteins, is in a symbiotic relationship with wasp species, virus particle does not contain the viral genome, which instead is integrated into the wasps' genomes, virus is transmitted together with insect eggs that are laid into living caterpillars, helps feed the wasp larvae by destroying caterpillars as food by means of the transported toxin genes, which resembles the mechanism of viral gene therapy vectors.

polymerase: Enzyme for synthesizing (polymerizing) biopolymers such as nucleic acids, required for replication of DNA or RNA in cells or viruses, the reverse transcriptase is specified as RNA-dependent DNA polymerase.

polymerase chain reaction: See PCR.

prebiotic: See probiotic.

primordial soup: Term coined by Charles Darwin, describes the environmental conditions on early earth in which the first biomolecules and life has emerged, may contain, according to John Sutherland, hydrogen cyanide, hydrogen sulfide, phosphate and water, that together with an energy source such as ultraviolet light, can give rise to the building blocks for life such as amino acids and nucleic acids and lipids.

prion: Fused word composed of "protein and infection", a proteinaceous infectious agent free of nucleic acids "protein only", prion proteins that are naturally present in the brain can become misfolded by environmental or genetic factors or infection with other misfolded prions, misfolding leads to aggregation and neurodegenerative diseases. The healthy prion proteins, called PrP, have different structures than the disease-causing folds (PrPSc after scrapie disease in sheep that is caused by misfolded prions), containing more beta sheets and less alpha turns. Nobel prize awarded to Stanley Prusiner in 1997, can infect humans by contaminated meat, caused "BSE" for bovine spongiform encephalopathy (mad cow disease), or kuru (formerly in tribes in Papua New Guinea due to cannibalism) or scrapie in sheep, general name for prion disease is TSE (transmissible spongiform encephalopathy).

probiotic: Often live bacteria in yogurt or dairy products introduced into the gut, in contrast, the prebiotics are specialized not digested plant fibers that act as fertilizer for the good bacteria which are already there.

prokaryote: Cell without a nucleus, bacteria and archaea, in contrast to eukaryotic cells that contain a nucleus.

prophage (DNA-): A DNA intermediate in the life cycle of DNA phages that is integrated into the DNA of the bacterial host.

protein: Composed of amino acids, main component of living creatures forming muscles, skin, organs, etc. — a steak or eggs in a pan for food is mainly protein composed of amino acids, short proteins are named peptides, proteins with catalytic activity are enzymes, only 2% of human genome codes for proteins.

protist: Unicellular microscopic eukaryotic organisms, including some algae, slime molds, sometimes including also certain multicellular and microbial eukaryotes, informal grouping of diverse eukaryotic organisms except for animals, fungi, and plants.

provirus (DNA-): DNA intermediate of retroviruses as essential step during replication, integrates into the DNA of host cells at random sites and is inherited during cell divisions, name relates to DNA prophage.

PrEP: Pre-exposure prophylaxis, protection of healthy individuals by anti-retroviral therapy to prevent infection by HIV-positive partners.

RISC: RNA-induced silencing complex with Argonaute (see above) cleaving RNA for silencing.

quasispecies: Population of RNA molecules, which are closely related but distinct, similar to swarms, typical of RNA-containing viruses with high mutation frequencies such as HIV and influenza viruses.

Raf: Oncogene of the rat fibrosarcoma virus, derived from cellular homologue c-Raf, a serine/threonine kinase, a new type of oncogene at the time of discovery (by K. Moelling), c-Raf is activated by upstream kinases and a component of the rather universal kinase signaling cascade Ras-Raf-MEK-ERK. In human cancers a partially activated form, B-Raf, is now target of several new drugs, frequent in malignant melanomas and other cancers, heterodimerizes with c-Raf. Activated c-Raf can lead to cancer but also to differentiation and growth arrest. This Janus-type behavior is also true for other oncogenes and a complication for therapies, which can result in the opposite effect, activating instead of inhibiting tumor growth.

RAG 1 and 2: Recombination-activating gene, leads to immunoglobulin diversity V(D)J by recombination, related to transposons and to RNases H cleavage enzymes, harboring the three hallmark acidic amino acids, DDE responsible for cleavage.

Ras: Oncogene of retrovirus for rat sarcoma derived from cellular homologue c-Ras, 20 kD in size, H-Ras (Harvey) and Ki-Ras (Kirsten) isolates, names of scientists, is normally a GTPase enzyme hydrolyzing GTP to GDP, if mutated (G12V, glycine to valine in codon 12) it is deregulated and constitutively active, turning "on" the downstream kinase signaling cascades (Ras-Raf-MEK-ERK) and cellular proliferation, most frequent oncogene in human tumors. More than 30% of human cancers are driven by oncogenic Ras. Difficult to treat because too many downstream effector molecules would be affected. 30 years of research did not lead to inhibitors, the National Cancer Institute, USA, launched a new research program in 2014, http://www.cancer.gov/research/key-initiatives/ras/the-problem.

Ras-Raf-MEK-ERK: Major signaling cascade consisting of protein kinases, also MAP kinase cascade (MAP for mitogen-activated protein, also described as Ras-MAPKKK-MAPKK-MAPK with K indicating kinase) which transfers phosphates from one kinase to the next kinase for downstream activation in response to extracellular stimuli via receptor tyrosine kinases, such as mitogens, stress, heat, proinflammatory cytokines, leading to pleiotropic effects such as proliferation, gene expression, differentiation, mitosis, cell survival, and apoptosis depending on cellular and environmental stimuli and factors. Cascades allow positive and negative feedback mechanisms and complex signaling. Ras has numerous downstream targets, Raf mainly only one, therefore useful target for intervention, transmitting signal from the outside of the cell to the nucleus for influencing gene expression, **MEK:** mitogen activated external kinase, **ERK:** extracellular signal regulated kinase leading to phosphorylation of transcription factors such as c-Myc, leading to gene regulation and cellular programs.

recombinant: Abbreviated as "r", means that a protein of one species is produced by gene technology in another organism, method to produce new therapeutics, used for new crops, regulated by law on genetically modified organisms, GMOs.

recombination: A natural mechanism of new combination of DNA, also under laboratory conditions, combination of DNA fragments by a procedure called cloning.

regulatory RNA: Often small non-coding ncRNA, regulates genes by activating methyltransferases which attach methyl groups to the DNA, or acetylate chromosomes, and change gene expression, often due to environmental factors leading to epigenetic effects, transient only for the lifetime of an organism and normally not inherited to the next generation, first discovered in multicolored maize by Barbara McClintock.

RELIK: Rabbit endogenous lentivirus type K, a complex retrovirus (like HIV) that has been stably integrated (endogenized) in the genome of rabbit hosts, described by Aris Katzourakis, minimum evolutionary age of 12 Mio years.

retrotransposon: Mobile genetic elements or jumping genes following copy-and-paste mechanism involving a reverse transcriptase step, related to retroviruses without coats which do not leave a cell, resulting in gene duplication, an important step during evolution of the cellular genomes, code for RT, RNase H, integrase, protease, flanked by LTRs, 8% of human genome corresponding to 450,000 copies are HERVs plus retrotransposons; retrotransposons without env rare in the human genome but abundant in plant, yeast and insect genomes", often fantasy names such as Gypsy or Copia are retrotransposon in *Drosophila* flies and Ty3 in yeast (or Ty1–Ty5), Sleeping Beauty or Mariner in fish, Lotus in plants, Tn in bacteria, DNA transposons and retrotransposons are so-called "selfish genetic elements" or "jumping genes" which need to be "tamed" or controlled by silencing by siRNA mechanisms with molecular scissors for cleavage, special control of TE in the germ cells occurs by piRNA (PIWI-interacting RNA). Number of retrotransposons can be high: 150,000 to 250,000 in the maize genome.

retrovirus: RNA virus with two identical viral plus-strand RNA genomes, replicates through an essential DNA intermediate, the DNA provirus, which is integrated normally into the host genomic DNA, coding for about 10 genes, gag (for structure), pol (for replication with RT, RNase H, Integrase, and Protease) and env (for coat proteins), and several auxiliary genes for regulation of replication mainly in HIV, can pick up cellular genes, which are transduced by the virus as oncogenes into other cells, the basis for cancer research, can also be manipulated for uptake of therapeutic genes as basis for gene therapy, have strong promoters, the long terminal repeats (LTR), to activate viral but also neighboring cellular genes.

reverse transcriptase or RT: RNA-dependent DNA polymerase, discovered as replication enzyme of retroviruses, but of almost universal importance, transcribing RNA into double-stranded DNA, which integrates into host genomes, is often fused to a nuclease, the ribonuclease H or RNase H, which removes the viral RNA after reverse transcription from RNA-DNA hybrids, and thereby allows first a single-stranded and then a double-stranded DNA to be produced, also used by pararetroviruses such as hepatitis B in humans or plant viruses such as cauliflower mosaic virus, and the monkey Foamy virus, very universal enzyme, present in bacteria, plankton, and yeast, with possibly essential role during evolution as generator or "inventor" of DNA, used by retrotransposons, is present in bacteria with thousands of copies with unknown functions, also present in a few phages, error-prone enzyme resulting in high mutagenic effect (quasispecies), which leads to innovation, also to escape from therapies with drug resistance and from vaccines, is targeted by several antiviral therapies. Most prevalent enzyme in the biosphere due to the large amount of transposable elements which code for reverse transcriptases, 13.5% of the collective genomes in samples from the ocean, 5% of the human genome mainly due to RTs of endogenous retroviruses and LINEs.

ribonuclease H: See also same RNase H, enzyme which specifically degrades the RNA strand of RNA-DNA hybrids during retroviral replication or retrotransposition, many related forms exist, PIWI, or Cas9.

ribosomal RNA: Component of ribosomes, the sequence is the basis for classification of different bacterial types, also defined as 16S/18S rRNA (ribosomal RNA) or at the DNA level rDNA, 16S in bacteria, 18S in eukaryotic cells.

ribosomes are ribozymes: Protein synthesis in ribosomes depends on the ancient ribozyme, slogan: "ribosomes are ribozymes!", catalytic RNA, active component that joins the amino acids.

ribozyme: Small enzymatically active circular RNA with hairpin-loop structure, sometimes inactive, also as viroids in plants, a single type is present in the liver of humans as hepatitis delta virus (HDV), also related to circular or circRNA.

RISC: RNA-induced silencing complex, molecular scissors in a protein complex to cleave RNA, generating small interfering RNA (siRNA or RNAi), RISC consists of the protein argonaute with the domains called PAZ and PIWI (related in structure and function to reverse transcriptase and RNase H of retroviruses).

RNA: Messenger (mRNA), transfer (tRNA), non-coding (ncRNA), small-interfering or silencer (siRNA), interfering (RNAi), long non-coding (lncRNA), long intergenic non-coding (lincRNA) from regions between genes, Piwi-interacting RNA (piRNA) or PIWI-RNA in germ cells, semen and embryos, circular (circRNA), micro RNA (miR) oncogene, repeat-associated siRNA (rasiRNA) controls retrotransposons in germ cells such as piRNA, ribosomal (rRNA), 16S rRNA for prokaryotes,18S rRNA for eukaryotes,

RNase H: Ribonuclease H, enzyme that cleaves and removes RNA in RNA-DNA hybrids during retroviral replication, molecular scissors, also removes RNA primers after initiation of DNA synthesis in eukaryotic cells, viral RNase H removes retroviral RNA after reverse transcription into DNA, RNase H1 and 2A,B,C in human cells, RNase HI und HII in bacteria, large family also comprises integrase, dicer, PIWI from argonaute in RISC with siRNA, RAG1/2 (recombination-activating genes 1/2) required for V(D)

J recombination for antibody production, RuvC endonuclease, Cas9 for cleavage of DNA in RNA-DNA hybrids (as an "DNase H"), or siDNase, or in Prp8 for splicing, most abundant protein fold in nature (see Moelling and Broecker, NYAS 2015), RNase H has five alpha helices and five beta sheets, and three amino acids binding a magnesium ion to keep the structure, only the triad DDE out of about a hundred AA are conserved; this triad is the hallmark of molecular scissors if one is searching for related family members but their distances apart vary.

RT: Abbreviation of reverse transcriptase.

SCID mice: Severe combined immune deficiency mice with a genetic defect, an animal model to analyze human tumor cells, which can be grown in these mice without being rejected, can carry human immune systems or human fetal liver cells which can be infected with HIV, as animal models.

SCID-NOD mice: SCID (severe combined immune deficient) and non-obese diabetic mice. Laboratory mouse model for studying foreign tissue, which is not rejected.

silencing siRNA: Shut-off (silencing) of gene expression in the RNA-induced silencing complex (RISC) which is responsible for cleavage of the targeted RNA, also known as RNA interference or RNAi. The cleavage complex RISC consists of the Argonaute protein and different classes of small non-coding RNAs. The small RNA, also called siRNA or silencer RNA, guide the enzyme Argonaute to complementary mRNAs and induces cleavage, which inhibits gene expression (no protein production by inhibition of translation), Argonaute contains a PIWI domain, an RNase H-folded molecular scissor, siRNA regulates cellular or viral RNA, antiviral defense, is part of immune systems also in plants, exists in *C. elegans* and is also a weak antiviral defense system in humans.

SINE: Small interspersed nuclear element, transposable element (TE), DNA element in the human genome, no RT, no LTRs, smaller than LINEs, often use the reverse transcriptase from LINEs to multiply in the genome, comprise Alu-Elements, 1.5 Mio copies in human genome (13%).

siRNA mechanism of silencing: Small regulatory or silencing RNA, arises from mRNA or viral double-stranded RNA intermediates by processing and cleavage by Drosha (molecular scissors), exported out of the nucleus, further processed by RISC to about 22 nucleotides, which removes one RNA of the double-strand, which binds to a target RNA with related sequence and cleaves the RNA, and eliminates translation and protein expression, and causing gene "silencing" or complete knock-out of a gene, two silencing effects: firstly transcriptional gene silencing (TGS), acting on the promoter by an RNA-induced methylation and thereby blocking the promoter and gene expression, and secondly by posttranscriptional gene silencing (PTGS), by binding to transcribed mRNAs of a gene and its cleavage leading to silencing of gene expression and stop of translation.

Sleeping Beauty: Name of a transposon reconstructed by genetic changes from fish, developed for gene therapy.

solo LTRs: Homologous recombination leads to deletion of the regions between the two LTRs leaving one behind, can activate neighboring genes. About 400,000 in the human genome.

splicing: Is a combination of various regions of an mRNA in eukaryotes, the exons by deleting intermediate sequences, often the non-protein-coding introns by lariat-type looped RNA exclusions, requires knot-free combination so that ribosomes can translate the new RNA sequences into proteins, important for increased genomic complexity and varieties, most important for the complexity of the human genome.

Src: First described oncogene, model, in RSV (Rous sarcoma virus) in chickens, sarcoma, not so frequent in human cancers.

stem cells, totipotent: Most versatile from *in vitro* fertilized eggs, can develop into any human cell type after 4 days, develop into **pluripotent** stem cells, which can also lead to all cell types but not to an entire organism, **multipotent** stem cells are less versatile, give rise to a limited range of cells, **adult** stem cells are multipotent, replace adult cells that have died or lost function, it is an undifferentiated cell in a differentiated tissue (such as brain, blood, muscle, etc.).

suicide: Driving HIV into suicide with a synthetic piece of DNA to form a hybrid, which activates the viral RNase H and destroys viral infectivity, designated also as siDNA (small interfering DNA, K. Moelling). It is a normal step during replication but can be activated prematurely also in viral particles. *Nature Biotechnology* editorial coined the word "suicide". Development as microbicide to prevent HIV transmission during sexual intercourse. The DNA is targeted to the polypurine tract, PPT, and forms a G-tetrad allowing viral uptake. The viral RNase H cannot be inhibited by drugs so far. This is an activation approach for therapy against HIV.

symbiosis: Co-existence of two or more species with long-lasting interdependence, prominent examples are mitochondria, former bacteria in mammalian cells, gain for both of them (mutualism), often gene reduction for the benefit of specialization. Some models suggest that archaea may have fused with bacteria by endosymbiosis leading finally to eukaryotic cells.

syncytin: Captured retroviral envelope protein associated with the unique placental structure of humans and other mammals, ruminants, cell layer consisting of syncytiotrophoblasts separates mother from fetus in humans, makes egg-laying obsolete, over 30 Mio years ago, fusogenic properties of the envelope protein is similar to HIV, where syncytia formation lead to giant cells by fusion, an indicator test for HIV, 18 syncytin family members are known in the animal world all derived from different retroviruses, protect growing embryos from immune-rejection by the mother.

Tara Oceans Expedition: Sailing exploration tour by European Molecular Biology Organization (EMBO), 2009 till 2013, through all oceans with 35,000 plankton sample collections from 200 stations, 40 Mio genes, most unknown, 5,000 viral populations in upper oceans, see *Science*, May 2015.

TE: Transposable elements of all kinds, also jumping genes.

telomerase: A telomeric reverse transcriptase (TERT) which carries its own short RNA as endogenous template and copies it up to thousandfold thereby elongating the DNA or chromosomal ends (the telomeres), active during embryogenesis and in tumors, in a normal cell telomerase activity is shut off and chromosomes are shortened at the ends with each round of cell division, a molecular clock, target for anticancer therapy or against aging, aging is not synchronous in different cells so telomeric length is no predictive marker for life expectancy of an organism.

temperate phage: Non-lytic phage, moderate, integrates its DNA into the DNA of the bacterial host as DNA prophage, can be induced by stress to become lytic by loss of a suppressor allowing replication of progeny and disrupting (lysing) the cell.

timeline: 14 Bio (billion years ago): Big Bang, 4.5 Bio: formation of earth, till 4 Bio: late heavy bombardments and giant impact collision, which generated the moon, 4 Bio: first biomolecules, 580 Mio (million y): complex multicellular life, 540 Mio: Cambrian explosion with diversification of life, 380 Mio: first vertebrate and animals, 230-65 Mio: dinosaurs, 65 Mio: cosmic impact by asteroid to Yucatán (present-day Mexico) with mass extinction and killing of the dinosaurs, 3.2 Mio: Lucy from Africa, 0.2 Mio: *Homo sapiens*, 200,000 to 40,000: Neanderthals.

Ti plasmid: Tumor-inducing double-stranded circular DNA plasmid from *Agrobacterium tumefaciens* used for gene therapy in plants, can transfer therapeutic genes.

transgenerational genetics: See also paramutation, inherited epigenetic chemical modifications of genes and chromatin without mutations, induced by environmental cues, longer-lasting than epigenetic changes for several or even many generations, memory in *C. elegans* is induced by piRNA for 30 generations, remembering ethanol and sex hormone smells, exists also in plants and mice — humans uncertain — research topic: how obese grandfathers have obese grandchildren.

transposon or transposable element (TE): Jumping gene, comprise all kinds of elements such as retroelements or DNA transposons, with mechanisms: "cut-and-paste" causing two genotoxic events, and "copy-and-paste", a retrotransposon, one genotoxic event with gene duplication, TE in human genomes almost 50%, maize 49 to 78%, wheat 90%, *C. elegans* 12%.

triplet: Three nucleotides of the mRNA form the codon, each codon is specific for one AA, are assembled on the ribosome with mRNA as matrix for protein synthesis, coding RNA leads to proteins, non-coding RNA (ncRNA) has regulatory functions.

tumor or T-antigen: Oncogene in small DNA viruses such as Simian Virus 40 (SV40) or Polyomavirus (Py), normally not relevant for human cancer but a cancer model system in the laboratory, cause cancer by integration of the virus.

tumor suppressor: Genes and proteins with antiproliferative function, their loss contributes to cancer, inhibit growth of a cell, counteracts the oncogenes, is often removed by oncogene proteins enhancing the tumorigenicity, loss of tumor suppressors is often a key event in tumor formation and its replacement by gene therapy is of some success against cancer. Viruses can titrate out tumor suppressor proteins for longer life of the cell and more viral progeny.

viroid: Small ribozyme, stable closed hairpin-looped RNA, today about 350 bases, often catalytically active, first biomolecule on our planet, formed 3.9 billion years ago, today mainly in plants, causing crop damage, hepatitis delta virus, only known human ribozyme. Naked virus-like element, no protein coding function but interacting with other nucleic acids, related to regulatory circRNA recently discovered and present in all human cells as chief regulator titrating out siRNAs. Viroids are discussed here as earliest biomolecules with properties like viruses, leading to the concept that viruses are our oldest ancestors, have no protein coats.

virome: All the viruses together present in a sample or ecosystem without knowledge of the individual viruses.

virophage: A virus infecting another virus in analogy to bacteriophage (virus of bacteria), present in giant viruses, making viruses to hosts of viruses and appearance of cell-like entities, names of virophages are Sputnik, or Ma virus in giant viruses, small, about 20,000 bp double-stranded DNA genomes, can destroy the giant virus hosts.

virus: Moving element, often composed of genes with or without proteins, "proteins only" are prions (not generally accepted as viruses, but one could include them), transport of genetic or structural information, horizontal and vertical transmission through somatic and/or germ cells, many unrelated replication mechanisms, highest genetic diversity on the planet, often highly variable by error-prone replication mechanisms, leading to quasispecies with many mutants existing simultaneously. DNA viruses are genetically more stable than RNA viruses, viruses lack ribosomes, except some in giant viruses, and depend on energy supply by cells. Such parasitism may be a late achievement during evolution. RNA molecules such as ribozymes, the viroids, are the most ancient biomolecules, non-coding for proteins (yet), can replicate without cells but need energy. Today viruses need cells for protein synthesis. Viruses are drivers of evolution and built our genomes, supplied in particular the antiviral defense (immune systems). About 50% of human genomes are related to retrovirus elements but may have been much more. They may represent the beginning of life and evolution. Also prions consisting only of proteins and ribozymes/viroids consisting only of RNA have virus-like properties, they replicate, mutate, evolve, can defend themselves, cause diseases(!) — so could be included into the classification of viruses.

virosphere: World of viruses, shown on a poster and listed by the International Committee on Taxonomy of viruses.

WHO: World Health Organization in Geneva, Switzerland, proclaims pandemics and emergency programs.

zoonosis, zoonoses (pl): Viruses transmitted from animal species to humans, most frequent source of viruses affecting humans, often causing disease in humans but not in the natural host animals, coevolution can abolish dangerous properties. Zoonotic transmissions were HIV, influenza, West Nile Fever virus, Ebola, SARS coronavirus, etc. Some of these viruses can spread from humans to humans, which increases the risk of infection.

References

General

Brockman, J., *The Edge*, Annual Question 2005: What do you believe is true even though you cannot prove it?

Blaser, M.J., *Missing Microbes* (Oneworld Publications, 2014).

Chin, J. (Ed.), *Control of Communicable Diseases Manual*, American Public Health Association, 2000.

Crawford, D.H., *Deadly Companions: How Microbes Shaped Our History* (Oxford University Press, New York, 2007).

Domingo, E., Parish, C., and Holland J. (Eds.), *Origin and Evolution of Viruses* (Academic Press, London, 2008), p. 477.

Dyson, F.J., *The Sun, the Genome and the Internet* (Oxford University Press, New York, 1999).

Dyson, F.J., The Scientist as Rebel, *The New York Review of Books*, 2006.

Dyson, F.J., *Origins of Life* (Cambridge University Press, 1986).

Eigen, M., *From Strange Simplicity to Complex Familiarity, A Treatise on Matter, Information, Life and Thought* (Oxford University Press, 2013).

Frebel, A., *Auf der Suche nach den ältesten Sternen* (S. Fischer Press, 2007).

Fischer, E.P., *Das Genom* (S. Fischer Press, 2004).

Flint, J. *et al.*, *Principles of Virology* (ASM Press, 2000).

Mahy, B.W.J., *The Dictionary of Virology* (Academic Press, 2009).

Moelling, K., Are viruses our oldest ancestors? *EMBO Reports*, 2012; 13:1033.

Moelling, K., What contemporary viruses tell us about evolution — a personal view, *Archives Virol.* 2013; 158:1833.

Napier, J., *Evolution* (McGraw-Hill, 2007).

Ryan, F., *Virolution* (Spektrum Akademic Press, 2010).

Science special issue: HIV and TB in South Africa, *Science* 2013; 339:873.

Schrödinger, E., *What Is Life?* (Cambridge University Press, 1967).

van Regenmortel, M. and Mahy, B., *Desk Encyclopedia of General Virology* (Academic Press, 2009).

Villarreal, L.P., *Viruses and the Evolution of Life* (American Society of Microbiology Press, 2005).

Watson, J.D., *The Double Helix* (Norton Critical Edition, 1980; Atheneum, 1980).

Wagener, Ch., *Molekulare Onkologie* (G. Thieme Verlag, 1999).

Witzany, G., A perspective on natural genetic engineering and natural genome editing. Introduction, *Ann. N. Y. Acad. Sci.* 2009; 1178: 1–5.

Witzany, G. (Ed.), *Viruses: Essential Agents of Life* (Springer, 2012).

Zimmer, C., *Parasite Rex: Inside the Bizarre World of Nature's Most Dangerous Creatures* (Free Press, 2000).

Zimmer, C., *A Planet of Viruses, 2nd ed.* (University of Chicago Press, 2015).

Chapter 1

Katzourakis, A., Gifford, R.J. *et al.*, Macroevolution of complex retroviruses, *Science* 2009; 325:1512.

Ziegler, Anna, *Photograph* 51, theatre play.

Maddox, B., *Rosalind Franklin: The Dark Lady of DNA* (HarperCollins, 2002).

Watson, J.D., and Crick, F., Molecular structure of nucleic acids: A structure for DNA. *Nature*, 1974; 248:765.

Crick, F., Central dogma of molecular biology, *Nature* 1970; 227:561.

Eigen, M., Error catastrophe and antiviral strategy, *PNAS* 2002; 99:13374.

Patel, B.H. *et al.*, Common origins of RNA, protein and lipid precursors in a protometabolism, *Nat. Chem.* 2015; 4:301.

Sweetlove, L., Number of species on Earth at 8.7 million, *Nature* 2011, doi:10.1038/news.2011498.

Splicing: Sharp, P.A., The discovery of split genes and RNA splicing, *Trends in Biochemical Sciences* 2005; 30:279.

Chapter 2

Influenza: Taubenberger, J.K., Influenza viruses: Breaking all the rules, *MBio.* 2013; 4: p.ii: e00365-13.

Fouchier, R.A.M., García-Sastre, A., and Kawaoka, Y., H5N1 virus: Transmission studies resume for avian flu, *Nature* 2013; 493:609.

Vaccination: Don't blame the CIA, *Nature* 2011; 265:475.

Bioterrorism: Lane, H.C., La Montagne, H.L., and Fauci, A.S., Bioterrorism: A clear and present danger, *Nature Med.* 2001; 7:1271.

HIV: Cohen, J., AIDS research. More woes for struggling HIV vaccine field, *Science* 2013; 340:667.

Naked DNA: Weber, R., Bossart, W. *et al.*, Moelling, K., Phase I clinical trial with HIV-1 gp160 plasmid vaccine in HIV-1-infected asymptomatic subjects, *Eur. J. Clin. Microbiol. Infect. Dis.* 2001; 11:800.

Lowrie, D.B. *et al.*, Moelling, K., Silva, C.L., Therapy of tuberculosis in mice by DNA vaccination, *Nature* 1999; 400:269.

Schlitz, J.G., Salzer, U., Mohajeri, M.H. *et al.*, Moelling, K., Antibodies from a DNA peptide vaccination decrease the brain amyloid burden in a mouse model of Alzheimer`s disease, *J. Mol. Med.* 2004; 82:706.

Microbicide: Haase, A.T., Early events in sexual transmission of HIV for interventions, *Annu. Rev. Med.* 2011; 62:127.

Matzen, K. *et al.*, Moelling, K., RNase H-mediated retrovirus destruction in vivo triggered by oligodeoxynucleotides, *Nat. Biotechnol.* 2007; 25:669; *Commentary*: Johnson, W.E., Assisted suicide for retroviruses, *Nat. Biotech.* 2007; 25:643.

Wittmer-Elzaouk, L. *et al.*, Moelling, K., Retroviral self-inactivation in the mouse vagina, *Antiviral Res.* 2009; 82:22.

HIV Future: Hütter, G., Transplantation of blood stem cells for HIV/AIDS?, *J. Int. AIDS Soc.* 2009; 12:10.

Faria, N.R. *et al.*, HIV eipdemiology. The early spread and ignition of HIV-1 in human, *Science* 2014; 346:56.

Chapter 3

Reverse transcriptase: Baltimore, D., RNA-dependent DNA polymerase in virions of RNA tumour viruses, *Nature* 1970; 226:1209.

Temin, H.M. and Mizutani, S., RNA-dependent DNA polymerase in virions of Rous sarcoma virus, *Nature* 1970; 226:1211.

Mölling, K. *et al.*, Association of viral RT with an enzyme degrading the RNA of RNA-DNA hybrids, *Nat. New. Biol.* 1971; 234:240.

Moelling, K., RT and RNase H: In a murine virus and in both subunits of an avian virus, *CSH SyQB* 1975; 39:269.

Moelling, K., Targeting the RNase H by rational drug design, *AIDS* 2012;26:1983.

Moelling, K. *et al.*, Relationship retroviral replication and RNA interference, *CSH SyQB* 2006; 71:365.

Simon, D.M. and Zimmerly, S., A diversity of uncharacterized RTs in bacteria, *NAR* 2008; 36:7219.

Lampson, B.C., Inouye, M., and Inouye, S., Retrons, msDNA, bacterial genome, *Cytogenet. Genome Res.* 2005; 110:491.

Moelling, K. and Broecker, F., The RT-RNase H: From viruses to antiviral defense, *Ann. N.Y. Acad. Sci.* 2015; 1341:126.

RNase H: Tisdale, M. *et al.*, Moelling, K., Mutations in the RNase H of HIV-1 infectivity, *J. Gen. Virol.* 1991; 72:59.

Broecker, F., Andrae, K., and Moelling, K., Activation of HIV RNase H, suicide a novel microbicide?, *ARHR* 2012; 28:1397.

Song, J.J. *et al.*, Structure of Argonaute and implications for RISC slicer, *Science* 2004; 305:1434.

Malik, H.S. and Eickbush, T.H., RNase H suggests a late origin of LTR from TE, *Genome Res.* 2001; 11:1187.

Crow, Y.J. *et al.*, Jackson, A.P., Mutations in genes encoding ribonuclease H2 subunits cause Aicardi-Goutière syndrome and mimic congenital viral brain infection. *Nat. Genet.* 2006; 38:910.

Cerritelli, S.M. and Crouch, R.J., Ribonuclease H: The enzymes in eukaryotes, *FEBS J.* 2009; 276:1494.

Nowotny, M., Retroviral integrase superfamily: The structural perspective, *EMBO Rep.* 2009; 10:144.

Arshan Nasir, A., Sun, F.-J., Kim, K.M., and Caetano-Anolles, G., Untangling the origin of viruses and their impact on cellular evolution, *Ann. N.Y. Acad. Sci.* 2014; 1341:61.

Telomerase: Greider, C.W. and Blackburn, E.H., A specific telomere terminal transferase, *Cell* 1985; 43:405.

Skloot, R., *The Immortal Life of Henrietta Lacks* (Crown Publishing Group, 2010).

Nandakumar, J. and Cech, T.R., Finding the end: Recruitment of telomerase, *Nat. Rev. Mol. Cell. Biol.* 2013; 14:69.

Budin, I. and Szostak, J.W., Transition from primitive to modern cell membranes, *PNAS* 2011; 108:5249.

Chapter 4

Sarcoma saga: Martin, G. S., The hunting of the Src, *Nat. Rev. Mol. Cell. Biol.* 2001; 2:467.

Yeatman, T.J. and Roskoski, R. Jr., A renaissance for SRC, *Nat. Rev. Cancer* 2004; 4: 470.

Oncoproteins: Donner, P., Greiser-Wilke, I. and Moelling, K., Nuclear localization and DNA binding of Myc, *Nature* 1982; 296:262 and Moelling, K. *et al.*, *Nature* 1984, 312:551, Myb, *Cell* 1985; 40:983.

Axel, R., Schlom, J., and Spiegelman, S., Presence in human breast cancer of RNA homologous to MMTV, *Nature* 1972; 235:32.

Raf-kinase: Moelling, K. *et al.*, Serine–threonine PK activities of Mil/Raf, *Nature* 1984; 312:558.

Zimmermann, S. and Moelling, K., Phosphorylation and regulation of Raf by Akt, *Science* 1999; 286:1741.

Rommel, C. *et al.*, Differentiation specific inhibition of Raf-MEK-ERK by Akt, *Science* 1999; 286:1738.

Zimmermann, S., Moelling K., and Radziwil, G., MEK1 mediates positive feedback on Raf, *Oncogene* 1997; 15:1503.

Dummer, R. and Flaherty, K.T., Resistance with tyrosine kinase inhibitors in melanoma, *Curr. Opin. Oncol.* 2012; 24:150.

Das Thakur, M., *et al.*, Modelling Vemurafenib resistance in melanoma, *Nature* 2013; 494:251.

Holderfield, M. *et al.*, Targeting RAF kinases for cancer therapy: BRAF-mutated melanoma and beyond, *Nature Rev. Cancer* 2014; 14:455.

McMahon, M., Parsing out the complexity of RAF inhibitor resistance, *Pigment Cell Melanoma Res.* 2011; 24:361.

Sun, C. *et al.*, Bernards, R., Reversible resistance to BRAF(V600E) inhibition in melanoma, *Nature* 2014; 508:118.

Myc: Liu, J. and Levens, D., Making Myc, *Curr. Top Microbiol. Immunol.* 2006; 302:1.

Tumor suppressor: Gateff E: Malignant neoplasms of genetic origin in Drosophila, *Science* 1978; 200:1448.

Sherr, C.J. and McCormick, F., The RB and p53 pathways in cancer. *Cancer Cell* 2002; 2:103.

Yi, L. *et al.*, Multiple roles of p53 in somatic and stem cell differentiation, *Cancer Res.* 2012; 72:563.

Wang, T. *et al.*, Retroviruses shape the transcriptional network of p53. *PNAS* 2007; 104:18613.

Metastases and cancer: Baumgartner, M. *et al.*, Moelling, K., SRC-migration and invasion by PDZ, *MCB* 2008; 28:642.

Broecker, F. *et al.*, Moelling, K., Transcription of C-terminal metastatic c-Src mutant, *FEBS J.* 2016; 283:1669.

Cancer general: Vogelstein, B. *et al.*, Cancer genome landscape, *Science* 2013; 339:1546.

Hanahan, D. and Weinberg, R.A., Hallmarks of cancer: The next generation, *Cell* 2011; 144:646.

zur Hausen, H., Papillomaviruses and cancer: From basics to clinical application, *Nat. Rev. Cancer* 2002; 2:342.

Cancer different: Han, Y.C. *et al.*, Ventura, A., miR-17-92-mutant mice, *Nat. Genet.* 2015; 47:766–75.

Prostata: Kearney, M. *et al.*, Coffin, J.M., Multiple sources of contamination in XMRV infection, *PLoS ONE* 2012; 7:e30889.

Bacteria and cancer: Salama, N. *et al.*, Life in the human stomach: Helicobacter pylori, *Nat. Rev. Microbiol.* 2013; 11:385.

Chapter 5

Phages and microbiome: Suttle, C.A., Viruses in the sea, *Nature* 2005; 437:356.

Reardon, S., News: Phage therapy: Phage therapy gets revitalized, *Nature* 2014; 510:15.

Young, R.Y. and Gill, J.J., Phage therapy redux — What is to be done?, *Science* 2015; 350:1163.

Delwart, E., A roadmap to the human virome, *PLoS Pathog.* 2013; 9:e1003146.

Zarowiecki, M., Metagenomics with guts, *Nat. Rev. Microbiol.* 2012; 10:674.

Turroni, F. *et al.*, Human gut microbiota and bifidobacteria, *A. van Leeuwenhoek* 2008; 94:35.

Gut: Turnbaugh, P.J. *et al.*, Gordon, J.I., The effect of diet on human gut microbiome, *Sci. Transl. Med.* 2009; 1:6ra14.

Qin, J. *et al.*, A human gut microbial gene catalogue by metagenomic sequencing, *Nature* 2010; 464:59.

Katsnelson, A. *et al.*, Twin study surveys genome for cause of multiple sclerosis, *Nature* 2010; 464: 1259.

Mokili, J.L., Rohwer, F. and Dutilh, B.E., Metagenomics and future virus discovery, *Curr. Opin. Virol.* 2012; 2:63.

Cesarean section: Dominguez-Bello, *et al.*, Microbiota of cesarean-born infants, *Nat. Med.* 2016; doi:10.1038/nm.4039.

Endogenous viruses: Weiss, R.A. and Stoye, J.P., Virology. Our viral inheritance, *Science* 2013; 340:820.

Feschotte, C. and Gilbert, C., Endogenous viruses: Viral evolution and host biology, *Nature Rev. Gen.* 2012; 13:283.

Plants global warming: Roossinck, M.J., The good viruses: Viral mutualistic symbioses. *Nat. Rev. Microbiol.* 2011; 9:99.

Roossinck, M.J., Lifestyles of plant viruses, *Philos. Trans. R. Soc. Lond. B. Biol. Sci.* 2010; 365:1899.

Placenta: Mi, S. *et al.*, Syncytin — a retroviral envelope in human placenta. *Nature* 2000; 403:785.

Polydnavirus: Bezier, A. *et al.*, Polydnaviruses of braconid wasps derive from an ancestral nudivirus, *Science* 2009; 323: 926.

Strand, M.R. and Burke, G.R., Polydnaviurses as symbionts and gene delivery systems, *PLoS* 2012; Pathog 8: e1002757.

Prionen: Mahal, S.P. *et al.*, Transfer of a prion strain leads to emergence of strain variants, *PNAS* 2010; 107:22653.

Chapter 6

Chlorella virus: Van Etten, J., Giant viruses; *American Scientist* 2011.

Van Etten, J., Lane, L.C. and Dunigan, D., DNA viruses: The really big ones (Giruses), *Ann. Rev. Microbiol.* 2010;64:83.

Broecker, F. *et al.*, Viral composition in the intestine after FT, *CSH Mol. Case Stud.* 2016; 2:a000448.

Moelling, K. (Ed.), Nutrition and microbiome, *Ann. N.Y. Acad. Sci.* (2016).

Mimivirus: Boyer, M. *et al.*, Mimivirus genome reduction after intraamoebal culture, *PNAS* 2011; 108:10296.

Raoult, D. and Forterre, P., Redefining viruses: Lessons from mimivirus, *Nat. Rev. Microbio.* 2006; 6:315.

Raoult, D. *et al.*, Claverie, J.M., The 1.2-megabase genome sequence of mimivirus, *Science* 2004; 306:1344.

Philippe, N. *et al.*, Pandoraviruses: Amoeba viruses with genomes up to 2.5 Mb, *Science* 2013; 341:281.

La Scola, B. *et al.*, Raoult, D., A giant virus in amoebae, *Science* 2003; 299:2033.

Mollivirus sibericus: Legendre, M. *et al.*, Mollivirus sib. 30,000y giant virus in Acanthamoea, *PNAS* 2015; 112:10795.

Sambavirus: Campos, R. K., Sambavirus: Mimivirus from rain forest, The Brazilian Amazon, *Virol. J.* 2014; 11:95.

Origin of giant viruses: Pennisi, E., Ever bigger viruses shake tree of life, *Science* 2013; 341:226.

Yutin, N., Wolf, Y.I., and Koonin, E.V., Origin of giant viruses from smaller DNA viruses, *Virology* 2014; 466–467:38.

Virophages, Sputnik: Yutin, N., Raoult, D. and Koonin, E., Virophages, polintons, and transpovirons: A complex evolutionary network of diverse selfish genetic elements with different reproduction strategies, *Virology J.* 2013; 1:15.

Fischer, M.G. and Suttle, C.A., A virophage at the origin of large DNA transposons, *Science* 2011; 332:231.

Forterre, P., The origin of viruses and their possible roles in major evolutionary transitions, *Virus Res.* 2006; 117:5.

Amoeba: Huber, H. *et al.*, A new phylum of archaea represented by a nanosized hyperthermophilic symbiont, *Nature* 2002; 417:63.

Slimani, M. *et al.*, Amoebae as battlefields for bacteria, giant viruses, and virophages. *J. Virol.* 2013; 87:4783.

Eyes: Yutin, N. and Koonin, E.V., Proteorhodopsin genes in giant viruses, *Biology Direct* 2012; 7:34.

Archaea: Stetter, K.O., A brief history of the discovery of hyperthermophilic life, *Biochem. Soc. Trans.* 2013; 41:416.

Podar, M. *et al.*, A genomic analysis of the archaeal system *Ignicoccus hospitalis–Nanoarchaeum equitans*, *Genome Biol.* 2008; 9: R158.

Mochizuki, T. *et al.*, Archaeal virus with exceptional architecture, largest ssDNA genome, *PNAS* 2012; 109:13386.

Prangishvili, D., Forterre, P. and Garrett, R.A., Viruses of the archaea: A unifying view, *Nat. Rev. Micro.* 2006; 4:837.

Chapter 7

Endogenous viruses: Weiss, R.A., The discovery of endogenous retroviruses, *Retrovirol.* 2006; 3:67.

Phoenix: Katzourakis, A. *et al.,* Discovery of first endogenous lentivirus, *PNAS* 2007; 10:6261.

Katzourakis, A. and Gifford, R.J., Endogenous viral elements in animal genomes, *PLoS Genet.* 2010: 6:e1001191.

Dewannieux, M. *et al.,* Identification of an infectious progenitor for the HERV-K, *Genome Res.* 2006; 16:1548.

Koalas: Tarlinton, R.E., Meers, J., and Young, P.R., Retroviral invasion of the koala genome, *Nature* 2006; 442:79.

Paleovirology: Belyi, V.A., Levine, A.J. and Skalka, A.M., Sequences from ancestral ssDNA viruses in vertebrate genomes: The parvoviridae and circoviridae are more than 40 to 50 mio years old, *J. Virol.* 2010; 84:12458.

Gifford, R.J., A transitional endogenous lentivirus from a basal primate for lentivirus evolution, PNAS 2008; 105:20362.

Katzourakis, A. *et al.,* Macroevolution of complex retroviruses, *Science* 2009; 32:1512.

Crippled viruses: Lander, E.S. *et al.,* Initial sequencing and analysis of the human genome, *Nature* 2001; 409:860.

Cordaux, R. and Batzer, M.A., The impact of retrotransposons on human genome evolution, *Nat. Rev. Genet.* 2009; 10:691.

SnapSot, Transposons, *Cell* 2008; 135:192.

Singer, T. *et al.,* LINE-1 retrotransposons in neuronal genomes? *Trends in Neurosciences,* 2010; 33:345.

Pollard, K.S. *et al.,* RNA expressed during cortical development evolved in humans, *Nature* 2006; 443:167.

Finnegan, D.J., Retrotransposons, *Current Biol.* 2012; 22: R432-7.

McClintock, B.: Harshey, R.M., The Mu story: How a maverick phage moved the field forword, *Mobile DNA* 2012; 3:21.

Sander, D.M. *et al.,* Intracisternal A-type retroviral particles in autoimmunity, *Microsc. Res. Tech.* 2005; 68:222.

Epigenetic Agouti mouse: Dolinoy, D.C. *et al.,* The Agouti mouse model: An epigenetic biosensor for nutritional and environmental alterations on the fetal epigenome, *Nutr. Rev.* 2008, 66 Suppl. 1:S7–11.

Rassoulzadegan, M., RNA-mediated non-Mendelian inheritance of an epigenetic change in the mouse, *Nature* 2006; 441:469.

Sleeping beauty: Luft, F.C. *et al.,* Sleeping Beauty jumps to new heights, *Mol. Med.* 2010; 88:641.

Fish: Amemiya, C.T. *et al.*, African coelacanth genome insights into tetrapod evolution, *Nature* 2013; 496:311.

Platypus: Warren, W.C. *et al.*, Genome analysis of the platypus unique evolution, *Nature* 2008; 453:175.

Introns: Morris, K. and Mattick, J.S., The rise of regulatory RNA, *Nat. Rev. Genet.* 2014; 15:423.

Chabannes, M. *et al.*, Three infectious viral species lying in wait in the banana genome, *J. Virol.* 2013; 87:8624.

Bartel, D.P., MicroRNAs target recognition and regulatory functions, *Cell* 2009; 136:215.

Zimmerly, S. *et al.*, GroupII intron mobility, *Cell* 1995; 82:545.

Beauregard, A., Curcio, M.J. and Belfort M., The take and give RT elements and hosts, *Ann. Rev. Genet.* 2008; 42:587.

Lambowitz, A.M. and Zimmerly S., Group II introns: Mobile ribozymes in DNA, *CSH Perspect. Biol.* 2011; 3:a003616.

Galej, W.P. *et al.*, Crystal structure of Prp8 reveals active site of slicosome, *Nature* 2013; 493:638.

Pena, V. *et al.*, Structure and function of an RNase H at the heart of the spliceosome, *EMBO J.* 2008; 27:2929.

ENCODE: Biemont, C. and Vieira, C., Junk DNA as an evolutionary force, *Nature* 2006; 443:521.

Venter, C., Multiple personal genomes await, *Nature* 2010, 464:676.

Lev, S. *et al.*, Venter, C., The diploid genome sequence of an individual human, *PLoS Biol.* 2007; 5:e254.

Fukai, E. *et al.*, Derepression of the plant Chromovirus LORE1 induces germline transposition in regenerated plants, *PLoS Genet.* 2010; 6(3):e1000868.

Chapter 8

RNA: Eigen, M., Error catastrophe and antiviral strategy, *PNAS* 2002; 99:13374.

Biebricher, C.K. and Eigen, M., What is a quasispecies, *Curr. Top Microbiol. Immunol.* 2006; 299:1.

Biebricher, C.K. and Eigen, M., The error threshold, *Virus Res.* 2005; 107:117.

Lincoln, T.A. and Joyce, G.F., Self-sustained replication of an RNA enzyme, *Science* 2009; 323:1229.

Doudna, J.A. and Szostak, J.W., RNA-catalysed synthesis of complementary-strand, *RNA Nature* 1989; 33:519.

Viroids: Steger, G. *et al.*, Structure of viroid replicative intermediates of PST viroid, *Nucl. Acids Res.* 1986; 14:9613.

Villarreal, L.P., The widespread evolutionary significance of viruses, *Viroids Cell Microbiol.* 2008;10:2168.

Cech, T.R. *et al.*, Hammerhead nailed down, *Nature* 1994; 372:39.

Koonin, E.V. and Dolja, V.V., A virocentric perspective on the evolution of life, *Curr. Opin. Virol.* 2013; 5:546.

Lambowitz, A.M. and Zimmerly, S., Mobile group II introns, *Ann. Rev. Genetics* 2004; 38:1.

Adamala, K., Engelhart, A.E., and Szostak, J.W., Generation of functional RNAs from inactive oligonucleotide complexes by non-enzymatic primer extension, *J. Am. Chem. Soc.* 2015; 137:483–489.

Forterre, P., Defining life: The virus viewpoint, *Origins of Life and Evolution of Biospheres* 2010; 40:51.

Moelling, K., Are viruses our oldest ancestors? *EMBO Reports* 2012; 13:1033.

Holmes, E.C., What does virus evolution tell us about virus origins? *Journal of Virology* 2011; 85:5247.

Plankton: Lescot, M. *et al.* and Ogata, H., Reverse transcriptase genes are highly abundant and transcriptionally active in marine plankton assemblages, *ISME Journal* 2015; 1–13.

Circular RNA: Memczak, S. *et al.*, Circular RNAs with regulatory potency, *Nature* 2013; 495:333.

Hansen, T.B. *et al.*, Natural RNA circles function as efficient microRNA sponges, *Nature* 2013; 495:384.

Hansen, T.B., Kjems, J., and Damgaard, C.K., Circular RNA and miR-7 in cancer, *Cancer Res.* 2013; 73:5609.

Ford, E. and Ares, M. Jr., Circular RNA using ribozymes from a T4 group I intron, *PNAS* 1994; 12; 91:3117.

Ribozymes and ribosomes: Wilusz, J.E. and Sharp, P.A., A circuitous route to non-coding RNA. *Science* 2013; 340:44.

Navarro, B. *et al.*, Viroids: Infect a host and cause disease without encoding proteins, *Biochemie* 2012; 94:1474.

Hammann, C. and Steger, G., Viroid-specific small RNA in plant disease, *RNA Biol.* 2012; 9:809.

Bartel, D.P., MicroRNAs target recognition and regulatory functions, *Cell* 2009; 136:215.

Eilus, J.E. and Sharp, P.A., A circuitous route to non-coding RNA, *Science* 2013; 340:440.

Proteins: Moore, P.B., and Steitz, T.A., The ribosome revealed, *Trends in Biochemical Sciences* 2005; 30:28.

Ma, B.G. *et al.*, Zhang, H.Y., Characters of very ancient proteins, *BBRC* 2008; 366:607.

Chaperone: Muller, G. *et al.*, NC protein of HIV-1 for increasing catalytic activity of a ribozyme, *J. Mol. Biol.* 1994; 242:422.

Clover leaf: Dreher, T.W., Viral tRNAs and tRNA-like structures, *Rev. RNA* 2010; 1:402.

Hammond, J.A., Comparison and functional implications of viral tRNA-like structures, *RNA* 2009; 15:294.

Witzany, G. (Editor), *Viruses: Essential Agents of Life* (Springer 2012), p. 414.

Plant viruses: Eickbush, D.G., Retrotransposon ribozyme and its self-cleavage site, *PLoS One* 2013; 8(9):e66441.

Webb, C.H. *et al.*, Widespread occurrence of self-cleaving ribozymes, *Science* 2009; 326:953.

Hammann, C. *et al.*, The ubiquitous hammerhead ribozyme, *RNA* 2012; 18:871.

Hepatitis delta virus: Braza, R. and Ganem, D., The HDAg may be of human origin, *Science* 1996; 274:90.

Taylor, J., and Pelchat, M., Origin of hepatitis delta virus, *Future Microbiol.* 2010; 5:393.

Flores, R., Ruiz-Ruiz, S. and Serra, P., Viroids and hepatitis delta virus, *Semin-Liver Dis.* 2012; 32:201.

Retrophage: Liu, M. *et al.*, Bordetella bacteriophages encoding RT-mediated tropism-switching, *J. Bacteriol.* 2004; 186:1503.

Doulatov, S. *et al.*, Tropism switching in Bordetella bacteriophage by diversity-generating retroelements, *Nature* 2004; 431:476.

Tobacco viruses: Buck, K.W., Replication of tobacco mosaic virus RNA, *Philos. Trans. R. Soc. Lond. B. Biol. Sci.* 1999; 354:613.

Beijerinck, M.W., Contagum vivum fluidum of tobacco leaves, Phytopathol Classics No. 7, ed. Johnson, J., Am. Phyto. Soc., 1898.

Dreher, T.W., Viral tRNAs and tRNA-like structures, *Rev. RNA* 2010; 1:402.

Hammond, J.A. *et al.*, 3D architectures of viral tRNA-like structures, *RNA* 2009; 15:294.

Apple tree: Mitrovic, J. *et al.*, Sequences of Stolbur phytoplasma from DNA, *Mol. Microbiol. Biotech.* 2014; 24:1.

Georgiades, K. *et al.*, Gene gain and loss events in *Rickettsia* and *Orientia* species, *Biol. Direct* 2011; 6:6.

Plants: Roossinck, M.J., Lifestyles of plant viruses, *Philos. Trans. R. Soc. Lond. B. Biol. Sci.* 2010; 365:1899.

Geminiviruses: Krupovic M. *et al.*, Geminiviruses: A tale of a plasmid becoming a virus, *BMC Evol. Biol.* 2009; 9:112.

Tulipomania: Lesnaw, J.A. and Ghabrial, S.A., Tulip breaking: Past, present and future, *Plant Disease* 2000; 84:1052.

Murray, J.D., How the leopard gets its spots, *Sci. Am.* 1988; 3:80–87.

Chapter 9

Interferon: McNab, F. *et al.,* O'Garra, A., Type Interferons in infectious disease, *Nat. Rev. Immunology* 2015; 15:87–103.

siRNA: Baulcombe, D.C. and Dean, C., Epigenetic regulation in plant to the environment, *CSH Pers. Biol.* 2014; 6(9):a019471.

Wilson, R.C. and Doudna, J.A., Molecular mechanisms of RNA interference, *Annu. Rev. Biophys.* 2013; 42:217.

Moelling K., Matskevich A., and Jung J.S., Relationship between retroviral replication and RNA interference, *CSH–SyQB* 2006; 71:365–8.

Matskevich, A. and Moelling, K., Dicer is involved in protection against influenza A virus infection, *J. Gen. Virol.* 2007; 88:2627–35.

Song J.J., *et al.,* The crystal structure of the Argonaute2 PAZ domain reveals an RNA binding motif in RNA effector complexes, *Nat. Struct. Biol.* 2003; 10:1026–32.

Nowotny, M., Retroviral integrase superfamily: The structural perspective. *EMBO Reports* 2009; 19:144–151.

Nasir, A. and Caetano-Anolles, G., Origin of viruses and their impact on cellular evolution, *Ann. N.Y. Acad. Sci.* 2015; 1341:61.

Moelling, K. and Broecker, F., The RT-RNase H: From viruses to antiviral defense, *Ann. N.Y. Acad. Sci.* 2015; 1341:126–35.

Orsay virus: Sterken, M.G. *et al.,* A heritable antiviral RNAi response limits Orsay virus infection, *PLoS One* 2014; 9(2).

CRISPR/Cas9: Doudna, J.A. and Charpentier, E., Genome editing. The new frontier of genome engineering with CRISPR-Cas9, *Science* 2014; 346 (6213):1258096.

Jinek, M. *et al.,* Structures of Cas9 endonucleases, *Science* 2014; 343, 1247997.

Hsu, P.D., Lander, E.S., and Zhang, F., Development and applications of CRISPR-Cas9 for genome engineering, *Cell* 2014; 157:1262.

Barrangou, R. and Marraffini, L.A., CRISPR-Cas systems: Procaryotes to adaptive immunity, *Review Mol. Cell.* 2014; 54:234.

Krupovic, M. *et al.,* Casposons: Self-synthesizing DNA transposons of CRISPR-Cas immunity, *BMC Biology* 2014; 12:36.

Horvath, P. and Barrangou R., CRISPR/Cas9, the immune system of bacteria and archaea, *Science* 2010; 327:167.

Swarts, D.C. *et al.,* DNA-guides DNA interference by a procaryotic Argonaute, *Nature* 2014; 507:258.

Moelling, K. *et al.,* Silencing of HIV by hairpin-loop DNA oligonucleotide (siDNA), *FEBS Letters* 2006; 580:3545.

Zhou, L. *et al.,* Transposition of hAT elements links TE and V(D)J recombination, *Nature* 2004; 432:995.

Bateman, A., Eddy, S.R., and Chothia, C., Members of the immunoglobulin superfamily in bacteria, *Protein Sci.*1996; 5:1939.

Beauregard, A., Curcio, M.J. and Belfort M., The take and give between TE, *Annu. Rev. Genet.* 2008; 42:587.

Antisense: Isselbacher, K.J., Retrospective: Paul C. Zamecnik (1912–2009), *Science* 2009; 326:1359.

Citron, M. and Schuster, H., The c4 repressors of bacteriophages P1 and P7 are antisense RNAs, *Cell* 1990; 62:591.

Chapter 10

Reardon, S., Phage therapy gets revitalized, *Nature* 2014; 510:15.

Young, Ry. and Gill, J.J., Phage therapy redux — What is to be done? *Science* 2015; 350:1163.

Viertel, T.M., Ritter, K. and Horz, H.P., Phage therapy against MDR pathogens, *J. Antimic. Chemother.* 2014; 69:2326–36.

Phagoburn: EU program: The main objective of Phagoburn is to assess the safety, effectiveness and pharmacodynamics of two therapeutic phage cocktails to treat either *E. coli* or *P. aeruginosa* burn wound infections, 2013–2016.

Vanessa, K. *et al.*, Gut microbiota from twins discordant for obesity, *Science* 2013; 341:1079.

Multiresistant bacteria: Nübel, U., MRSA transmission on a neonatal intensive care unit, *PLoS One.* 2013;8(1):e54898.

Stower, H., Medical genetics: Narrowing down obesity genes, *Nat. Med.* 2014;20:349.

Lepage, P. *et al.*, Dore, A., Metagenomic insight into our gut`s microbiome, *Gut* 2013; 62:146.

Smemo, S., Obesity-associated variants within FTO and with IRX3, *Nature* 2014; 507:371.

Blaser, M. and Bork, P. *et al.*, The microbiome explored: Recent insights and future challenges, *Nat. Rev. Microbiol.* 2013;11:213.

Transgenerational epigenetics: Tobie, E.W. *et al.*, DNA methylation to prenatal famine, *Hu. Mol. Gen.* 2009; 18;2:4046.

Bygren, L.O. *et al.*, Grandmothers' early food supply influenced mortality of female grandchildren, *BMC Genet.* 2014; 15:12.

Norman, J.M. *et al.* Alterations in the enteric virome in inflammatory bowel disease, *Cell* 2015;160: 447.

Grossniklaus, U. *et al.*, Transgenerational epigenetic inheritance: How important is it?, *Nat. Rev. Genet.* 2013; 14:228.

Pembrey, M.E. *et al.*, Sex-specific, male-line transgenerational responses in humans, *Eur. J. Hum. Genet.* 2006 ; 14:159.

Arabidopsis Genome Initiative, Analysis of the genome sequence of *Arabidopsis thaliana*, *Nature* 2006; 441:469.

Waterland, R.A. and Jirtle, R.L.,TE targets for early nutritional effects on epigenetic gene regulation. *MCB* 2003; 23:5293.

Slotkin, R.K. and Martienssen, R., TE and the epigenetic regulation of the genome, *Nat. Rev. Genet.* 2007; 8:2.

Pfeifer, A. *et al.*, Verma, I.M., Lentiviral vectors: Lack of gene silencing in embryos, *PNAS* 2002; 99:2140.

Dutch famine: Lumey, L.H. *et al.*, The Dutch hunger winter families study, *Int. J. Epidemiol.* 2007; 36:1196.

Roseboom, T., de Rooij, S. and Painter, R., The Dutch famine and consequences for adult health, *Early Hum. Dev.* 2006; 82:485.

Cao-Lei, L. *et al.*, DNA methylation triggered by prenatal stress exposure to ice storm, *PLoS One* 2014; 9:9(e107653).

Ost, A. *et al.*, Paternal diet defines offspring chromatin state and intergenerational obesity, *Cell* 2014; 159:1352.

Phage therapy: Vandenheuvel, D., Lavigen, R. and Brüssow, H., Bacteriophage therapy, *Annu. Rev. Virol.* 2015; 2:599.

Gut-brain axis: Bercik, P. *et al.*, Intestinal microbiota affect BDNF, *Gastroenterology* 2011; 141:599.

Obesity gene counts: Le Chatelier *et al.*, Richness of human gut microbiome and metabolic markers, *Nature* 2013;500:541.

Gut microbiota: Qin, J. *et al.*, Wang, J., A metagenome-wide study of gut microbiota in diabetes, *Nature* 2012; 490:55.

Cotillard, A. *et al.*, Ehrlich, S.D., Dietary intervention impact on gut microbial gene richness, *Nature* 2013; 500:585.

Ackerman, D. and Gems, D., The mystery of *C. elegans* aging: An emerging role for fat, *Bioessays* 2012; 34:466.

Fecal transfer: Broecker F., Rogler, G., and Moelling, K., Intestinal microbiome of *C. diff.* with FT, *Digestion* 2013;88:243.

Reyes, A. *et al.*, Viruses in the faecal microbiota of monozygotic twins and their mothers, *Nature* 2010; 466:344.

Gut virome: Norman, J.M. *et al.*, Alterations in the enteric virome in inflammatory bowel disease, *Cell* 2015;160:447.

Broecker, F. *et al.*, Bacterial and viral compositions of *C. diff.* patient after fecal transplant, *CSH Molecular Case Stud.* 2015; 2:a000448.

Elba worms: McCutcheon, J.P. and Moran, N.A., Genome reduction in symbiotic bacteria, *Nat. Rev. Micr.* 2012; 10:13.

Dubilier, N., Bergin, C. and Lott, C., Symbiotic diversity in marine animals, *Nat. Rev. Micro.* 2008; 6:725.

Chapter 11

Viruses: Baltimore, D., Gene therapy. Intracellular immunization, *Nature* 1988; 335:39.

Ott, M.G. *et al.*, Grez, M., Correction of X-linked chronic granulomatosis by gene therapy, *Nat. Med.* 2006; 12:401.

Rossi, J.J., June, C.H. and Kohn, D.B., Genetic therapies against HIV, *Nat. Biotechnol.* 2007; 25:144.

Pachuk, C.J. *et al.*, Selective cleavage of bcr-abl chimeric RNAs by a ribozyme, *Nucl. Acids Res.* 1994; 22:301.

Lipizzan horses, malignant melanoma: Heinzerling, L. *et al.*, IL-12 DNA into melanoma patients and white horses, *Hu. Gene. Ther.* 2005; 16:35, *J. Mol. Med.* 2001; 78:692.

Mosquito vaccination: Bian, G. *et al.*, Wolbachia in Anopheles against plasmodium infection, *Science* 2013; 340:748.

Plants gene therapy: Fresco, L.O., The GMO stalemate in Europe, *Science* 2012; 33:883.

Pappas, K.M., Cell-cell signaling and the Agrobacterium tumefaciens Ti plasmid, *Plasmid* 2008; 60:89.

Thomson, H., Plant science: The chestnut resurrection, *Nature* 2012; 490:22–3.

Choi, G.H. and Nuss, D.L., Hypovirulence of chestnut blight fungus by an infectious viral cDNA, *Science* 1992; 257:800–803.

D'Hont, *et al.*, The banana (*Mus acuminata*) genome and the evolution of mono-cotyledonous plants, *Nature* 2012;488:213.

Schnable, P.S. *et al.*, The B73 maize genome: Complexity, diversity, and dynamics, *Science* 2009; 326:1112.

Yeast/fungi: Goffeau, A. *et al.*, Life with 6000 genes, *Science* 1996; 274:546.

Goffeau, A., Genomics: Multiple moulds, *Nature* 2005; 438:1092.

Roossinck, M.J., The good viruses: Viral mutualistic symbiose, *Nat. Rev. Microbiol.* 2011; 9:99.

Márquez, L.M. *et al.*, A virus in a fungus in a plant: Symbioses for thermal toler-ance, *Science* 2007; 315:513.

Esteban, R. and Fujimura,T., Yeast 23S RNA narnavirus shows signals for replica-tion, *PNAS* 2015; 100, 5:2568.

Stem cells: Qian, L. *et al.*, In vivo reprogramming cardiac fibroblasts into cardio-myocytes, *Nature* 2012; 485:593.

Hydra: Chapman, *et al.*, The dynamic genome of hydra, *Nature* 2010; 464:592.

Renard, E. *et al.*, Origin of the neuro-sensory system: In sponges, *Integr. Zoolog.* 2009; 4:294–308.

Smelick, C. and Ahmed S., Achieving immortality in the *C. elegans* germline, *Ageing Res. Rev.* 2005; 4:67–82.

Third tooth: Arany, P.R., Photoactivtion of TGF-β1 directs stem cell for regeneration, *Sci. Transl. Med.* 2014; 6(238)238.

Chapter 12

Synthetic biology: Hutchinson, C.A. *et al.*, Design and synthesis of a minimal bacterial genome, *Science* 2016; 351:1414.

Glass, J.I. *et al.*, Venter C: Essential genes of a minimum bacterium, *PNAS* 2006; 103:425.

Gil, R. *et al.*, A determination of the core of a minimal bacterial gene set, *Microbiol. Mol. Rev.* 2004; 68:518.

Annaluru, N. *et al.*, Total synthesis of a functional designer eukaryotic chromosome, *Science* 2014; 344:55–8.

Nystedt, B. *et al.*, The Norway spruce genome sequence and conifer genome evolution, *Nature* 2013; 497:579.

Moreira, D. and Lopez-Garcia, P., Ten reasons to exclude viruses from the tree of life, *Nat. Rev. Microbiol* 2009; 7:306.

Lu, T. and Tauxe, W., Cocktail maker proteins and semiconductors, *Nature* 2015; 528:S14.

Forterre, P., The virocell concept and environmental microbiology, *ISME J.* 2013; 7:233.

TARA Ocean and gut: Sunagawa, S. *et al.*, Structure and function of the global ocean microbiome, *Science* 2015; 348:1–9.

TARA virome: Brum, J.R. *et al.*, Seasonal time bombs: Dominant temperate viruses, *ISME J.* 2016; 10:437–49.

TARA Ocean plankton: Lescot, M. *et al.*, RT genes in marine plankton, *ISME J.* 2015; doi:10.1038/ismej.2015.

Moelling, K. and Broecker, F., The RT-RNase H: From viruses to antiviral defense, *Ann. N.Y. Acad. Sci.* 2015; 1341:126.

Retrons: Farzadfard, F. and Lu, T.K., Genomically encoded analog memory, *Science* 2014;346 1256272. doi: 10.1126/.

Front-runners and late-progressors: Moelling, K., Are viruses our oldest ancestors? *EMBO Reports.* 2012; 13:1033.

Koonin, E., Senkevich T. and Dolja, V., The ancient virus world and evolution of cells, *Biology Direct* 2006;1:29.

Koonin, E.V. *et al.*, Reasons why viruses are relevant for the origin of cells, *Nat. Rev. Micro.* 2009; 7:615.

Villarreal, L.P., The widespread evolutionary significance of viruses, in Domingo, E., Parish, C. and Holland, J. (Eds.), *Origin and Evolution of Viruses* (Academic Press, 2008), pp. 477–516.

Dolja, V.V. and Koonin, E.V., Common origins and diversity of plant and animal viromes, *Curr. Opin. Virol.* 2011; 5:322.

Monster: Spiegelman, S. *et al.*, Synthesis of a self-propagating and infectious NA with an enzyme, *PNAS* 1965; 54:919.

Oehlenschläger, F. and Eigen, M., 30 years later: A new approach to Sol Spiegelman's and Leglie Orgel's in vitro, *Origins of Life and Evolution of Biospheres*, 1997; 2:437.

Kacian, D.L. *et al.*, Replicating RNA for extracellular evolution and replication, *PNAS* 1972; 69:3038.

Lucky: Cano, R.J. and Borucki, M.K., Revival and identification of bacterial spores in 25- to 40-million-year-old Dominican amber, *Science* 1995; 268:1060.

Body size: Katzourakis, A. and Gifford R., Larger body size leads to lower retroviral activity, *PloS Pathog.* 2014, 10 (7):e100414.

Obesity and cancer: Luo, C. and Puigserver, P., Dietary fat promotes intestinal dysregulation, *Nature* 531,42–43.

Transgenerational inheritance: Rassoulzadegan, M. *et al.*, Cuzin, F., RNA-mediated non-Mendelian inheritance of an epigenetic change in the mouse, *Nature* 2006; 441:469-74; and *Essays Biochem.* 2010;48(1):101–6.

Seth, M., The *C. elegans* CSR-1 Argonaute pathway counteracts epigenetic silencing, *Cell* 2013; 27:656.

Simon, M. *et al.*, Reduced insulin/IGF-1 signaling restores germ cell immortality to *C. elegans* Piwi mutants, *Cell Rep.* 2014; 7:762.

Weick, E.M. and Miska, E., piRNAs: From biogenesis to function, *Development* 2014; 141:3458.

Sarkies, P., and Miska, E.A., Small RNAs break out: Mobile small RNAs, *Nat. Rev. Mol. Cell. Biol.* 2014; 15:525.

Shirayama, M. *et al.*, piRNAs initiate an epigenetic memory of nonself RNA in *C. elegans*, *Cell* 2012; 150:65.

Ashe, A. *et al.*, Miska, E.A., piRNA can trigger a multigenerational epigenetic memory, *Cell* 2012; 150:88.

Ng, S.Y. *et al.*, Long ncRNAs in development and disease of CNS, *Trends Genet.* 2013; 29:461.

Orlando, L. and Willerslev, E., An epigenetic window into the past, *Science* 2014; 345:511.

Finnegan, D.J., Retrotransposon, *Current Biol.* 2012; 22: R432-7.

Pembrey, M. *et al.*, Human transgenerational responses to early-life experience, *J. Med. Genet.* 2014;51:56.

Ost, A. *et al.*, Paternal diet defines offspring chromatin and intergenerational obesity, *Cell* 2014; 159:1352.

Alleman, M., *et al.*, An RNA-dependent RNA polymerase is required for paramutation in maize, *Nature* 2006; 442:295.

Lolle, S.J., Non-Mendelian inheritance of extra-genomic information in Arabidopsis, *Nature* 2005; 434:505.

Color patterns: Murray, J. D., How the leopard gets its spots, *Sci. Am.* 1988; 3: 80–87.

Honey bees: Miklos, G. L., and Maleszka, R., Epigenetic comunication in human and honey bees, *Horm. Behav.* 2011; 59:399.

Kohl, J. V., Nutrient dependent/pheromone evolution: A model, *Socioaff. Neurosci. Psychol.* 2013; 3:20553.

Ganko, E.W. *et al.*, Contribution of LTR retrotransposons to *C. elegans* gene evolution, *Mol. Bio. Evol.* 2003; 20:1925.

Illustrations Credit

Pages 2, 37 and 137: With permission of Prof. Dr. Hans Gelderblom, Robert Koch Institute, Bundesgesundheitsamt, Nordufer 20, 13353 Berlin, Germany

Page 10: Courtesy of P. Rona, OAR/NOAA Photo Library, NURP (National Undersea Research Program); NOAA (National Oceanic and Atmospheric Administration), NOAA Library, http://www.photolib.noaa.gov/htmls/nur04506.htm Freedomdefinded.org

Page 113: *Top*: Courtesy of Prof. Dr. Curtis Suttle, Department of Earth and Ocean Sciences, University of British Columbia, Laboratory of Marine Virology and Microbiology, Vancouver, Canada.
Bottom left: Red tide, www.whoi.edu/redtide/Woods Hole Oceanographic Institutions, red tide algal bloom at Leigh, Cape Rodney, New Zealand, photo by M. Godfrey is gratefully acknowledged.
Bottom right: NEODAAS (NERC (National Centre for Earth Observation), Earth Observation Data Acquisition and Analysis Service), Cornwall coast, effect of *E. hux virus lysis*, Plymouth Marine Laboratory, UK, courtesy of Steve Groom, USGS Landsat image courtesy of https://upload.wikimedia.org/wikipedia/commons/8/86/Cwall99_lg.jpg hohe pix

Page 137: *Top left*: Retroviruses courtesy of Hans Gelderblom, see p. 2; *Middle left*: Phage hand-drawn by KM; *Bottom left*: Two archaea viruses: Haring, M., Rachel, R., Peng, X. Garrett, R.A., and Prangishvili, D., Viral diversity in hot springs of Pozzuoli, Italy, and characterization of a unique archaeal virus. Acidianus bottle-shaped virus from a new family, the *Ampullaeviridae*: Fig. 6 in *J. Virol.* 2005; 79:9904 (bought by KM from publisher), American Society of Microbiology; *Middle and right*: two mimiviruses, *Megavirus chilensis* (*middle*) and pandoravirus (*right*): Philippe N., Legendre M.,

Doutre G., Couté Y., Poiret O., Lescot M., Arslan D., Seltzer V., Bertaux L., Bruley C., Carin J., Claverie J.M., and Abergel C., Pandoraviruses: amoebaeviruses with genomes up to 2.5 Mb reaching that of parasitic eukaryotes. Fig. 1 and Fig. 2 from: *Science* 2013; 341:281. With permission from Copyright Clearance Centre, copyright@marketing.copyright.com to KM. The permission by Chantal Abergel, IGS (Information Genomique et Structurale), 13288 Marseille, France, for the pictures of the giant viruses *Megavirus chilensis* and Pandoravirus is gratefully acknowledged.

Page 138: *Left*: *Emiliania huxleyi* algae, Natural History Museum of London, UK, commons.wikimedia.org, science photo library; *Right*: Chalk coastal line, island of Ruegen, Northern Germany, commons. wikimedia

Page 163: *Bat*: Commons.wikimedia.org; *koala*: Commons.wikimedia.org

Page 167: Modified from Cordaux, R. and Batzer, M.A., The impact of retrotransposons on human genome evolution, *Nat. Rev. Genet.* 2009; 10:691, see also: Lander, E.S. *et al.*, Initial sequencing and analysis of the human genome, *Nature* 2010; 409:860

Page 180: *Agouti mouse*: Courtesy of Dana Dolinoy, Department of Environmental Health Sciences, School of Public Health, University of Michigan, Ann Arbor, Michigan, USA

Page 277: *Elba worm*: Courtesy of C. Lott/Hydra/ Max Planck Institute for Marine Microbiology, 28359 Bremen, Germany, courtesy of Prof. Dr. Nicole Dubilier

Page 308: *Hydra*: Courtesy of Dr. Michael Plewka, plingfactory, 45525 Hattingen, Germany, plingfactory@plingfactory.de, www.plingfactory.de, OK 21.4.2016

Note: All graphs and other photographs are produced or owned by the author.

Index